ROGER CARAS'
TREASURY OF CLASSIC
NATURE TALES

ROGER CARAS'
TREASURY OF CLASSIC
NATURE TALES

T·T

TRUMAN TALLEY BOOKS
DUTTON
NEW YORK

DUTTON

Published by the Penguin Group
Penguin Books USA Inc., 375 Hudson Street,
New York, New York 10014, U.S.A.
Penguin Books Ltd, 27 Wrights Lane,
London W8 5TZ, England
Penguin Books Australia Ltd, Ringwood,
Victoria, Australia
Penguin Books Canada Ltd, 10 Alcorn Avenue,
Toronto, Ontario, Canada M4V 3B2
Penguin Books (N.Z.) Ltd, 182–190 Wairau Road,
Auckland 10, New Zealand

Penguin Books Ltd, Registered Offices:
Harmondsworth, Middlesex, England

First published by Truman Talley Books/Dutton, an imprint of
New American Library, a division of Penguin Books USA Inc.
Distributed in Canada by McClelland & Stewart Inc.

First Printing, December, 1992
1 3 5 7 9 10 8 6 4 2
Copyright © Roger Caras and Martin H. Greenberg, 1992
All rights reserved.

Henry Beston: From The Outermost House: A Year of Life on the Great Beach of Cape Cod *by Henry Beston. Copyright 1928, 1949,* © *1956 by Henry Beston. Reprinted by permission of Henry Holt and Company, Inc.*
Thomas Bledsoe: From Brown Bear Summer *by Thomas Bledsoe. Copyright* © *1987 by Thomas Bledsoe. Used by permission of Truman Talley Books/Dutton, an imprint of New American Library, a division of Penguin Books USA Inc.*
Edward Abbey: From Desert Solitaire *by Edward Abbey. Copyright* © *1981 by Edward Abbey. Reprinted by permission of Don Congdon Associates, Inc.*
Stephen J. Gould: "The Misnamed, Mistreated, and Misunderstood Irish Elk." From Ever Since Darwin: Reflections in Natural History, *by Stephen Jay Gould, by permission of W. W. Norton & Company, Inc. Copyright* © *1977 by Stephen Jay Gould.*
Lee Wulff: From The Compleat Lee Wulff. *Copyright* © *1989 by Lee Wulff and reprinted by permission of Truman Talley Books/Dutton, a division of New American Library and Penguin Books USA Inc.*
Joseph Wood Krutch: From The Great Chain of Life *by Joseph Wood Krutch. Copyright* © *1956 by Joseph Wood Krutch. Reprinted by permission of Houghton Mifflin Company. All Rights Reserved.*

REGISTERED TRADEMARK—MARCA REGISTRADA
LIBRARY OF CONGRESS CATALOGING IN PUBLICATION DATA
Caras, Roger A.
[Treasury of classic nature tales]
Roger Caras' treasury of classic nature tales / Roger Caras.
p.   cm.
ISBN 0-525-93422-7
1. Natural history.   I. Title.
QH81.C354   1992
508—dc20   92-52887
CIP

Printed in the United States of America
Set in Centaur and Garamond No. 3

CONTENTS

Introduction
ROGER CARAS
XI

PART ONE
THE AMERICAN WILDERNESS:
The Nineteenth-Century Observers

Travel in Concord
HENRY DAVID THOREAU
3

CONTENTS

Passenger Pigeon
JOHN JAMES AUDUBON
30

My First Summer in the Sierra
JOHN MUIR
40

Travels in Florida
WILLIAM BARTRAM
54

The Canyon
JOHN W. POWELL
73

Buffalo Country
GEORGE CATLIN
93

Roughing It
MARK TWAIN
108

The Range of Light
JOHN MUIR
113

Raggylug: The Story of a Cottontail Rabbit
ERNEST THOMPSON SETON
129

Birch Browsings
JOHN BURROUGHS
154

CONTENTS

PART TWO
CREATURES GREAT AND SMALL:
The Twentieth-Century Naturalists

The Land of Little Rain
MARY AUSTIN
169

The Bird and the Machine
LOREN EISELEY
178

The Endless Migrations
ROGER CARAS
189

On the Edge of the Abyss
JACK RUDLOE
217

The Twilight Seas
SALLY CARRIGHAR
238

The Mysterious Lands
ANN HAYMOND ZWINGER
279

Summer's End
RACHEL L. CARSON
299

Night on the Great Beach
HENRY BESTON
309

Johnny Bear
ERNEST THOMPSON SETON
323

ix

CONTENTS

Brown Bear Summer
THOMAS BLEDSOE
344

The Predators
RICHARD ELLIS
387

The Serpents of Paradise
EDWARD ABBEY
395

The Misnamed, Mistreated, and Misunderstood Irish Elk
STEPHEN J. GOULD
404

Chicago Spiders
CHARLES D. STEWART
415

Antarctic Penguins
G. MURRAY LEVICK
429

The Atlantic Salmon
LEE WULFF
455

The Great Chain of Life
JOSEPH WOOD KRUTCH
465

Roger Caras

INTRODUCTION

The writer who seeks to observe and record nature as it really exists is a hunter. He or she stalks the prey, takes all of the risks of the hunter who consumes (more risk, in fact, for there is usually no gun should a miscalculation occur), but comes away with impressions and knowledge, for that was the goal. That there is no blood, no carcass, no trophy head or filet makes the quest no less of a hunt. The animals, the places or the plants observed, or all of them have given up a different kind of trophy. They have surrendered both secrets and perspective.

What makes a piece of nature writing a classic? *Classic* is one of the easiest words in our language to use (we apply it to everything from wine to cars, from postage stamps to music) but is one of the most difficult to define or, more important, justify. When it is applied to nature writing it means, I think, that a unique perspective is revealed in the shared experience of writer and reader. Surely John Muir's record of his first summer in the Sierra is unique. It is *his* record of *his* impressions and reactions that were achieved through *his* personal experiences. Neither Sally Carrighar's nor Rachel Carson's perspective could possibly be the same as any other observer's perspective now or then or ever. Classic nature writing offers a unique perspective, not something second or third hand, not something canned or rehashed, but rather first person and immediate as well as personal. The experience is enriched by the observer, not altered; it is made better for the quality of the observer's mind and expertise.

Style is of great significance as well. Memorable nature writing has style, movement, and it is evocative. Since it doesn't have the built-in audience that science-fiction writing and spy/crime thrillers can depend on, nature writing has been kept a secret from a great many otherwise literate people, the most important secret being that it is often very good writing indeed. The kind of writing in this book, then, is both unique in viewpoint and is an example of skilled craftsmanship and art.

The word *evocative* is worth considering on its own. The French say evocative stories have ambiance. I call it texture, and every place in the world has its own. With our five acknowledged senses we gain an overall impression of a place, how it looks, what it sounds like, what it smells like, what its tastes are and, if we can touch it, how it feels. For each observer, of course, the total sensual experience is again something unique. No one can possibly know what you yourself hear when you hear a tree frog or a Mozart concerto. When you and your companion agree that a leaf is red or green or gold or all three, you still do not really know what the other person is seeing. The names of colors are words only, but when they exist as experiences they are the individual property of each person.

The nature writer of merit, then, takes a place or a moment and re-creates it for the reader so cleverly and with such insight that it evokes for the reader the actual experience. It is almost like being there yourself.

There is always the danger that a nature writer will become so self-indulgent (I think poets often do this) that communication fails or is sharply curtailed, as the writer goes off simply doing his own thing, in today's vernacular. But that is not what the exercise is all about. Nature writing in simple terms is show-and-tell, it is communication. In fact it must be that or it is something of a sham.

If a passage or a volume of nature writing carries you away to a different place, evokes for you the world the writer saw through his or her eyes, tells you something that is perhaps new and different and offers you all this in a form that is art, chances are that you are onto something good, perhaps even classic. It should be an adventure if not in the conventional sense then an adventure of the mind and soul.

There is an important, greater aspect to all of this. Nature writing has led our collective conscience to new levels of awareness. We usually encounter nature writing in our children's literature first, and as we grow we find the well is never dry. At every level of our intellectual growth there is a wealth of literature about the natural world waiting for us. Then, eventually, many of us encounter the natural world as voyagers and as voters. I think we do better on both accounts for the wondrous exposure we have in our reading. Loren Eiseley awoke a lot of stirrings in us; John Muir certainly gave us pause and made us search our souls; Rachel Carson described the shoreline so no beach could be the same for us ever again. Stephen J. Gould continues to evoke the wonder of nature as well as anyone ever has; he is the king of "gee-whiz" in the best possible sense. Henry David Thoreau brought it all together and gave it meaning, Ernest Thompson Seton popularized it, John Burroughs deepened it, and Joseph Wood Krutch acted as not only a literary man but as a kind and loving philosopher and guide.

All these people are part of our ecological awareness today, and while they were awakening the sleeping giant of human concern, they were entertaining us. Anybody who can do that simultaneously has to be considered a positive force in the affairs of man and nature.

This wide-ranging anthology is as up-to-date as we can make it, "we" representing Martin Greenberg, master anthologist, and myself. In our era of growing ecological concern for the diversity of living things, this book warns of man's continuing encroachment on Earth's last wild places. Yet it does so by celebrating creatures great and small and the myriad ways they live and die in the last great ecosystems of the natural world.

July 15,1992
Thistle Hill Farm
Freeland, Maryland

PART ONE

THE AMERICAN WILDERNESS:
The Nineteenth-Century Observers

Henry David Thoreau

TRAVEL
IN
CONCORD

I think that no experience which I have today comes up to, or is comparable with the experiences of my boyhood. And not only this is true, but as far back as I can remember I have unconsciously referred to the experiences of a previous state of existence, "For life is a forgetting," etc. Formerly, methought, nature developed as I developed, and grew up with me. My life was ecstasy. In youth, before I lost any of my senses, I can remember that I was all alive, and inhabited my body with inexpressible satisfaction; both its weariness and its refreshment were sweet to me. This earth was the most glorious instrument, and I was audience to its strains. To have such sweet impressions made on us, such ecstasies begotten of the breezes! I can remember how I was astonished. I said to myself—I said to others—"There comes into my mind such an indescribable, infinite, all-absorbing, divine, heavenly pleasure, a sense of elevation and expansion, and [I] have had nought to do with it. I perceive that I am dealt with by superior powers. This is a pleasure, a joy, an existence which I have not procured myself. I speak as a witness on the stand, and tell what I have perceived." The morning and the evening were sweet to me, and I led a life aloof from society of men. I wondered if a mortal had ever known what I knew. I looked in books for some recognition of a kindred experience, but strange to say, I found none. Indeed, I was slow to discover that other men had had this experience, for it had been possible to read books and to associate with men on other grounds. The Maker of me was improving me. When I detected this interference I was profoundly moved. For years I marched as to a music in comparison with which the military music of the streets is noise and discord. I was daily intoxicated, and yet no man could call me intemperate. With all your science can you tell how it is, and whence it is that light comes into the soul?

There is little or nothing to be remembered written on the subject of getting an honest living. Neither the New Testament nor Poor Richard speaks to our condition. I cannot think of a single page which entertains, much less answers the questions which I put to myself on this subject. How to make the getting our living poetic! for if it is not poetic, it is not life but death that we get. Is it that men are too disgusted with their experience to speak of it? or that commonly they do not question the common modes? The most practically important of all questions, it seems to me, is how shall I get my living, and yet I find little or nothing said to the purpose in any book. Those who are living on the interest of money inherited, or dishonestly—i.e. by false methods—acquired, are of course incompetent to answer it. I consider that society with all its arts has done nothing for us in this respect. One would think from looking at literature that this question had never disturbed a solitary individual's musings. Cold and hunger seem more friendly to my nature than those methods which men have adopted and advise to ward them off. If it were not that I desire to do something here—accomplish some work—I should certainly prefer to suffer and die rather than be at the pains to get a living by the modes men propose.

While the Republic has already acquired a history world-wide, America is still unsettled and unexplored. Like the English in New Holland, we live only on the shores of a continent even yet, and hardly know where the rivers come from which float our navy. The very timber and boards and shingles of which our houses are made grew but yesterday in a wilderness where the Indian still hunts and the moose runs wild. New York has her wilderness within her own borders; and though the sailors of Europe are familiar with the soundings of her Hudson, and Fulton long since invented the steamboat on its waters, an Indian is still necessary to guide her scientific men to its head-waters in the Adirondack country.

We have advanced by leaps to the Pacific, and left many a lesser Oregon and California unexplored behind us. Though the railroad and the telegraph have been established on the shores of Maine, the Indian still looks out from her interior mountains over all these to the sea. There stands the city of Bangor, fifty miles up the Penobscot, at the head of navigation for vessels of the largest class, the principal lumber depot on this continent, with a population of twelve thousand, like a star on the edge of night, still hewing at the forests of which it is built, already overflowing with the luxuries and refinement of

Europe, and sending its vessels to Spain, to England, and to the West Indies for its groceries—and yet only a few axmen have gone "up river," into the howling wilderness which feeds it. The bear and deer are still found within its limits; and the moose, as he swims the Penobscot, is entangled amid its shipping, and taken by foreign sailors in its harbor. Twelve miles in the rear, twelve miles of railroad, are Orono and the Indian Island, the home of the Penobscot tribe; and then commence the batteau and the canoe, and the military road; and sixty miles above, the country is virtually unmapped and unexplored, and there still waves the virgin forest of the New World.

Maine, perhaps, will soon be where Massachusetts is. A good part of her territory is already as bare and commonplace as much of our neighborhood, and her villages generally are not so well shaded as ours.

And what are we coming to in our Middlesex towns? A bald, staring townhouse, or meetinghouse, and a bare liberty pole, as leafless as it is fruitless, for all I can see. We shall be obliged to import the timber for the last, hereafter, or splice such sticks as we have. And our ideas of liberty are equally mean with these. The very willow-rows lopped every three years for fuel or powder, and every sizable pine and oak, or other forest tree, cut down within the memory of man! As if individual speculators were to be allowed to export the clouds out of the sky, or the stars out of the firmament, one by one. We shall be reduced to gnaw the very crust of the earth for nutriment.

The kings of England formerly had their forests "to hold the king's game," for sport or food, sometimes destroying villages to create or extend them; and I think that they were impelled by a true instinct. Why should not we, who have renounced the king's authority, have our national preserves, where no villages need be destroyed, in which the bear and panther, and some even of the hunter race, may still exist, and not be "civilized off the face of the earth"—our forests, not to hold the king's game merely, but to hold and preserve the king himself also, the lord of creation—not for idle sport or food, but for inspiration and our own true recreation? Or shall we, like the villains, grub them all up, poaching on our own national domains?

Would it not be a luxury to stand up to one's chin in some retired swamp for a whole summer's day, scenting the sweetfern and bilberry blows, and lulled by the minstrelsy of gnats and mosquitoes? . . . Say twelve hours of genial and familiar converse with the leopard frog. The sun to rise behind alder and dogwood, and climb buoyantly to his meridian of three hands' breadth, and finally sink to rest behind some bold western hummock. To hear the evening chant of the mosquito from a thousand green chapels, and the bittern begin to boom from his concealed fort like a sunset gun! Surely, one may as profitably be soaked in the juices of a marsh for one day, as pick his way dryshod over sand. Cold and damp—are they not as rich experience as warmth and dryness?

Visited my nighthawk on her nest. Could hardly believe my eyes when I stood within seven feet and beheld her sitting on her eggs, her head to me. She looked so Saturnian, so one with the earth, so sphinx-like, a relic of the reign of Saturn which Jupiter did not destroy, a riddle that might well cause a man to go dash his head against a stone. It was not an actual living creature, far less a winged creature of the air, but a figure in stone or bronze, a fanciful production of art, like the gryphon or phoenix. In fact, with its breast toward me, and owing to its color or size no bill perceptible, it looked like the end [of] a brand, such as are common in a clearing, its breast mottled or alternately waved with dark brown and gray, its flat, grayish, weather-beaten crown, its eyes nearly closed, purposely, lest those bright beads should betray it, with the stony cunning of the sphinx. A fanciful work in bronze to ornament a mantel. It was enough to fill one with awe. The sight of this creature sitting on its eggs impressed me with the venerableness of the globe. There was nothing novel about it. All the while, this seemingly sleeping bronze sphinx, as motionless as the earth, was watching me with intense anxiety through those narrow slits in its eyelids. Another step, and it fluttered down the hill close to the ground, with a wabbling motion, as if touching the ground now with the tip of one wing, now with the other, so ten rods to the water, which [it] skimmed close over a few rods, then rose and soared in the air above me. Wonderful creature, which sits motionless on its eggs on the barest, most exposed hills, through pelting storms of rain or hail, as if it were a rock or a part of the earth itself, the outside of the globe, with its eyes shut and its wings folded, and, after the two days' storm, when you think it has become a fit symbol of the rheumatism, it suddenly rises into the air a bird, one of the most aerial, supple, and graceful of creatures, without stiffness in its wings or joints! It was a fit prelude to meeting Prometheus bound to his rock on Caucasus.

Suddenly, looking down the river, I saw a fox some sixty rods off, making across to the hills on my left. As the snow lay five inches deep, he made but slow progress, but it was no impediment to me. So, yielding to the instinct of the chase, I tossed my head aloft and bounded away, snuffing the air like a foxhound, and spurning the world and the Humane Society at each bound. It seemed the woods rang with the hunter's horn, and Diana and all the satyrs joined in the chase and cheered me on. Olympian and Elean youths were waving palms on the hills. In the meanwhile I gained rapidly on the fox; but he showed a remarkable presence of mind, for, instead of keeping up the face of the hill, which was steep and unwooded in that part, he kept along the slope in the direction of the forest, though he lost ground by it. Notwithstanding his fright, he took no step which was not beautiful. The course on his part was a series of most graceful curves. It was a sort of leopard canter, I should say, as if he were nowise impeded by the snow, but was husbanding his strength all the while. When he doubled I wheeled and cut him off, bounding with fresh vigor, and Antaeus-like, recovering my strength each time I touched the snow. Having got near enough for a fair view, just as he was slipping into the wood, I gracefully yielded him the palm. He ran as though there were not a bone in his back, occasionally dropping his muzzle to the snow for a rod or two, and then tossing his head aloft when satisfied of his course. When he came to a declivity he put his forefeet together and slid down it like a cat. He trod so softly that you could not have heard it from any nearness, and yet with such expression that it would not have been quite inaudible at any distance. So, hoping this experience would prove a useful lesson to him, I returned to the village by the highway of the river.

A neat herd of cows approached, of unusually fair proportions and smooth, clean skins, evidently petted by their owner, who must have carefully selected them. One more confiding heifer, the fairest of the herd, did by degrees approach as if to take some morsel from our hands, while our hearts leaped to our mouths with expectation and delight. She by degrees drew near with her fair limbs progressive, making pretense of browsing; nearer and nearer, till there was wafted toward us the bovine fragrance—cream of all the dairies that ever were or will be—and then she raised her gentle muzzle toward us, and snuffed an honest recognition within hand's reach. I saw 'twas possible for his herd to inspire with love the herdsman. She was as delicately featured as a hind. Her hide was mingled white and fawn-color, and on her muzzle's tip there was a white spot not bigger than a daisy, and on her side toward me the map of Asia plain to see.

Farewell, dear heifer! Though thou forgettest me, my prayer to heaven shall be that thou may'st not forget thyself. There was a whole bucolic in her snuff. I saw her name was Sumach. And by the kindred spots I knew her mother, more sedate and matronly, with full-grown bag; and on her sides was Asia, great and small, the plains of Tartary, even to the pole, while on her daughter it was Asia Minor. She was not disposed to wanton with the herdsman.

And as I walked, she followed me, and took an apple from my hand, and seemed to care more for the hand than apple. So innocent a face as I have rarely seen on any creature, and I have looked in [the] face of many heifers. And as she took the apple from my hand, I caught the apple of her eye. She smelled as sweet as the clethra blossom. There was no sinister expression. And for horns, though she had them, they were so well disposed in the right place, bent neither up nor down, I do not now remember she had any. No horn was held toward me.

A hen-hawk sails away from the wood southward. I get a very fair sight of it sailing overhead. What a perfectly regular and neat outline it presents! an easily recognized figure anywhere. Yet I never see it represented in any books. The exact correspondence of the marks on one side to those on the other, as the black or dark tip of one wing to the other, and the dark line midway the wing. I have no idea that one can get as correct an idea of the form and color of the under side of a hen-hawk's wings by spreading those of a dead specimen in his study as by looking up at a free and living hawk soaring above him in the fields. The penalty for obtaining a petty knowledge thus dishonestly is that it is less interesting to men generally, as it is less significant. Some, seeing and admiring the neat figure of the hawk sailing two or three hundred feet above their heads, wish to get nearer and hold it in their hands, perchance, not realizing that they can see it best at this distance, better now, perhaps, than ever they will again. What is an eagle in captivity!—screaming in a courtyard! I am not the wiser respecting eagles for having seen one there. I do not wish to know the length of its entrails.

I spend a considerable portion of my time observing the habits of the wild animals, my brute neighbors. By their various movements and migrations they fetch the year about to me. Very significant are the flight of geese and the migration of suckers, etc., etc. But when I consider that the nobler animals have been exterminated here—the cougar, panther, lynx, wolverine, wolf, bear, moose, deer, the beaver, the turkey, etc., etc.—I cannot but feel as if I lived in a tamed and, as it were, emasculated country. Would not the motions of those larger and wilder animals have been more significant still? Is it not a maimed and imperfect nature that I am conversant with? As if I were to study a tribe of Indians that had lost all its warriors. Do not the forest and the meadow now lack expression, now that I never see nor think of the moose with a lesser forest on his head in the one, nor of the beaver in the other? When I think what were the various sounds and notes, the migrations and works, and changes of fur and plumage which ushered in the spring and marked the other seasons of the year, I am reminded that this my life in nature, this particular round of natural phenomena which I call a year, is lamentably incomplete. I listen to [a] concert in which so many parts are wanting. The whole civilized country is to some extent turned into a city, and I am that citizen whom I pity. Many of those animal migrations and other phenomena by which the Indians marked the season are no longer to be observed. I seek acquaintance with Nature—to know her moods and manners. Primitive Nature is the most interesting to me. I take infinite pains to know all the phenomena of the spring, for instance, thinking that I have here the entire poem, and then, to my chagrin, I hear that it is but an imperfect copy that I possess and have read, that my ancestors have torn out many of the first leaves and grandest passages, and mutilated it in many places. I should not like to think that some demigod had come before me and picked out some of the best of the stars. I wish to know an entire heaven and an entire earth.

The simplest and most lumpish fungus has a peculiar interest to us, compared with a mere mass of earth, because it is so obviously organic and related to ourselves, however mute. It is the expression of an idea; growth according to a law; matter not dormant, not raw, but inspired, appropriated by spirit. If I take up a handful of earth, however separately interesting the particles may be, their relation to one another appears to be that of mere juxtaposition generally. I might have thrown them together thus. But the humblest fungus betrays a life akin to my own. It is a successful poem in its kind. There is suggested something superior to any particle of matter, in the idea or mind which uses and arranges the particles.

I cannot but see still in my mind's eye those little striped breams poised in Walden's glaucous water. They balance all the rest of the world in my estimation at present, for this is the bream that I have just found, and for the time I neglect all its brethren and am ready to kill the fatted calf on its account. For more than two centuries have men fished here, and have not distinguished this permanent settler of the township. It is not like a new bird, a transient visitor that may not be seen again for years, but there it dwells and has dwelt permanently, who can tell how long? When my eyes first rested on Walden the striped bream was poised in it, though I did not see it; and when

Tahatawan paddled his canoe there. How wild it makes the pond and the township, to find a new fish in it! America renews her youth here. But in my account of this bream I cannot go a hair's breadth beyond the mere statement that it exists—the miracle of its existence, my contemporary and neighbor, yet so different from me! I can only poise my thought there by its side and try to think like a bream for a moment. I can only think of precious jewels, of music, poetry, beauty, and the mystery of life. I only see the bream in its orbit, as I see a star, but I care not to measure its distance or weight. The bream, appreciated, floats in the pond as the center of the system, another image of God. Its life no man can explain more than he can his own. I want you to perceive the mystery of the bream. I have a contemporary in Walden. It has fins where I have legs and arms. I have a friend among the fishes, at least a new acquaintance. Its character will interest me, I trust, not its clothes and anatomy. I do not want it to eat. Acquaintance with it is to make my life more rich and eventful. It is as if a poet or an anchorite had moved into the town, whom I can see from time to time and think of yet oftener. Perhaps there are a thousand of these striped bream which no one had thought of in that pond—not their mere impressions in stone, but in the full tide of the bream life.

Though science may sometimes compare herself to a child picking up pebbles on the seashore, that is a rare mood with her; ordinarily her practical belief is that it is only a few pebbles which are *not* known, weighed and measured. A new species of fish signifies hardly more than a new name. See what is contributed in the scientific reports. One counts the fin rays; another measures the intestines; a third daguerreotypes a scale, etc., etc.; otherwise there's nothing to be said. As if all but this were done, and these were very rich and generous contributions to science. Her votaries may be seen wandering along the shore of the ocean of truth, with their backs to that ocean, ready to seize on the shells which are cast up. You would say that the scientific bodies were terribly put to it for objects and subjects. A dead specimen of an animal, if it is only well preserved in alcohol, is just as good for science as a living one preserved in its native element.

What is the amount of my discovery to me? It is not that I have got one in a bottle, that it has got a name in a book, but that I have a little fishy friend in the pond. How was it when the youth first discovered fishes? Was it the number of their fin rays or their arrangement, or the place of the fish in some system that made the boy dream of them? Is it these things that interest mankind in the fish, the inhabitant of the water? No, but a faint recognition of a living contemporary, a provoking mystery. One boy thinks of fishes and goes a-fishing from the same motive that his brother searches the poets for rare lines. It is the poetry of fishes which is their chief use; their flesh is their lowest use. The beauty of the fish, that is what it is best worth the while to measure. Its place in our systems is of comparatively little importance. Generally the boy loses some of his perception and his interest in the fish; he degenerates into a fisherman or an ichthyologist.

Shad are still taken in the basin of Concord River, at Lowell, where they are said to be a month earlier than the Merrimack shad, on account of the warmth of the water. Still patiently, almost pathetically, with instinct not to be discouraged, not to be *reasoned* with, revisiting their old haunts, as if their stern fates would relent, and still met by the Corporation with its dam. Poor shad! where is thy redress? When Nature gave thee instinct, gave she thee the heart to bear thy fate? Still wandering the sea in thy scaly armor to inquire humbly at the mouths of rivers if man has perchance left them free for thee to enter. By countless shoals loitering uncertain meanwhile, merely stemming the tide there, in danger from sea foes in spite of thy bright armor, awaiting new instructions, until the sands, until the water itself, tell thee if it be so or not. Thus by whole migrating nations, full of instinct, which is thy faith, in this backward spring, turned adrift; and perchance knowest not where men do *not* dwell, where there are *not* factories, in these days. Armed with no sword, no electric shock, but mere shad, armed only with innocence and a just cause, with tender dumb mouth only forward, and scales easy to be detached.—I for one am with thee, and who knows what may avail a crowbar against that Billerica dam?—Not despairing when whole myriads have gone to feed those sea-monsters during thy suspense, but still brave, indifferent, on easy fin there, like shad reserved for higher destinies. Willing to be decimated for man's behoof after the spawning season. Away with the superficial and selfish phil-*anthropy* of men—who knows what admirable virtue of fishes may be below low-water mark, bearing up against a hard destiny, not admired by that fellow creature who alone can appreciate it! Who hears the fishes when they cry? It will not be forgotten by some memory that we were contemporaries. Thou shalt ere long have thy way up the rivers, up all the rivers of the globe, if I am not mistaken. Yea, even thy dull watery dream shall be more than realized. If it were not so, but thou wert to be overlooked at first and at last, then would not I take their heaven. Yes, I say so, who think I know better than thou canst. Keep a stiff fin, then, and stem all the tides thou mayst meet.

I thrive best on solitude. If I have had a companion only one day in a week—unless it were one or two I could name—I find that the value of the week to me has been seriously affected. It dissipates my days, and often it takes me another week to get over it. As the Esquimaux of Smith's Strait in North Greenland laughed when Kane warned them of their utter extermination, cut off as they were by ice on all sides from their race, unless they attempted in season to cross the glacier southward, so do I laugh when you tell me of the danger of impoverishing myself by isolation. It is here that the walrus and the seal, and the white bear, and the eider ducks and auks on which I batten, most abound.

A man asked me the other night whether such and such persons were not as happy as anybody; being conscious, as I perceived, of much unhappiness himself and not aspiring to much more than an animal content. "Why!" said I, speaking to his condition, "the stones are happy, Concord River is happy, and I am happy too. When I took up a fragment of a Walnut shell this morning, I saw by its very grain and composition, its form and color, etc., that it was made for happiness. The most brutish and inanimate objects that are made suggest an everlasting and thorough satisfaction; they are the homes of content. Wood, earth, mold, etc., exist for joy. Do you think that Concord River would have continued to flow these million of years by Clamshell Hill and round Hunt's Island, if it had not been happy—if it had been miserable in its channel, tired of existence, and cursing its Maker and the hour that it sprang?"

The catechism says that the chief end of man is to glorify God and enjoy him forever, which of course is applicable mainly to God as seen in his works. Yet the only account of its beautiful insects—butterflies, etc.—which God has made and set before us which the State ever thinks of spending any money on is the account of those which are injurious to vegetation! This is the way we glorify God and enjoy him forever. Come out here and behold a thousand painted butterflies and other beautiful insects which people the air; then go to the libraries and see what kind of prayer and glorification of God is there recorded. Massachusetts has published her report on "Insects Injurious to Vegetation," and our neighbor, the "Noxious Insects of New

York." We have attended to the evil and said nothing about the good. This is looking a gift horse in the mouth with a vengeance. Children are attracted by the beauty of butterflies, but their parents and legislators deem it an idle pursuit. The parents remind me of the devil, but the children, of God. Though God may have pronounced his work good, we ask, "Is it not poisonous?"

Here have been three ultra-reformers, lecturers on Slavery, Temperance, the Church, etc., in and about our house and Mrs. Brooks's the last three or four days—A. D. Foss, once a Baptist minister in Hopkinton, N.H.; Loring Moody, a sort of traveling pattern-working chaplain; and H. C. Wright, who shocks all the old women with his infidel writings. Though Foss was a stranger to the others, you would have thought them old and familiar cronies. (They happened here together by accident.) They addressed each other constantly by their Christian names, and rubbed you continually with the greasy cheeks of their kindness. They would not keep their distance, but cuddle up and lie spoon-fashion with you, no matter how hot the weather nor how narrow the bed—chiefly——. I was awfully pestered with his benignity; feared I should get greased all over with it past restoration; tried to keep some starch in my clothes. He wrote a book called A Kiss for a Blow, and he behaved as if there were no alternative between these, or as if I had given him a blow. I would have preferred the blow, but he was bent on giving me the kiss, when there was neither quarrel nor agreement between us. I wanted that he should straighten his back, smooth out those ogling wrinkles of benignity about his eyes, and, with a healthy reserve, pronounce something in a downright manner. It was difficult to keep clear of his slimy benignity, with which he sought to cover you before he swallowed you and took you fairly into his bowels. It would have been far worse than the fate of Jonah. I do not wish to get any

nearer to a man's bowels than usual. They lick you as a cow
her calf. They would fain wrap you about with their bowels.
——addressed me as "Henry" within one minute from the time
I first laid eyes on him, and when I spoke, he said with drawling,
sultry sympathy, "Henry, I know all you would say; I understand you perfectly; you need not explain anything to me;" and
to another, "I am going to dive into Henry's inmost depths."
I said, "I trust you will not strike your head against the bottom."
He could tell in a dark room, with his eyes blinded and in
perfect stillness, if there was one there whom he loved. One
of the most attractive things about the flowers is their beautiful
reserve.

This afternoon, being on Fair Haven Hill, I heard the sound
of a saw, and soon after from the Cliff saw two men sawing
down a noble pine beneath, about forty rods off. I resolved to
watch it till it fell, the last of a dozen or more which were left
when the forest was cut and for fifteen years have waved in
solitary majesty over the sprout-land. I saw them like beavers
or insects gnawing at the trunk of this noble tree, the diminutive
manikins with their cross-cut saw which could scarcely span it.
It towered up a hundred feet, as I afterward found by mea-
surement, one of the tallest, probably, in the township and
straight as an arrow, but slanting a little toward the hillside, its
top seen against the frozen river and the hills of Conantum. I
watch closely to see when it begins to move. Now the sawers
stop, and with an ax open it a little on the side toward which
it leans, that it may break the faster. And now their saw goes
again. Now surely it is going; it is inclined one-quarter of the
quadrant, and breathless, I expect its crashing fall. But no, I
was mistaken; it has not moved an inch; it stands at the same
angle as at first. It is fifteen minutes yet to its fall. Still its
branches wave in the wind, as if it were destined to stand for
a century, and the wind soughs through its needles as of yore;

it is still a forest tree, the most majestic tree that waves over Musketaquid. The silvery sheen of the sunlight is reflected from its needles; it still affords an inaccessible crotch for the squirrel's nest; not a lichen has forsaken its mast-like stem, its raking mast—the hill is the hulk. Now, now's the moment! The manikins at its base are fleeing from their crime. They have dropped the guilty saw and ax. How slowly and majestically it starts! as if it were only swayed by a summer breeze, and would return without a sigh to its location in the air. And now it fans the hillside with its fall; and it lies down to its bed in the valley, from which it is never to rise, as softly as a feather, folding its green mantle about it like a warrior, as if, tired of standing, it embraced the earth with silent joy, returning its elements to the dust again.

I went down and measured it. It was about four feet in diameter where it was sawed, about one hundred feet long. Before I had reached it the axmen had already half divested it of its branches. Its gracefully spreading top was a perfect wreck on the hillside, as if it had been made of glass; and the tender cones of one year's growth upon its summit appealed in vain and too late to the mercy of the chopper. Already he has measured it with his ax, and marked off the mill-logs it will make. And the space it occupied in upper air is vacant for the next two centuries. It is lumber. He has laid waste the air. When the fish hawk in the spring revisits the banks of the Musketaquid, he will circle in vain to find his accustomed perch, and the hen-hawk will mourn for the pines lofty enough to protect her brood. A plant which it has taken two centuries to perfect, rising by slow stages into the heavens, has this afternoon ceased to exist. Its sapling top had expanded to this January thaw as the forerunner of summers to come. Why does not the village bell sound a knell? I hear no knell tolled. I see no procession of mourners in the streets, or the woodland aisles. The squirrel has leaped to another tree; the hawk has circled farther off, and has now settled upon a new eyrie; but the woodman is preparing [to] lay his ax at the root of that also.

Today it snows again, covering the ground. To get the value of the storm we must be out a long time and travel far in it, so that it may fairly penetrate our skin, and we be, as it were, turned inside-out to it, and there be no part in us but is wet or weatherbeaten—so that we become storm men instead of fair-weather men. Some men speak of having been wetted to the skin once as a memorable event in their lives, which, notwithstanding the croakers, they survived.

The snow is finally turned to a drenching rain.

Found amid the sphagnum on the dry bank on the south side of the Turnpike, just below Everett's meadow, a rare and remarkable fungus, such as I have heard of but never seen before. The whole height six and three-quarters inches, two-thirds of it being buried in the sphagnum. It may be divided into three parts, pileus, stem, and base—or scrotum, for it is a perfect phallus. One of those fungi named *impudicus*, I think. In all respects a most disgusting object, yet very suggestive.

It was as offensive to the eye as to the scent, the cap rapidly melting and defiling what it touched with a fetid, olivaceous, semiliquid matter. In an hour or two the plant scented the whole house, wherever placed, so that it could not be endured. I was afraid to sleep in my chamber, where it had lain, until the room had been well ventilated. It smelled like a dead rat in the ceiling, in all the ceilings of the house. Pray, what was Nature thinking of when she made this? She almost puts herself on a level with those who draw in privies.

The thin snow now driving from the north and lodging on my coat consists of those beautiful star crystals, not cottony and chubby spokes, as on the 13th December, but thin and partly transparent crystals. How full of the creative genius is the air in which these are generated! I should hardly admire more if real stars fell and lodged on my coat. Nature is full of genius, full of the divinity; so that not a snowflake escapes its fashioning hand.

A divinity must have stirred within them before the crystals did thus shoot and set. Wheels of the storm-chariots. The same law that shapes the earth-star shapes the snow-star. As surely as the petals of a flower are fixed, each of these countless snow-stars comes whirling to earth, pronouncing thus, with emphasis, the number six, Order, *kóomos*.

On the Saskatchewan, when no man of science is there to behold, still down they come, and not the less fulfill their destiny, perchance melt at once on the Indian's face. What a world we live in! where myriads of these little disks, so beautiful to the most prying eye, are whirled down on every traveler's coat, the observant and the unobservant, and on the restless squirrel's fur, and on the far-stretching fields and forests, the wooded dells, and the mountaintops. Far far away from the haunts of man, they roll down some little slope, fall over and come to their bearings, and melt or lose their beauty in the mass, ready anon to swell some little rill with their contribution, and so at last, the universal ocean from which they came. There they lie, like the wreck of chariot wheels after a battle in the skies. Meanwhile the meadow mouse shoves them aside in his gallery, the schoolboy casts them in his snowball, or the wood-man's sled glides smoothly over them, these glorious spangles, the sweeping of heaven's floor. And they all sing, melting as they sing of the mysteries of the number six—six, six, six, He takes up the water of the sea in his hand, leaving the salt; He disperses it in mist through the skies; He re-collects and sprinkles it like grain in six-rayed snowy stars over the earth, there to lie till He dissolves its bonds again.

Very little evidence of God or men did I see just then, and life not as rich and inviting an enterprise as it should be, when my attention was caught by a snowflake on my coat sleeve. It was one of those perfect, crystalline, star-shaped ones, six-rayed, like a flat wheel with six spokes, only the spokes were perfect little pine trees in shape, arranged around a central spangle. This little object, which, with many of its fellows, rested unmelting on my coat, so perfect and beautiful, reminded me that Nature had not lost her pristine vigor yet, and why should man lose heart? . . . I may say that the Maker of the world exhausts his skill with each snowflake and dewdrop that he sends down. We think that the one mechanically coheres, and that the other simply flows together and falls, but in truth they are the produce of *enthusiasm*, the children of an ecstasy, finished with the artist's utmost skill.

In the fall the loon (*Colymbus glacialis*) came, as usual, to moult and bathe in the pond, making the woods ring with his wild laughter before I had risen. At rumor of his arrival all the Mill-dam sportsmen are on the alert, in gigs and on foot, two-by-two and three-by-three, with patent rifles and conical balls and spyglasses. They come rustling through the woods like autumn leaves, at least ten men to one loon. Some station themselves on this side of the pond, some on that, for the poor bird cannot be omnipresent; if he dive here he must come up there. But now the kind October wind rises, rustling the leaves and rippling the surface of the water, so that no loon can be heard or seen, though his foes sweep the pond with spyglasses, and make the woods resound with their discharges. The waves generously rise and dash angrily, taking sides with all waterfowl, and our sportsmen must beat a retreat to town and shop and unfinished jobs. But they were too often successful. When I went to get a pail of water early in the morning I frequently saw this stately bird sailing out of my cove within a few rods. If I endeavored

to overtake him in a boat, in order to see how he would maneuver, he would dive and be completely lost, so that I did not
discover him again, sometimes, till the latter part of the day.
But I was more than a match for him on the surface. He commonly went off in a rain.

As I was paddling along the north shore one very calm
October afternoon—for such days especially they settle on to
the lakes, like the milkweed down—having looked in vain over
the pond for a loon, suddenly one sailing out from the shore
toward the middle a few rods in front of me set up his wild
laugh and betrayed himself. I pursued with a paddle and he
dived, but when he came up I was nearer than before. He dived
again, but I miscalculated the direction he would take, and we
were fifty rods apart when he came to the surface this time,
for I had helped to widen the interval; and again he laughed
long and loud, and with more reason than before. He maneu-
vered so cunningly that I could not get within half-a-dozen rods
of him. Each time, when he came to the surface, turning his
head this way and that, he coolly surveyed the water and the
land, and apparently chose his course so that he might come
up where there was the widest expanse of water and at the
greatest distance from the boat. It was surprising how quickly
he made up his mind and put his resolve into execution. He
led me at once to the widest part of the pond, and could not
be driven from it. While he was thinking one thing in his brain,
I was endeavoring to divine his thought in mine. It was a pretty
game, played on the smooth surface of the pond, a man against
a loon. Suddenly your adversary's checker disappears beneath
the board, and the problem is to place yours nearest to where
his will appear again. Sometimes he would come up unex-
pectedly on the opposite side of me, having apparently passed
directly under the boat. So longwinded was he and so un-
weariable, that when he had swum farthest he would immediately plunge again, nevertheless; and then no wit could divine

23

where in the deep pond, beneath the smooth surface, he might be speeding his way like a fish, for he had time and ability to visit the bottom of the pond in its deepest part. It is said that loons have been caught in the New York lakes eighty feet beneath the surface, with hooks set for trout—though Walden is deeper than that. How surprised must the fishes be to see this ungainly visitor from another sphere speeding his way amid their schools! Yet he appeared to know his course as surely under water as on the surface, and swam much faster there. Once or twice I saw a ripple where he approached the surface, just put his head out to reconnoiter, and instantly dived again. I found that it was as well for me to rest on my oars and wait his reappearing, as to endeavor to calculate where he would rise; for again and again, when I was straining my eyes over the surface one way, I would suddenly be startled by his unearthly laugh behind me. But why, after displaying so much cunning, did he invariably betray himself the moment he came up, by that loud laugh? Did not his white breast enough betray him? He was indeed a silly loon, I thought. I could commonly hear the plash of the water when he came up, and so also detect him. But after an hour he seemed as fresh as ever, dived as willingly, and swam yet farther than at first. It was surprising to see how serenely he sailed off with unruffled breast when he came to the surface, doing all the work with his webbed feet beneath. His usual note was this demoniac laughter, yet somewhat like that of a waterfowl; but occasionally, when he had balked me most successfully and come up a long way off, he uttered a long-drawn unearthly howl, probably more like that of a wolf than any bird; as when a beast puts his muzzle to the ground and deliberately howls. This was his looning,— perhaps the wildest sound that is ever heard here, making the woods ring far and wide. I concluded that he laughed in derision of my efforts, confident of his own resources. Though the sky was by this time overcast, the pond was so smooth that I could see where he broke the surface when I did not hear him. His white breast, the stillness of the air, and the smoothness of the water were all against him. At length, having come up fifty rods off, he uttered one of those prolonged howls, as if calling on the god of loons to aid him, and immediately there came a wind from the east and rippled the surface, and filled the whole air with misty rain; and I was impressed as if it were the prayer of the loon answered, and his god was angry with me; and so I left him disappearing far away on the tumultuous surface.

As I turned round the corner of Hubbard's Grove, saw a wood-chuck, the first of the season, in the middle of the field, six or seven rods from the fence which bounds the wood, and twenty rods distant. I ran along the fence and cut him off, or rather overtook him, though he started at the same time. When I was only a rod and a half off, he stopped, and I did the same; then he ran again, and I ran up within three feet of him, when he stopped again, the fence being between us. I squatted down and surveyed him at my leisure. His eyes were dull black and rather inobvious, with a faint chestnut (?) iris, with but little expression and that more of resignation than of anger. The general aspect was a coarse grayish brown, a sort of grisel (?). A lighter brown next the skin, then black or very dark brown and tipped with whitish rather loosely. The head between a squirrel and a bear, flat on the top and dark brown, and darker still or black on the tip of the nose. The whiskers black, two inches long. The ears very small and roundish, set far back and nearly buried in the fur. Black feet, with long and slender claws for digging. It appeared to tremble, or perchance shivered with cold. When I moved, it gritted its teeth quite loud, sometimes striking the under jaw against the other chatteringly, sometimes grinding one jaw on the other, yet as if more from instinct than anger. Whichever way I turned, that way it headed. I took a twig a foot long and touched its snout, at which it started

forward and bit the stick, lessening the distance between us to
two feet, and still it held all the ground it gained. I played with
it tenderly awhile with the stick, trying to open its gritting jaws.
Even its long incisors, two above and two below, were presented. But I thought it would go to sleep if I stayed long
enough. It did not sit upright as sometimes, but *standing* on
its forefeet with its head down, i.e. half-sitting, half-standing.
We sat looking at one another about half an hour, till we began
to feel mesmeric influences. When I was tired, I moved away,
wishing to see him run, but I could not start him. He would
not stir as long as I was looking at him or could see him. I
walked round him; he turned as fast and fronted me still. I sat
down by his side within a foot. I talked to him quasi forest
lingo, baby talk, at any rate in a conciliatory tone, and thought
that I had some influence on him. He gritted his teeth less. I
chewed checkerberry leaves and presented them to his nose at
last without a grit; though I saw that by so much gritting of the
teeth he had worn them rapidly and they were covered with a
fine white powder, which, if you measured it thus, would have
made his anger terrible. He did not mind any noise I might
make. With a little stick I lifted one of his paws to examine it,
and held it up at pleasure. I turned him over to see what color
he was beneath (darker or more purely brown), though he
turned himself back again sooner than I could have wished.
His tail was also all brown, though not very dark, rattail-like,
with loose hairs standing out on all sides like a caterpillar brush.
He had a rather mild look. I spoke kindly to him. I reached
checkerberry leaves to his mouth. I stretched my hands over
him, though he turned up his head and still gritted a little. I
laid my hand on him, but immediately took it off again, instinct
not being wholly overcome. If I had had a few fresh bean leaves,
thus in advance of the season, I am sure I should have tamed
him completely. It was a frizzly tail. His is a humble, terrestrial
color like the partridge's, well concealed where dead wiry grass

rises above darker brown or chestnut dead leaves—a modest color. If I had had some food, I should have ended with stroking him at my leisure. Could easily have wrapped him in my handkerchief. He was not fat nor particularly lean. I finally had to leave him without seeing him move from the place. A large, clumsy, burrowing squirrel. *Arctomys*, bear-mouse. I respect him as one of the natives. He lies there, by his color and habits so naturalized amid the dry leaves, the withered grass, and the bushes. A sound nap, too, he has enjoyed in his native fields, the past winter. I think I might learn some wisdom of him. His ancestors have lived here longer than mine. He is more thoroughly acclimated and naturalized than I. Bean leaves the red man raised for him, but he can do without them.

There is a period in the history of the individual, as of the race, when the hunters are the "best men," as the Algonquins called them. We cannot but pity the boy who has never fired a gun; he is no more humane, while his education has been sadly neglected. This was my answer with respect to those youths who were bent on this pursuit, trusting that they would soon outgrow it. No humane being, past the thoughtless age of boyhood, will wantonly murder any creature which holds its life by the same tenure that he does. The hare in its extremity cries like a child. I warn you, mothers, that my sympathies do not always make the usual phil-*anthropic* distinctions.

Such is oftenest the young man's introduction to the forest, and the most original part of himself. He goes thither at first as a hunter and fisher, until at last, if he has the seeds of a better life in him, he distinguishes his proper objects—as a poet or naturalist it may be—and leaves the gun and fish pole behind. The mass of men are still and always young in this respect. In some countries a hunting parson is no uncommon sight. Such a one might make a good shepherd's dog, but is far from being the Good Shepherd. I have been surprised to consider that the only obvious employment, except wood chopping, ice cutting, or the like business, which ever to my knowledge detained at Walden Pond for a whole half-day any of my fellow citizens, whether fathers or children of the town, with just one exception, was fishing. Commonly they did not think that they were lucky, or well paid for their time, unless they got a long string of fish, though they had the opportunity of seeing the pond all the while. They might go there a thousand times before the sediment of fishing would sink to the bottom and leave their purpose pure; but no doubt such a clarifying process would be going on all the while. The Governor and his Council faintly remember the pond, for they went a-fishing there when they were boys; but now they are too old and dignified to go a-fishing, and so they know it no more forever. Yet even they expect to go to heaven at last. If the legislature regards it, it is chiefly to regulate the number of hooks to be used there; but they know nothing about the hook of hooks with which to angle for the pond itself, impaling the legislature for a bait. Thus, even in civilized communities, the embryo man passes through the hunter stage of development.

I never in all my walks came across a man engaged in so simple and natural an occupation as building his house. We belong to the community. It is not the tailor alone who is the ninth part of a man; it is as much the preacher, and the merchant, and the farmer. Where is this division of labor to end? and what object does it finally serve? No doubt another *may* also think of me; but it is not therefore desirable that he should do so to the exclusion of my thinking for myself.

I left the woods for as good a reason as I went there. Perhaps it seemed to me that I had several more lives to live, and could not spare any more time for that one. It is remarkable how easily and insensibly we fall into a particular route, and make a beaten track for ourselves. I had not lived there a week before my feet wore a path from my door to the pond-side; and though it is five or six years since I trod it, it is still quite distinct. It is true, I fear, that others may have fallen into it, and so helped to keep it open. The surface of the earth is soft and impressible by the feet of men; and so with the paths which the mind travels. How worn and dusty, then, must be the highways of the world, how deep the ruts of tradition and conformity! I did not wish to take a cabin passage, but rather to go before the mast and on the deck of the world, for there I could best see the moon-light amid the mountains. I do not wish to go below now.

John James Audubon

PASSENGER
PIGEON

I shall begin my description of the Passenger Pigeon with an account of its flight, because the most important facts connected with its habits relate to its migrations. It migrates for food rather than with a view to escaping the severity of the northern weather, or to seek a southern climate for breeding. Consequently such flights do not take place at any fixed period or season of the year. Indeed it sometimes happens that a continuance of a supply of food in one district will keep these birds away from any other place for years. I know, at least, to a certainty, that in Kentucky they remained for several years constantly, and were nowhere else to be found. They all suddenly disappeared one season when the mast (or acorns, nuts and tree pods) was exhausted, and they did not return for a long period. Similar facts have been observed in other States.

Their great power of flight enables them to survey and pass over an astonishing extent of country in a very short time, as proved by well-known facts. Pigeons have been killed in the neighborhood of New York with their crops full of rice, which they must have collected in the fields of Georgia and Carolina. These districts are the nearest possible for a supply of such food. Their power of digestion is so remarkable that they can entirely assimilate food in twelve hours; therefore, they must have travelled between three and four hundred miles in six hours, showing their speed to be at an average of about one mile a minute. A velocity such as this would enable one of these birds to visit the European continent in less than three days, were it so inclined.

This great power of flight is seconded by as great a power of vision, which enables them to inspect the country below as they travel at that swift rate, and to discover their food with facility. Thus they obtain the object of their journey. I have also proved another point by my observation that they fly high and in an extended front when passing over sterile or poor country. This enables them to survey hundreds of acres at once. On the other hand, when the land is richly covered with food, or the trees abundantly hung with mast, they fly low in order to discover the part most plentifully supplied.

Their body is formed in an elongated oval, which they steer with a long, well-plumed tail and propel with well-set wings, the muscles of which are very large and powerful for the size of the bird. A bird seen gliding through the woods and close to the observer passes like a thought; and the eye tries in vain to see it again, but the bird is gone. It propels itself with extreme rapidity by repeated flapping of the wings, which it brings more or less close to its body, according to the degree of velocity required. Like the domestic Pigeon, it often flies in a circling manner during the love season, supporting itself with both wings elevated at an angle, and keeping them in that position until it is about to alight. Now and then during these circular flights, the tips of the primary quills of each wing are made to strike against each other, producing a smart rap which may be heard thirty or forty yards away. Before alighting, the Wild Pigeon, like the Carolina Parrot and a few other species of birds, breaks the force of its flight by repeated flapping, as if apprehensive of injury from too sudden contact with the branch or spot of ground where it intends to settle.

The multitudes of Wild Pigeons in our American woods are astonishing. Indeed, after having viewed them so often and under so many circumstances, I now feel inclined even to pause and reassure myself that what I am going to relate is fact. Yet I have seen it all, and in the company, too, of persons who like myself were struck with amazement.

In the autumn of 1813 I left my house at Henderson on the banks of the Ohio, on my way to Louisville ninety miles distant. In passing over the Kentucky barrens a few miles beyond Hardinsburg, I observed the Passenger Pigeons flying from northeast to southwest in greater numbers than I had ever seen them before, it seemed to me. Feeling an inclination to count the flocks that might pass within the reach of my eye in one hour, I dismounted, seated myself on an eminence, and began to mark a dot with my pencil for every flock that passed. In a short time, finding this task impracticable because the birds were pouring by in countless multitudes, I arose. But before I travelled on, I counted the dots that I had put down and found that one hundred and sixty flocks had been recorded in twenty-one minutes. I met still more, farther on. The air was literally filled with Pigeons, and the noon-day light was obscured as by an eclipse. The dung fell in spots not unlike melting flakes of snow; and the continuous buzz of wings tended to lull my senses.

While waiting for dinner at Young's Inn at the confluence of Salt River with the Ohio, I saw, at my leisure, immense legions still going by. Their front reached far beyond the Ohio on the west, and the beechwood forests directly east of me. Not a single bird alighted, for not a nut or acorn was that year to be seen in the neighborhood. Consequently they were flying so high that different attempts to reach them with a capital rifle proved ineffectual; nor did the reports disturb them in the least. I cannot describe to you the extreme beauty of their aerial evolutions when a Hawk chanced to press upon the rear of a flock. At once, like a torrent, and with a noise like thunder, they rushed in a compact mass, pressing upon each other towards the center. In these almost solid masses they darted forward in undulating and angular lines, descended to the earth and swept close over it with inconceivable velocity. Then they mounted perpendicularly so as to resemble a vast column, and, when high, they were seen wheeling and twisting within their continued lines, which resembled the coils of a gigantic serpent.

Before sunset I reached Louisville, fifty-five miles from Hardinsburg. The Pigeons were still passing in undiminished numbers. They continued to do so for three days in succession. The people were all in arms, and the banks of the Ohio were crowded with men and boys incessantly shooting at the pilgrims, which flew lower as they passed the river. Multitudes were thus destroyed. For a week or more, the population fed on no other flesh than that of Pigeons, and talked of nothing but Pigeons.

It is extremely interesting to see flock after flock performing exactly the same evolutions which a preceding flock has traced in the air. Thus should a Hawk charge on a group at a certain point, the angles, curves and undulations described by the birds in their efforts to escape the dreaded talons of the plunderer are undeviatingly followed by the next flock that comes up. Should the bystander happen to witness one of these affrays and be struck with the rapidity and elegance of the motions, and desire to see them repeated, his wishes will be gratified if he but remain in the same place until the next flock of Pigeons comes along.

As soon as the Pigeons discover a sufficiency of food to entice them to alight, they fly around in circles, reviewing the countryside below. During these evolutions the dense mass which they form presents a beautiful spectacle, as it changes its direction, turning from a glistening sheet of azure, as the backs of the birds come simultaneously into view, to a suddenly presented, rich deep purple. After that they pass lower, over the woods, and for a moment are lost among the foliage. Again they emerge and glide aloft. They may now alight, but the next moment take to wing as if suddenly alarmed, the flapping of their wings producing a noise like the roar of distant thunder, as they sweep through the forests to see if danger is near. However, hunger soon brings them to the ground. On alighting they industriously throw aside the withered leaves in quest of the fallen mast. The rear ranks continually rise, passing over the main body and alighting in front, and in such rapid succession that the whole flock seems still on the wing. The quantity of ground swept in this way is astonishing. So completely has it been cleared that the gleaner who might follow in the rear of the flock would find his labor completely lost. While feeding, their avidity is at times so great that, in attempting to swallow a large acorn or nut, they may be seen to gasp for a long while as if in the agonies of suffocation.

When the woods are filled with these Pigeons, they are killed in immense numbers, although no apparent diminution comes of it. About mid-day, after their repast is finished, they settle on the trees to enjoy rest and digest their food. On the ground and on the branches they walk with ease, frequently jerking their beautiful tails and moving their necks backward and forward in the most graceful manner. As the sun begins to sink beneath the horizon, they depart *en masse* for the roosting place which, not infrequently, is hundreds of miles away, a fact ascertained by persons who have kept track of their arrivals and departures.

Let us inspect their place of nightly rendezvous. One of these curious roosting places on the banks of the Green River in Kentucky I repeatedly visited. As always, it was in a part of the forest where the trees were huge and where there was little underbrush. I rode through it for more than forty miles, and on crossing it in different parts I found it rather more than three miles wide on average. My first view of it was at nearly two hours before sunset, about two weeks before the coming of the Pigeons. Few of these birds were then to be seen, but a great gathering of persons with horses and wagons, guns and ammunition had pitched camp on the edge of the forest.

Two farmers from the vicinity of Russellville, more than a hundred miles distant, had driven more than three hundred hogs to be fattened on the Pigeons they hoped to slaughter. Here and there, people were busy plucking and salting birds already killed, and they sat amid large piles of them. The dung lay several inches deep, covering the whole roosting place. I noticed that many trees two feet in diameter were broken off at no great distance from the ground; and the branches of many of the largest and tallest had given way. It was as if the forest had been swept by a tornado, proving to me that the number of birds must be immense beyond conception.

As the time of the arrival of the Passenger Pigeons approached, their foes anxiously prepared to receive them. Some persons were ready with iron pots containing sulphur, others with torches of pine knots; many had poles, and the rest, guns.

The sun went down, yet not a Pigeon had arrived. However, everything was ready, and all eyes were fixed on the clear sky which could be glimpsed amid the tall tree-tops.

Suddenly a general cry burst forth, "Here they come!" The noise they made, even though still distant, reminded me of a hard gale at sea, passing through the rigging of a close-reefed vessel. As the birds arrived and passed over me, I felt a current of air that surprised me. Thousands of the Pigeons were soon knocked down by the pole-men, while more continued to pour in. The fires were lighted, then a magnificent, wonderful, and almost terrifying sight presented itself. The Pigeons, arriving by the thousands, alighted everywhere, one above another, until solid masses were formed on the branches all around. Here and there the perches gave way with a crash under the weight, and fell to the ground, destroying hundreds of birds beneath, and forcing down the dense groups of them with which every stick was loaded. The scene was one of uproar and confusion. I found it quite useless to speak, or even to shout, to those persons nearest to me. Even the gun reports were seldom heard, and I was made aware of the firing only by seeing the shooters reloading.

No one dared venture nearer the devastation. Meanwhile, the hogs had been penned up. The picking up of the dead and wounded birds was put off till morning. The Pigeons were constantly coming, and it was past midnight before I noticed any decrease in the' number of those arriving. The uproar continued the whole night. I was anxious to know how far away the sound could be heard, so I sent off a man used to roaming the forest, who returned in two hours with the information that he had heard it distinctly three miles from the roosting place.

Towards the approach of day, the noise somewhat subsided. Long before I could distinguish them plainly, the Pigeons began to move off in a direction quite different from the one in which they flew when they arrived the evening before. By sunrise all that were able to fly had disappeared. The howling of the wolves now reached our ears, and the foxes, lynxes, cougars, bears, raccoons, opossums and polecats were sneaking off. Eagles and Hawks, accompanied by a crowd of Vultures, took their place and enjoyed their share of the spoils.

Then the authors of all this devastation began to move among the dead, the dying, and the mangled, picking up the Pigeons and piling them in heaps. When each man had as many as he could possibly dispose of, the hogs were let loose to feed on the remainder.

Persons unacquainted with these birds might naturally conclude that such dreadful havoc would soon put an end to the species. But I have satisfied myself by long observation that nothing but the gradual diminution of our forests can accomplish their decrease. They not infrequently quadruple their number yearly, and always at least double it. In 1805 I saw schooners loaded with Pigeons caught up the Hudson River, coming into the wharf at New York where the birds sold for a cent apiece. I knew a man in Pennsylvania who caught and killed more than five hundred dozen in a clap-net in one day. Sometimes the net took twenty dozens or more at a single haul. In March, 1830, the Passenger Pigeon was so abundant in the New York markets that piles of them met the eye in every direction. I have seen the Negroes grow weary of killing them as the birds alighted for weeks at a time to drink the water from the pipes at the saltworks of Shawnee Town, on the Tennessee-Kentucky border.

The places they choose for breeding are of interest; though, as I have said, the season varies. Food is most plentiful and most attainable, and water is always at a convenient distance. The tallest trees of the forest are those in which the Pigeons nest. To them countless pairs resort and prepare to fulfil one of the great laws of Nature. At this period the note of the Pigeon is a soft *coo—coo—coo—coo*, much shorter than that of the domestic variety. The common notes resemble *kee —kee—kee—kee*, the first of these being the loudest, and the others gradually diminishing in power. With his tail spread and his wings drooping, the male, whether on the ground or on the branches, follows the female with a pompous demeanor. His body is elevated, his throat swells, his eyes sparkle, as he continues his cooing. Now and then he rises on the wing and flies a few yards towards the fugitive and timorous female. Like the domestic Pigeon and others, they caress each other by billing, an action in which the bill of the one is placed transversely in that of the other. Both birds alternately disgorge the contents of their crops by repeated efforts. After these preliminaries, the Pigeons begin building their nest in peace and harmony, crossing a few dry twigs in the fork of some branches.

Sometimes fifty to a hundred nests may be seen in the same tree. Were I not anxious that you should not feel disposed to refer my account of the Wild Pigeons to the marvelous, however wonderful, I might estimate a much greater number than one hundred.

There are two, broadly elliptical, pure white eggs. The male keeps the female supplied with food during incubation. Indeed the tenderness and affection shown by these birds towards their mates are striking in the highest degree. It is a remarkable fact that each brood hatched usually consists of a male and female.

Here again, the tyrant of creation, man, interferes, disturbing the harmony of this peaceful scene. As the young birds grow up, their enemies, armed with axes, reach the spot and seize and destroy all they can. The trees are felled, and are made to fall in such a way that the cutting of one causes the overthrow of another. Or the crash shakes the trees near by so much that the young Pigeons, or squabs, are violently tossed to the ground. In this way immense quantities are destroyed.

The young are fed by the parents as described above. In other words, the old bird puts its bill in the mouth of the young one crosswise, or with the back of the two parts of the bill opposite the young bird's open mouth and disgorges the contents of its crop as food. As soon as the young are able to shift for themselves, they leave their parents and continue apart until they reach maturity. By the end of six months they are capable of reproducing their species.

The flesh of the Wild Pigeon is of a dark color, but affords tolerable eating. That of young birds is much esteemed. The skin is covered with small white filmy scales. The feathers fall off at the least touch, like those of the Carolina Mourning Dove. Like other Pigeons, it immerses its head up to the eyes while drinking.

In March, 1830, I bought about three hundred and fifty of these birds in the New York market at four cents apiece, and carried most of them alive to England. I distributed them among several noblemen, giving some to the Zoological Society also. A curious change of habits has taken place in those which I presented to the Earl of Derby in 1830. That nobleman has assured me that ever since they began breeding in his aviaries near Liverpool they have laid only one egg. My noble friend has raised many and distributed them freely. It is therefore not surprising that some which have escaped confinement have been shot. But that the Passenger Pigeon should have a natural claim to be admitted into the British fauna appears to me very doubtful.

This bird wanders continually in search of food throughout all parts of North America, and is wonderfully abundant at times in particular districts.

John Muir

MY FIRST SUMMER
IN THE SIERRA

THROUGH THE FOOTHILLS
WITH A FLOCK OF SHEEP

In the great Central Valley of California there are only two
seasons—spring and summer. The spring begins with the first
rainstorm, which usually falls in November. In a few months
the wonderful flowery vegetation is in full bloom, and by the
end of May it is dead and dry and crisp, as if every plant had
been roasted in an oven.

Then the lolling, panting flocks and herds are driven to

the high, cool, green pastures of the Sierra. I was longing for the mountains about this time, but money was scarce and I couldn't see how a bread supply was to be kept up. While I was anxiously brooding on the bread problem, so troublesome to wanderers, and trying to believe that I might learn to live like the wild animals, gleaning nourishment here and there from seeds, berries, etc., sauntering and climbing in joyful independence of money or baggage, Mr. Delaney, a sheep owner for whom I had worked a few weeks, called on me and offered to engage me to go with his shepherd and flock to the headwaters of the Merced and Tuolumne Rivers—the very region I had most in mind. I was in the mood to accept work of any kind that would take me into the mountains whose treasures I had tasted last summer in the Yosemite region. The flock, he explained, would be moved gradually higher through the successive forest belts as the snow melted, stopping for a few weeks at the best places we came to. These I thought would be good centers of observation from which I might be able to make many telling excursions within a radius of eight or ten miles of the camps to learn something of the plants, animals, and rocks; for he assured me that I should be left perfectly free to follow my studies. I judged, however, that I was in no way the right man for the place, and freely explained my shortcomings, confessing that I was wholly unacquainted with the topography of the upper mountains, the streams that would have to be crossed, and the wild sheep-eating animals, etc.; in short that, what with bears, coyotes, rivers, canyons, and thorny, bewildering chaparral. I feared that half or more of his flock would be lost. Fortunately these shortcomings seemed insignificant to Mr. Delaney. The main thing, he said, was to have a man about the camp whom he could trust to see that the shepherd did his duty; and he assured me that the difficulties that seemed so formidable at a distance would vanish as we went on, encouraging me further by saying that the shepherd would do all the herding; that I could study plants and rocks and scenery as much as I liked; and that he would himself accompany us to the first main camp and make occasional visits to our higher ones to replenish our store of provisions and see how we prospered. Therefore I concluded to go, though still fearing, when I saw the silly sheep bouncing one by one through the narrow gate of the home corral to be counted, that of the two thousand and fifty many would never return.

I was fortunate in getting a fine St. Bernard dog for a companion. His master, a hunter with whom I was slightly acquainted, came to me as soon as he heard that I was going to spend the summer in the Sierra and begged me to take his favorite dog, Carlo, with me, for he feared that if he were compelled to stay all summer on the plains the fierce heat might be the death of him. "I think I can trust you to be kind to him," he said, "and I am sure he will be good to you. He knows all about the mountain animals, will guard the camp, assist in managing the sheep, and in every way be found able and faithful." Carlo knew we were talking about him, watched our faces, and listened so attentively that I fancied he understood us. Calling him by name, I asked him if he was willing to go with me. He looked me in the face with eyes expressing wonderful intelligence, then turned to his master, and after permission was given by a wave of the hand toward me and a farewell patting caress, he quietly followed me as if he perfectly understood all that had been said and had known me always.

June 3, 1869. This morning provisions, camp-kettles, blankets, plant-press, etc., were packed on two horses, the flock headed for the tawny foothills, and away we sauntered in a cloud of dust; Mr. Delaney, bony and tall, with sharply hacked profile like Don Quixote, leading the pack-horses; Billy, the proud shepherd, a Chinaman and a Digger Indian to assist in driving for the first few days in the brushy foothills; and myself with notebook tied to my belt.

IN CAMP ON THE NORTH FORK OF THE MERCED

June 8. The sheep, now grassy and good-natured, slowly nibbled their way down into the valley of the North Fork of the Merced at the foot of Pilot Peak Ridge to the place selected by the Don for our first central camp, a picturesque hopper-shaped hollow formed by converging hill slopes at a bend of the river. Here racks for dishes and provisions were made in the shade of the riverbank trees, and beds of fern fronds, cedar plumes, and various flowers, each to the taste of its owner, and a corral back on the open flat for the wool.

June 9. How deep our sleep last night in the mountain's heart, beneath the trees and stars, hushed by solemn-sounding waterfalls and many small soothing voices in sweet accord whispering peace! And our first pure mountain day, warm, calm, cloudless—how immeasurable it seems, how serenely wild! I can scarcely remember its beginning. Along the river, over the hills, in the ground, in the sky, spring work is going on with joyful enthusiasm, new life, new beauty, unfolding, unrolling in glorious exuberant extravagance—new birds in their nests, new winged creatures in the air, and new leaves, new flowers, spreading, shining, rejoicing everywhere.

June 19. Pure sunshine all day. How beautiful a rock is made by leaf shadows! Those of the live oak are particularly clear and distinct, and beyond all art in grace and delicacy, now still as if painted on stone, now gliding softly as if afraid of noise, now dancing, waltzing in swift, merry swirls, or jumping on

43

and off sunny rocks in quick dashes like wave embroidery on seashore cliffs. How true and substantial is this shadow beauty, and with what sublime extravagance is beauty thus multiplied! The big orange lilies are now arrayed in all their glory of leaf and flower. Noble plants, in perfect health, Nature's darlings.

June 20. Some of the silly sheep got caught fast in a tangle of chaparral this morning, like flies in a spider's web, and had to be helped out. Carlo found them and tried to drive them from the trap by the easiest way. How far above sheep are intelligent dogs! No friend and helper can be more affectionate and constant than Carlo. The noble St. Bernard is an honor to his race.

The air is distinctly fragrant with balsam and resin and mint—every breath of it a gift we may well thank God for. Who could ever guess that so rough a wilderness should yet be so fine, so full of good things. One seems to be in a majestic domed pavilion in which a grand play is being acted with scenery and music and incense—all the furniture and action so interesting we are in no danger of being called on to endure one dull moment. God himself seems to be always doing his best here, working like a man in a glow of enthusiasm.

June 23. Oh, these vast, calm, measureless mountain days, inciting at once to work and rest! Days in whose light everything seems equally divine, opening a thousand windows to show us God. Nevermore, however weary, should one faint by the way who gains the blessings of one mountain day; whatever his fate, long life, short life, stormy or calm, he is rich forever.

July 8. Now away we go toward the topmost mountains. Many still, small voices, as well as the noon thunder, are calling, "Come higher." Farewell, blessed dell, woods, gardens, streams, birds, squirrels, lizards, and a thousand others. Farewell. Farewell.

THE YOSEMITE

July 15. Followed the Mono Trail up the eastern rim of the basin nearly to its summit, then turned off southward to a small shallow valley that extends to the edge of the Yosemite, which we reached about noon, and encamped. After luncheon I made haste to high ground, and from the top of the ridge on the west side of Indian Canyon gained the noblest view of the summit peaks I have ever yet enjoyed. Nearly all the upper basin of the Merced was displayed, with its sublime domes and canyons, dark up-sweeping forests, and glorious array of white peaks deep in the sky, every feature glowing, radiating beauty that pours into our flesh and bones like heat rays from fire. Sunshine over all; no breath of wind to stir the brooding calm. Never before had I seen so glorious a landscape, so boundless an affluence of sublime mountain beauty. The most extravagant description I might give of this view to anyone who has not seen similar landscapes with his own eyes would not so much as hint its grandeur and the spiritual glow that covered it. I shouted and gesticulated in a wild burst of ecstasy, much to the astonishment of St. Bernard Carlo, who came running up to me, manifesting in his intelligent eyes a puzzled concern that was very ludicrous, which had the effect of bringing me to my senses. A brown bear, too, it would seem, had been a spectator of the show I had made of myself, for I had gone but a few yards when I started one from a thicket of brush. He evidently considered me dangerous, for he ran away very fast, tumbling over the tops of the tangled manzanita bushes in his haste. Carlo drew back, with his ears depressed as if afraid, and kept looking me in the face, as if expecting me to pursue and shoot, for he had seen many a bear battle in his day.

Following the ridge, which made a gradual descent to the south, I came at length to the brow of that massive cliff that stands between Indian Canyon and Yosemite Falls, and here the far-famed valley came suddenly into view throughout almost its whole extent. The noble walls—sculptured into endless variety of domes and gables, spires and battlements and plain mural precipices—all atremble with the thunder tones of the falling water. The level bottom seemed to be dressed like a garden—sunny meadows here and there, and groves of pine and oak; the river of Mercy sweeping in majesty through the midst of them and flashing back the sunbeams. The great Tissiack, or Half-Dome, rising at the upper end of the valley to a height of nearly a mile, is nobly proportioned and lifelike, the most impressive of all the rocks, holding the eye in devout admiration, calling it back again and again from falls or meadows, or even the mountains beyond—marvelous cliffs, marvelous in sheer dizzy depth and sculpture, types of endurance. Thousands of years have they stood in the sky exposed to rain, snow, frost, earthquake and avalanche, yet they still wear the bloom of youth.

July 19. Watching the daybreak and sunrise. The pale rose and purple sky changing softly to daffodil yellow and white, sunbeams pouring through the passes between the peaks and over the Yosemite domes, making their edges burn; the silver firs in the middle ground catching the glow on their spiry tops, and our camp grove fills and thrills with the glorious light. Everything awakening alert and joyful; the birds begin to stir and innumerable insect people. Deer quietly withdraw into leafy hiding-places in the chaparral; the dew vanishes, flowers spread their petals, every pulse beats high, every life cell rejoices, the very rocks seem to thrill with life. The whole landscape glows like a human face in a glory of enthusiasm, and the blue sky, pale around the horizon, bends peacefully down over all like one vast flower.

About noon, as usual, big bossy cumuli began to grow above the forest, and the rainstorm pouring from them is the most imposing I have yet seen. The silvery zigzag lightning lances are longer than usual, and the thunder gloriously impressive, keen, crashing, intensely concentrated, speaking with such tremendous energy it would seem that an entire mountain is being shattered at every stroke, but probably only a few trees are being shattered, many of which I have seen on my walks hereabouts strewing the ground. At last the clear ringing strokes are succeeded by deep low tones that grow gradually fainter as they roll afar into the recesses of the echoing mountains, where they seem to be welcomed home. Then another and another peal, or rather crashing, splintering stroke, follows in quick succession, perchance splitting some giant pine or fir from top to bottom into long rails and slivers, and scattering them to all points of the compass. Now comes the rain, with corresponding extravagant grandeur, covering the ground high and low with a sheet of flowing water, a transparent film fitted like a skin upon the rugged anatomy of the landscape, making the rocks glitter and glow, gathering in the ravines, flooding the streams, and making them shout and boom in reply to the thunder.

How interesting to trace the history of a single raindrop! It is not long, geologically speaking, as we have seen, since the first raindrops fell on the newborn leafless Sierra landscapes. How different the lot of these falling now! Happy the showers that fall on so fair a wilderness—scarce a single drop can fail to find a beautiful spot—on the tops of the peaks, on the shining glacier pavements, on the great smooth domes, on forests and gardens and brushy moraines, plashing, glinting, pattering, laving. Some go to the high snowy fountains to swell their well-saved stores; some into the lakes, washing the mountain windows, patting their smooth glassy levels, making dimples and bubbles and spray; some into the waterfalls and cascades, as if eager to join in their dance and song and beat their foam yet finer; good luck and good work for the happy mountain raindrops, each one of them a high waterfall in itself, descending from the cliffs and hollows of the clouds to the cliffs and hollows of the rocks, out of the sky-thunder into the thunder of the falling rivers. Some, falling on meadows and bogs, creep silently out of sight to the grass roots, hiding softly as in a nest, slipping, oozing hither, thither, seeking and finding their appointed work. Some, descending through the spires of the woods, sift spray through the shining needles, whispering peace and good cheer to each one of them. Some drops with happy aim glint on the sides of crystals—quartz, hornblende, garnet, zircon, tourmaline, feldspar—patter on grains of gold and heavy way-worn nuggets; some, with blunt plap-plap and low bass drumming, fall on the broad leaves of Veratrum, Saxifrage, Cypripedium. Some happy drops fall straight into the cups of flowers, kissing the lips of lilies. How far they have to go, how many cups to fill, great and small, cells too small to be seen, cups holding half a drop as well as lake basins between the hills, each replenished with equal care; every drop in all the blessed throng a silvery newborn star with lake and river, garden and grove, valley and mountain, all that the landscape holds reflected in its crystal depths, God's messenger, angel of love sent on its way with majesty and pomp and display of power that makes man's greatest shows ridiculous.

Now the storm is over, the sky is clear, the last rolling thunder-wave is spent on the peaks, and where are the raindrops now—what has become of all the shining throng? In winged vapor rising some are already hastening back to the sky, some have gone into the plants, creeping through invisible doors into the round rooms of cells, some are locked in crystals of ice, some in rock crystals, some in porous moraines to keep their small springs flowing, some have gone journeying on in the rivers to join the larger raindrop of the ocean. From form to form, beauty to beauty, ever changing, never resting, all are speeding on with love's enthusiasm, singing with the stars the eternal song of creation.

July 20. Fine calm morning; air tense and clear; not the slightest breeze astir; everything shining, the rocks with wet crystals, the plants with dew, each receiving its portion of irised dew-drops and sunshine like living creatures getting their breakfast, their dew manna coming down from the starry sky like swarms of smaller stars. How wondrous fine are the particles in showers of dew, thousands required for a single drop, growing in the dark as silently as the grass! What pains are taken to keep this wilderness in health—showers of snow, showers of rain, show-ers of dew, floods of light, floods of invisible vapor, clouds, winds, all sorts of weather, interaction of plant on plant, animal on animal, etc., beyond thought! How fine Nature's methods! How deeply with beauty is beauty overlaid! the ground covered with crystals, the crystals with mosses and lichens and low-spreading grasses and flowers, these with larger plant leaf over leaf with ever-changing color and form, the broad palms of the first outspread over these, the azure dome over all like a bell-flower, and star above star.

Yonder stands the South Dome, its crown high above our camp, though its base is four thousand feet below us; a most noble rock, it seems full of thought, clothed with living light, no sense of dead stone about it, all spiritualized, neither heavy-looking nor light, steadfast in serene strength like a god.

Our shepherd is a queer character and hard to place in this wilderness. His bed is a hollow made in red dry-rot punky dust beside a log which forms a portion of the south wall of the corral. Here he lies with his wonderful everlasting clothing on, wrapped in a red blanket, breathing not only the dust of the decayed wood but also that of the corral, as if determined to take ammoniacal snuff all night after chewing tobacco all day. Following the sheep he carries a heavy six-shooter swung from his belt on one side, and his luncheon on the other. The ancient cloth in which the meat, fresh from the frying pan, is tied, serves as a filter through which the clear fat and gravy juices drop down on his right hip and leg in clustering stalactites. This oleaginous formation is soon broken up, however, and diffused and rubbed evenly into his scanty apparel, by sitting down, rolling over, crossing his legs while resting on logs, etc., making shirt and trousers water-tight and shiny. His trousers, in particular, have become so adhesive with the mixed fat and resin that pine needles, thin flakes and fibers of bark, hair, mica scales and minute grains of quartz, hornblende, etc., feathers, seed wings, moth and butterfly wings, legs and antennae of innumerable insects, or even whole insects such as the small beetles, moths and mosquitoes, with flower petals, pollen dust and indeed bits of all plants, animals, and minerals of the region adhere to them and are safely imbedded, so that though far from being a naturalist he collects fragmentary specimens of everything and becomes richer than he knows. His specimens are kept passably fresh, too, by the purity of the air and the resiny bituminous beds into which they are pressed. Man is a microcosm, at least our shepherd is, or rather his trousers. These precious overalls are never taken off, and nobody knows how old they are, though one may guess by their thickness and concentric structure. Instead of wearing thin they wear thick, and in their stratification have no small geological significance.

Besides herding the sheep, Billy is the butcher, while I have agreed to wash the few iron and tin utensils and make the bread. Then, these small duties done, by the time the sun is fairly above the mountaintops I am beyond the flock, free to rove and revel in the wilderness all the big immortal days.

Sketching on the North Dome. It commands views of nearly all the valley besides a few of the high mountains. I would fain draw everything in sight—rock, tree, and leaf. But little can I do beyond mere outlines—marks with meanings like words, readable only to myself—yet I sharpen my pencils and work on as if others might possibly be benefited. Whether these picture-sheets are to vanish like fallen leaves or go to friends like letters, matters not much; for little can they tell to those who have not themselves seen similar wildness, and like a language have learned it. No pain here, no dull empty hours, no fear of the past, no fear of the future. These blessed mountains are so compactly filled with God's beauty, no petty personal hope or experience has room to be. Drinking this champagne water is pure pleasure, so is breathing the living air, and every movement of limbs is pleasure, while the whole body seems to feel beauty when exposed to it as it feels the campfire or sunshine, entering not by the eyes alone, but equally through all one's flesh like radiant heat, making a passionate ecstatic pleasure-glow not explainable. One's body then seems homogeneous throughout, sound as a crystal.

Perched like a fly on this Yosemite dome, I gaze and sketch and bask, oftentimes settling down into dumb admiration without definite hope of ever learning much, yet with the longing, unresting effort that lies at the door of hope, humbly prostrate before the vast display of God's power, and eager to offer self-denial and renunciation with eternal toil to learn any lesson in the divine manuscript.

It is easier to feel than to realize, or in any way explain, Yosemite grandeur. The magnitudes of the rocks and trees and streams are so delicately harmonized they are mostly hidden. Sheer precipices three thousand feet high are fringed with tall trees growing close like grass on the brow of a lowland hill, and extending along the feet of these precipices a ribbon of meadow a mile wide and seven or eight long, that seems like a strip a farmer might mow in less than a day. Waterfalls, five hundred to one or two thousand feet high, are so subordinated to the mighty cliffs over which they pour that they seem like wisps of smoke, gentle as floating clouds, though their voices fill the valley and make the rocks tremble. The mountains, too, along the eastern sky, and the domes in front of them, and the succession of smooth rounded waves between, swelling higher, higher, with dark woods in their hollows, serene in massive exuberant bulk and beauty, tend yet more to hide the grandeur of the Yosemite temple and make it appear as a subdued subordinate feature of the vast harmonious landscape. Thus every attempt to appreciate any one feature is beaten down by the overwhelming influence of all the others. And, as if this were not enough, lo! in the sky arises another mountain range with topography as rugged and substantial-looking as the one beneath it—snowy peaks and domes and shadowy Yosemite valleys—another version of the snowy Sierra, a new creation heralded by a thunderstorm. How fiercely, devoutly wild is Nature in the midst of her beauty-loving tenderness!—painting lilies, watering them, caressing them with gentle hand, going from flower to flower like a gardener while building rock mountains and cloud mountains full of lightning and rain. Gladly we run for shelter beneath an overhanging cliff and examine the reassuring ferns and mosses, gentle love tokens growing in cracks and chinks. Daisies, too, and ivesias, confiding wild children of light, too small to fear. To these one's heart goes home, and the voices of the storm become gentle. Now the sun breaks forth and fragrant steam arises. The birds are out singing on the edges of the groves. The west is flaming in gold and purple, ready for the ceremony of the sunset, and back I go to camp with my notes and pictures, the best of them printed in my mind as dreams. A fruitful day, without measured beginning or ending. A terrestrial eternity. A gift of good God.

July 23. Another midday cloudland, displaying power and beauty that one never wearies in beholding, but hopelessly unsketchable and untellable. What can poor mortals say about clouds? While a description of their huge glowing domes and ridges, shadowy gulfs and canyons, and feather-edged ravines is being tried, they vanish, leaving no visible ruins. Nevertheless, these fleeting sky mountains are as substantial and significant as the more lasting upheavals of granite beneath them. Both alike are built up and die, and in God's calendar difference of duration is nothing. We can only dream about them in wondering, worshiping admiration, happier than we dare tell even to friends who see farthest in sympathy, glad to know that not a crystal or vapor particle of them, hard or soft, is lost; that they sink and vanish only to rise again and again in higher and higher beauty. As to our own work, duty, influence, etc., concerning which so much fussy pother is made, it will not fail of its due effect, though, like a lichen on a stone, we keep silent.

William Bartram

TRAVELS IN FLORIDA

The Indian not returning this morning, I set sail alone. The coasts on each side had much the same appearance as already described. The Palm trees here seem to be of a different species from the Cabbage tree; their strait trunks are sixty, eighty or ninety feet high, with a beautiful taper of a bright ash colour, until within six or seven feet of the top, where it is a fine green colour, crowned with an orb of rich green plumed leaves: I have measured the stem of these plumes fifteen feet in length, besides the plume, which is nearly of the same length.

The little lake, which is an expansion of the river, now

appeared in view; on the East side are extensive marshes, and on the other high forests and Orange groves, and then a bay, lined with vast Cypress swamps, both coasts gradually approaching each other, to the opening of the river again, which is in this place about three hundred yards wide; evening now drawing on, I was anxious to reach some high bank of the river, where I intended to lodge, and agreeably to my wishes, I soon after discovered on the West shore, a little promontory, at the turning of the river, contracting it here to about one hundred and fifty yards in width. This promontory is a peninsula, containing about three acres of high ground, and is one entire Orange grove, with a few Live Oaks, Magnolias and Palms. Upon doubling the point, I arrived at the landing, which is a circular harbour, at the foot of the bluff, the top of which is about twelve feet high; and back of it is a large Cypress swamp, that spreads each way, the right wing forming the West coast of the little lake, and the left stretching up the river many miles, and encompassing a vast space of low grassy marshes. From this promontory, looking Eastward across the river, we behold a landscape of low country, unparalleled as I think; on the left is the East coast of the little lake, which I had just passed, and from the Orange bluff at the lower end, the high forests begin, and increase in breadth from the shore of the lake, making a circular sweep to the right, and contain many hundred thousand acres of meadow, and this grand sweep of high forests encircles, as I apprehend, at least twenty miles of these green fields, interspersed with hommocks or islets of evergreen trees, where the sovereign Magnolia and lordly Palm stand conspicuous. The islets are high shelly knolls, on the sides of creeks or branches of the river, which wind about and drain off the super-abundant waters that cover these meadows, during the winter season.

The evening was temperately cool and calm. The crocodiles began to roar and appear in uncommon numbers along the shores and in the river. I fixed my camp in an open plain, near the utmost projection of the promontory, under the shelter of a large Live Oak, which stood on the highest part of the ground and but a few yards from my boat. From this open, high situation, I had a free prospect of the river, which was a matter of no trivial consideration to me, having good reason to dread the subtle attacks of the alligators, who were crowding about my harbour. Having collected a good quantity of wood for the purpose of keeping up a light and smoke during the night, I began to think of preparing my supper, when, upon examining my stores, I found but a scanty provision, I thereupon determined, as the most expeditious way of supplying my necessities, to take my bob and try for some trout. About one hundred yards above my harbour, began a cove or bay of the river, out of which opened a large lagoon. The mouth or entrance from the river to it was narrow, but the waters soon after spread and formed a little lake, extending into the marshes, its entrance and shores within I observed to be verged with floating lawns of the Pistia and Nymphea and other aquatic plants; these I knew were excellent haunts for trout.

The verges and islets of the lagoon were elegantly embellished with flowering plants and shrubs; the laughing coots with wings half spread were tripping over the little coves and hiding themselves in the tufts of grass; young broods of the painted summer teal, skimming the still surface of the waters, and following the watchful parent unconscious of danger, were frequently surprised by the voracious trout, and he in turn, as often by the subtle, greedy alligator. Behold him rushing forth from the flags and reeds. His enormous body swells. His plaited tail brandished high, floats upon the lake. The waters like a cataract descend from his opening jaws. Clouds of smoke issue from his dilated nostrils. The earth trembles with his thunder. When immediately from the opposite coast of the lagoon, emerges from the deep his rival champion. They suddenly dart upon each other. The boiling surface of the lake marks their rapid course, and a terrific conflict commences. They now sink to the bottom folded together in horrid wreaths. The water becomes thick and discoloured. Again they rise, their jaws clap together, re-echoing through the deep surrounding forests. Again they sink, when the contest ends at the muddy bottom of the lake, and the vanquished makes a hazardous escape, hiding himself in the muddy turbulent waters and sedge on a distant shore. The proud victor exulting returns to the place of action. The shores and forests resound his dreadful roar, together with the triumphing shouts of the plaited tribes around, witnesses of the horrid combat.

My apprehensions were highly alarmed after being a spectator of so dreadful a battle; it was obvious that every delay would but tend to increase my dangers and difficulties, as the sun was near setting, and the alligators gathered around my harbour from all quarters; from these considerations I concluded to be expeditious in my trip to the lagoon, in order to take some fish. Not thinking it prudent to take my fusee with me, lest I might lose it overboard in case of a battle, which I had every reason to dread before my return, I therefore furnished myself with a club for my defence, went on board, and penetrating the first line of those which surrounded my harbour, they gave way; but being pursued by several very large ones, I kept strictly on the watch, and paddled with all my might towards the entrance of the lagoon, hoping to be sheltered there from the multitude of my assailants; but ere I had half-way reached the place, I was attacked on all sides, several endeavouring to overset the canoe. My situation now became precarious to the last degree: two very large ones attacked me closely, at the same instant, rushing up with their heads and part of their bodies above the water, roaring terribly and belching floods of water over me. They struck their jaws together so close to my ears, as almost to stun me, and I expected every moment to be dragged out of the boat and instantly devoured, but I applied my weapons so effectually about me, though at random, that I was so successful as to beat them off a little; when, finding that they designed to renew the battle, I made for the shore, as the only means left me for my preservation, for, by keeping close to it, I should have my enemies on one side of me only, whereas I was before surrounded by them, and there was a probability, if pushed to the last extremity, of saving myself, by jumping out of the canoe on shore, as it is easy to outwalk them on land, although comparatively as swift as lightning in the water. I found this last expedient alone could fully answer my expectations, for as soon as I gained the shore they drew off and kept aloof. This was a happy relief, as my confidence was, in some degree, recovered by it. On recollecting myself, I discovered that I had almost reached the entrance of the lagoon, and determined to venture in, if possible to take a few fish and then return to my harbour, while day-light continued; for I could now, with caution and resolution, make my way with safety along shore, and indeed there was no other way to regain my camp, without leaving my boat and making my retreat through the marshes and reeds, which, if I could even effect, would have been in a manner throwing myself away, for then there would have been no hopes of ever recovering my bark, and returning in safety to any settlements of men. I accordingly proceeded and made good my entrance into the lagoon, though not without opposition from the alligators, who formed a line across the entrance, but did not pursue me into it, nor was I molested by any there, though there were some very large ones in a cove at the upper end. I soon caught more trout than I had present occasion for, and the air was too hot and sultry to admit of their being kept for many hours, even though salted or barbecued. I now prepared for my return to camp, which I succeeded in with but little trouble, by keeping close to the shore, yet I was opposed upon re-entering the river out of the lagoon, and pursued near to my landing (though not closely attacked) particularly by an old daring one, about twelve feet in length, who kept close after me, and when I stepped on shore and turned about, in order to draw up my canoe, he rushed up near my feet and lay there for some time, looking me in the face, his head and shoulders out of water; I resolved he should pay for his temerity, and having a heavy load in my fusee, I ran to my camp, and returning with my piece, found him with his foot on the gunwale of the boat, in search of fish. On my coming up he withdrew sullenly and slowly into the water, but soon returned and placed himself in his former position, looking at me and seeming neither fearful nor any way disturbed. I soon dispatched him by lodging the contents of my gun in his head, and then proceeded to cleanse and prepare my fish for supper, and accordingly took them out of the boat, laid them down on the sand close to the water, and began to scale them, when, raising my head, I saw before me, through the clear water, the head and shoulders of a very large alligator, moving slowly towards me; I instantly stepped back, when, with a sweep of his tail, he brushed off several of my fish. It was certainly most providential that I looked up at that instant, as the monster would probably, in less than a minute, have seized and dragged me into the river. This incredible boldness of the animal disturbed me greatly, supposing there could now be no reasonable safety for me during the night, but by keeping continually on the watch; I therefore, as soon as I had prepared the fish, proceeded to secure myself and effects in the best manner I could: in the first place, I hauled my bark upon the shore, almost clear out of the water, to prevent their oversetting or sinking her. After this every moveable was taken out and carried to my camp, which was but a few yards off; then ranging some dry wood in such order as was the most convenient, cleared the ground round about it, that there might be no impediment in my way, in case of an attack in the night, either from the water or the land; for I discovered by this time, that this small isthmus, from its remote situation and fruitfulness, was resorted to by bears and wolves. Having prepared myself in the best manner I could, I charged my gun and proceeded to reconnoitre my camp and the adjacent grounds; when I discovered that the peninsula and grove, at the distance of about two hundred yards from my encampment, on the land side, were invested by a Cypress swamp, covered with water, which below was joined to the shore of the little lake, and above to the marshes surrounding the lagoon, so that I was confined to an islet exceedingly circumscribed, and I found there was no other retreat for me, in case of an attack, but by either ascending one of the large Oaks, or pushing off with my boat.

It was by this time dusk, and the alligators had nearly ceased their roar, when I was again alarmed by a tumultuous noise that seemed to be in my harbour, and therefore engaged my immediate attention. Returning to my camp I found it undisturbed, and then continued on to the extreme point of the promontory, where I saw a scene, new and surprising, which at first threw my senses into such a tumult, that it was some time before I could comprehend what was the matter; however, I soon accounted for the prodigious assemblage of crocodiles at this place, which exceeded every thing of the kind I had ever heard of.

How shall I express myself so as to convey an adequate idea of it to the reader, and at the same time avoid raising suspicions of my want of veracity. Should I say, that the river (in this place) from shore to shore, and perhaps near half a mile above and below me, appeared to be one solid bank of fish, of various kinds, pushing through this narrow pass of St. Juans into the little lake, on their return down the river, and that the alligators were in such incredible numbers, and so close together from shore to shore, that it would have been easy to have walked across on their heads, had the animals been harmless. What expressions can sufficiently declare the shocking scene that for some minutes continued, whilst this mighty army of fish were forcing the pass? During this attempt, thousands, I may say hundreds of thousands of them were caught and swallowed by the devouring alligators. I have seen an alligator take up out of the water several great fish at a time, and just squeeze them betwixt his jaws, while the tails of the great trout flapped about his eyes and lips, ere he had swallowed them. The horrid noise of their closing jaws, their plunging amidst the broken banks of fish, and rising with their prey some feet upright above the water, the floods of water and blood rushing out of their mouths, and the clouds of vapour issuing from their wide nostrils, were truly frightful. This scene continued at intervals during the night, as the fish came to the pass. After this sight, shocking and tremendous as it was, I found myself somewhat easier and more reconciled to my situation, being convinced that their extraordinary assemblage here was owing to this annual feast of fish, and that they were so well employed in their own enjoyment that I had little occasion to fear their paying any attention to me. As it was almost night, I returned to my camp; whilst I was revising the notes of my past day's journey, I was suddenly roused with a noise behind me toward the main land; I sprang up on my feet, and listening, I distinctly heard some creature wading in the water of the isthmus; I seized my gun and went cautiously from my camp, directing my steps towards the noise; when I advanced about thirty yards, I halted behind a coppice of Orange trees, and soon perceived two very large bears, which had made their way through the water, and had landed in the grove, about one hundred yards distance from me, and were advancing towards me. I waited until they were within thirty yards of me. They there began to snuff and look towards my camp, I snapped my piece, but it flashed, on which they both turned about and galloped off, plunging through the water and swamp, never halting as I suppose, until they reached fast land, as I could hear them leaping and plunging a long time; they did not presume to return again, nor was I molested by any other creature, except being occasionally awakened by the whooping of owls, screaming of bitterns, or the wood-rats running amongst the leaves.

The wood-rat is a very curious animal, they are not half the size of the domestic rat; of a dark brown or black colour; their tail slender and shorter in proportion, and covered thinly with short hair; they are singular with respect to their ingenuity and great labour in the construction of their habitations, which are conical pyramids about three or four feet high, constructed with dry branches, which they collect with great labour and perseverance, and pile up without any apparent order, yet they are so interwoven with one another, that it would take a bear or wild-cat some time to pull one of these castles to pieces, and allow the animals sufficient time to secure a retreat with their young.

The noise of the crocodiles kept me awake the greater part of the night, but when I arose in the morning, contrary to my expectations, there was perfect peace; very few of them to be seen, and those were asleep on the shore, yet I was not able to suppress my fears and apprehensions of being attacked by them in future; and indeed yesterday's combat with them, notwithstanding I came off in a manner victorious, or at least made a safe retreat, had left sufficient impression on my mind to damp my courage, and it seemed too much for one of my strength, being alone in a very small boat to encounter such collected danger. To pursue my voyage up the river, and be obliged every evening to pass such dangerous defiles, appeared to me as perilous as running the gauntlet betwixt two rows of Indians armed with knives and fire brands; I however resolved to continue my voyage one day longer, if I possibly could with safety, and then return down the river, should I find the like difficulties to oppose. Accordingly I got every thing on board, charged my gun, and set sail cautiously along shore; as I passed by Battle lagoon, I began to tremble and keep a good look out, when suddenly a huge alligator rushed out of the reeds, and with a tremendous roar, came up, and darted as swift as an arrow under my boat, emerging upright on my lea quarter, with open jaws, and belching water and smoke that fell upon me like rain in a hurricane; I laid soundly about his head with my club and beat him off, and after plunging and darting about my boat, he went off on a strait line through the water, seemingly with the rapidity of lightning, and entered the cape of the lagoon; I now employed my time to the very best advantage in paddling close along shore, but could not forbear looking now and then behind me, and presently perceived one of them coming up again; the water of the river hereabouts was shoal and very clear. The monster came up with the usual roar and menaces, and passed close by the side of my boat, when I could distinctly see a young brood of alligators to the number of one hundred or more, following after her in a long train. They kept close together in a column without straggling off to the one side or the other, the young appeared to be of an equal size, about fifteen inches in length, almost black, with pale yellow transverse waved clouds or blotches, much like rattle snakes in colour. I now lost sight of my enemy again.

Still keeping close along shore, on turning a point or projection of the river bank, at once I beheld a great number of hillocks or small pyramids, resembling hay cocks, ranged like an encampment along the banks, they stood fifteen or twenty yards distant from the water, on a high marsh, about four feet perpendicular above the water; I knew them to be the nests of the crocodile, having had a description of them before, and now expected a furious and general attack, as I saw several large crocodiles swimming abreast of these buildings. These nests being so great a curiosity to me, I was determined at all events immediately to land and examine them. Accordingly I ran my bark on shore at one of their landing places, which was a sort of nick or little dock, from which ascended a sloping path or road up to the edge of the meadow, where their nests were, most of them deserted, and the great thick whitish egg-shells lay broken and scattered upon the ground round about them.

The nests or hillocks are of the form of an obtuse cone, four feet high and four or five feet in diameter at their bases; they are constructed with mud, grass and herbage: at first they lay a floor of this kind of tempered mortar on the ground, upon which they deposit a layer of eggs, and upon this a stratum of mortar seven or eight inches in thickness, and then another layer of eggs, and in this manner one stratum upon another, nearly to the top: I believe they commonly lay from one to two hundred eggs in a nest: these are hatched I suppose by the heat of the sun, and perhaps the vegetable substances mixed with the earth, being acted upon by the sun, may cause a small degree of fermentation, and so increase the heat in those hillocks. The ground for several acres about these nests shewed evident marks of a continual resort of alligators; the grass was every where beaten down, hardly a blade or straw was left standing; whereas, all about, at a distance, it was five or six feet high, and as thick as it could grow together. The female, as I imagine, carefully watches her own nest of eggs until they are all hatched, or perhaps while she is attending her own brood, she takes under her care and protection, as many as she can get at one time, either from her own particular nest or others: but certain it is, that the young are not left to shift for themselves, having had frequent opportunities of seeing the female alligator, leading about the shores her train of young ones, just like a hen does her brood of chickens, and she is equally assiduous and courageous in defending the young, which are under their care, and providing for their subsistence; and when she is basking upon the warm banks, with her brood around her, you may hear the young ones continually whining and barking, like young puppies. I believe but few of a brood live to the years of full growth and magnitude, as the old feed on the young as long as they can make prey of them.

The alligator when full grown is a very large and terrible creature, and of prodigious strength, activity and swiftness in the water. I have seen them twenty feet in length, and some are supposed to be twenty-two or twenty-three feet; their body is as large as that of a horse; their shape exactly resembles that of a lizard, except their tail, which is flat or cuniform, being compressed on each side, and gradually diminishing from the abdomen to the extremity, which, with the whole body is covered with horny plates or squamae, impenetrable when on the body of the live animal, even to a rifle ball, except about their head and just behind their forelegs or arms, where it is said they are only vulnerable. The head of a full grown one is about three feet, and the mouth opens nearly the same length, the eyes are small in proportion and seem sunk deep in the head, by means of the prominency of the brows; the nostrils are large, inflated and prominent on the top, so that the head in the water, resembles, at a distance, a great chunk of wood floating about. Only the upper jaw moves, which they raise almost perpendicular, so as to form a right angle with the lower one. In the fore part of the upper jaw, on each side, just under the nostrils, are two very large, thick, strong teeth or tusks, not very sharp, but rather the shape of a cone, these are as white as the finest polished ivory, and are not covered by any skin or lips, and always in sight, which gives the creature a frightful appearance; in the lower jaw are holes opposite to these teeth, to receive them; when they clap their jaws together it causes a surprising noise, like that which is made by forcing a heavy plank with violence upon the ground, and may be heard at a great distance. But what is yet more surprising to a stranger, is the incredible loud and terrifying roar, which they are capable of making, especially in the spring season, their breeding time; it most resembles very heavy distant thunder, not only shaking the air and waters, but causing the earth to tremble; and when hundreds and thousands are roaring at the same time, you can scarcely be persuaded, but that the whole globe is violently and dangerously agitated.

An old champion, who is perhaps absolute sovereign of a little lake or lagoon (when fifty less than himself are obliged to content themselves with swelling and roaring in little coves round about) darts forth from the reedy coverts all at once, on the surface of the waters, in a right line; at first seemingly as rapid as lightning, but gradually more slowly until he arrives at the center of the lake, when he stops; he now swells himself by drawing in wind and water through his mouth, which causes a loud sonorous rattling in the throat for near a minute, but it is immediately forced out again through his mouth and nostrils, with a loud noise, brandishing his tail in the air, and the vapour ascending from his nostrils like smoke. At other times, when swollen to an extent ready to burst, his head and tail lifted up, he spins or twirls round on the surface of the water. He acts his part like an Indian chief when rehearsing his feats of war, and then retiring, the exhibition is continued by others who dare to step forth, and strive to excel each other, to gain the attention of the favourite female.

Having gratified my curiosity at this general breeding place and nursery of crocodiles, I continued my voyage up the river without being greatly disturbed by them: in my way I observed islets or floating fields of the bright green Pistia, decorated with other amphibious plants, as Senecio Jacobea, Persicaria amphibia, Coreopsis bidens, Hydrocotile fluitans, and many others of less note.

The swamps on the banks and islands of the river, are generally three or four feet above the surface of the water, and very level; the timber large and growing thinly, more so than what is observed to be in the swamps below Lake George; the black, rich earth is covered with moderately tall and very succulent tender grass, which when chewed is sweet and agreeable to the taste, somewhat like young sugar-cane: it is a jointed decumbent grass, sending out radiculae at the joints into the earth, and so spreads itself, by creeping over its surface.

The large timber trees, which possess the low lands, are Acer rubrum, Ac. nigundo, Ac. glaucum, Ulmus sylvatica, Fraxinus excelsior, Frax. aquatica, Ulmus suberifer, Gleditsia monosperma, Gledit. triacanthus, Diospyros Virginica, Nyssa aquatica, Nyssa sylvatica, Juglans cinerea, Quercus dentata, Quercus phillos, Hopea tinctoria, Corypha palma, Morus rubra, and many more. The Palm grows on the edges of the banks, where they are raised higher than the adjacent level ground, by the accumulation of sand, river-shells, &c. I passed along several miles by those rich swamps. The channels of the river which encircle the several fertile islands, I had passed, now uniting, formed one deep channel near three hundred yards over. The banks of the river on each side began to rise and present shelly bluffs, adorned by beautiful Orange groves, Laurels and Live Oaks. And now appeared in sight a tree that claimed my whole attention: it was the Carica papaya, both male and female, which were in flower; and the latter both in flower and fruit, some of which were ripe, as large, and of the form of a pear, and of a most charming appearance.

This admirable tree is certainly the most beautiful of any vegetable production I know of; the towering Laurel Magnolia, and exalted Palm, indeed exceed it in grandeur and magnificence, but not in elegance, delicacy and gracefulness; it rises erect, with a perfectly strait tapering stem, to the height of fifteen or twenty feet, which is smooth and polished, of a bright ash colour, resembling leaf silver, curiously inscribed with the footsteps of the fallen leaves, and these vestiges are placed in a very regular uniform imbricated order, which has a fine effect, as if the little column were elegantly carved all over. Its perfectly spherical top is formed of very large lobesinuate leaves, supported on very long footstalks; the lower leaves are the largest as well as their petioles the longest, and make a graceful sweep or flourish, like the long ∫ or the branches of a sconce candlestick. The ripe and green fruit are placed round about the stem or trunk, from the lowermost leaves, where the ripe fruit are, and upwards almost to the top; the heart or inmost pithy part of the trunk is in a manner hollow, or at best consists of very thin porous medullae or membranes; the tree very seldom branches or divides into limbs, I believe never unless the top is by accident broken off when very young: I saw one which had two tops or heads, the stem of which divided near the earth. It is always green, ornamented at the same time with flowers and fruit, which like figs come out singly from the trunk or stem.

After resting and refreshing myself in these delightful shades, I left them with reluctance, embarking again after the fervid heats of the meridian sun were abated, for some time I passed by broken ridges of shelly high land, covered with groves of Live Oak, Palm, Olea Americana, and Orange trees; frequently observing floating islets and green fields of the Pistia near the shores of the river and lagoons.

There is in this river and in the waters all over Florida, a very curious and handsome bird, the people call them Snake Birds. I think I have seen paintings of them on the Chinese screens and other India pictures: they seem to be a species of cormorant or loon (Colymbus cauda elongata) but far more beautiful and delicately formed than any other species that I have ever seen. The head and neck of this bird are extremely small and slender, the latter very long indeed, almost out of all proportion, the bill long, strait and slender, tapering from its base to a sharp point, all the upper side, the abdomen and thighs, are as black and glossy as a raven's, covered with feathers so firm and elastic, that they in some degree resemble fish-scales, the breast and upper part of the belly are covered with feathers of a cream colour, the tail is very long, of a deep black, and tipped with a silvery white, and when spread, represents an unfurled fan. They delight to sit in little peaceable communities, on the dry limbs of trees, hanging over the still waters, with their wings and tails expanded, I suppose to cool and air themselves, when at the same time they behold their images in the watery mirror: at such times, when we approach them, they drop off the limbs into the water as if dead, and for a minute or two are not to be seen; when on a sudden at a vast distance, their long slender head and neck only appear, and have very much the appearance of a snake, and no other part of them is to be seen when swimming in the water, except sometimes the tip end of their tail. In the heat of the day they are seen in great numbers, sailing very high in the air, over lakes and rivers.

I doubt not but if this bird had been an inhabitant of the Tiber in Ovid's days, it would have furnished him with a subject, for some beautiful and entertaining metamorphoses. I believe they feed entirely on fish, for their flesh smells and tastes intolerably strong of it, it is scarcely to be eaten unless constrained by insufferable hunger.

I had now swamps and marshes on both sides of me, and evening coming on apace, I began to look out for high land to encamp on, but the extensive marshes seemed to have no bounds; and it was almost dark when I found a tolerable suitable place, and at last was constrained to take up on a narrow strip of high shelly bank, on the West side. Great numbers of crocodiles were in sight on both shores; I ran my bark on shore at a perpendicular bank four or five feet above the water, just by the roots and under the spreading limbs of a great Live Oak: this appeared to have been an ancient camping place by Indians and strolling adventurers, from ash heaps and old rotten fire brands, and chunks, scattered about on the surface of the ground; but was now evidently the harbour and landing place of some sovereign alligator: there led up from it a deep beaten path or road, and was a convenient ascent.

I did not approve of my intended habitation from these circumstances; and no sooner had I landed and moored my canoe to the roots of the tree, than I saw a huge crocodile rising up from the bottom close by me, who, when he perceived that I saw him, plunged down again under my vessel; this determined me to be on my guard, and in time to provide against a troublesome night: I took out of my boat every moveable, which I carried upon the bank, then chose my lodging close to my canoe, under the spreading Oak; as hereabouts only, the ground was open and clear of high grass and bushes, and consequently I had some room to stir and look round about. I then proceeded to collect firewood which I found difficult to procure. Here were standing a few Orange trees. As for provisions, I had saved one or two barbecued trout; the remains of my last evening's collection in tolerable good order, though the sultry heats of the day had injured them; yet by stewing them up afresh with the lively juice of Oranges, they served well enough for my supper: having by this time but little relish or appetite for my victuals; for constant watching at night against the attacks of alligators, stinging of mosquitoes and sultry heats of the day; together, with the fatigues of working my bark, had almost deprived me of every desire but that of ending my troubles as speedy as possible. I had the good fortune to collect together a sufficiency of dry sticks, to keep up a light and smoke, which I laid by me, and then spread my skins and blankets upon the ground, kindled up a little fire and supped before it was quite dark. The evening was however, extremely pleasant, a brisk cool breeze sprang up, and the skies were perfectly serene, the stars twinkling with uncommon brilliancy. I stretched myself along before my fire; having the river, my little harbour and the stern of my vessel in view, and now through fatigue and weariness I fell asleep, but this happy temporary release from cares and troubles I enjoyed but a few moments, when I was awakened and greatly surprised by the terrifying screams of Owls in the deep swamps around me, and what increased my extreme misery was the difficulty of getting quite awake, and yet hearing at the same time such screaming and shouting, which increased and spread every way for miles around, in dreadful peals vibrating through the dark extensive forests, meadows and lakes. I could not after this surprise recover the former peaceable state and tranquility of mind and repose, during the long night, and I believe it was happy for me that I was awakened, for at that moment the crocodile was dashing my canoe against the roots of the tree, endeavouring to get into her for the fish, which I however prevented. Another time in the night I believe I narrowly escaped being dragged into the river by him, for when again through excessive fatigue I had fallen asleep, but was again awakened by the screaming owl, I found the monster on the top of the bank, his head towards me not above two yards distant, when starting up and seizing my fusee well loaded, which I always kept under my head in the night time, he drew back and plunged into the water. After this I roused up my fire, and kept a light during the remaining part of the night, being determined not to be caught napping so again, indeed the mosquitoes alone would have been abundantly sufficient to keep any creature awake that possessed their perfect senses, but I was overcome, and stupefied with incessant watching and labour: as soon as I discovered the first signs of day-light, I arose, got all my effects and implements on board and set sail, proceeding upwards, hoping to give the mosquitoes the slip, who were now, by the cool morning dews and breezes, driven to their shelter and hiding places; I was mistaken however in these conjectures, for great numbers of them, which had concealed themselves in my boat, as soon as the sun arose, began to revive, and sting me on my legs, which obliged me to land in order to get bushes to beat them out of their quarters.

John W. Powell

THE CANYON

August 13. We are now ready to start on our way down the Great Unknown. Our boats, tied to a common stake, are chafing each other, as they are tossed by the fretful river. They ride high and buoyant, for their loads are lighter than we could desire. We have but a month's rations remaining. The flour has been resifted through the mosquito-net sieve; the spoiled bacon has been dried, and the worst of it boiled; the few pounds of dried apples have been spread in the sun, and reshrunken to their normal bulk; the sugar has all melted, and gone on its way down the river; but we have a large sack of coffee. The lightening of the boats has this advantage; they will ride the waves better, and we shall have but little to carry when we make a portage.

We are three quarters of a mile in the depths of the earth, and the great river shrinks into insignificance, as it dashes its angry waves against the walls and cliffs, that rise to the world above; they are but puny ripples, and we but pigmies, running up and down the sands, or lost among the boulders.

We have an unknown distance yet to run; an unknown river yet to explore. What falls there are, we know not; what rocks beset the channel we know not; what walls rise over the river, we know not. Ah, well! we may conjecture many things. The men talk as cheerfully as ever; jests are bandied about freely this morning; but to me the cheer is somber and the jests are ghastly.

With some eagerness, and some anxiety, and some misgiving, we enter the canyon below, and are carried along by the swift water through walls which rise from its very edge. They have the same structure as we noticed yesterday—tiers of irregular shelves below, and, above these, steep slopes to the foot of marble cliffs. We run six miles in a little more than half an hour, and emerge into a more open portion of the canyon, where high hills and ledges of rock intervene between the river and the distant walls. Just at the head of this open place the river runs across a dike; that is, a fissure in the rocks, open to depths below, has been filled with eruptive matter, and this, on cooling, was harder than the rocks through which the crevice was made, and, when these were washed away, the harder volcanic matter remained as a wall, and the river has cut a gateway through it several hundred feet high, and as many wide. As it crosses the wall, there is a fall below, and a bad rapid, filled with boulders of trap; so we stop to make a portage. Then we go, gliding by hills and ledges, with distant walls in view; sweeping past sharp angles of rock; stopping at a few points to examine rapids, which we find can be run, until we have made another five miles, when we land for dinner.

Then we let down with lines, over a long rapid, and start again. Once more the walls close in, and we find ourselves in a narrow gorge, the water again filling the channel, and very swift. With great care, and constant watchfulness, we proceed, making about four miles this afternoon, and camp in a cave.

August 14. At daybreak we walk down the bank of the river, on a little sandy beach, to take a view of a new feature in the canyon. Heretofore, hard rocks have given us bad river; soft rocks, smooth water; and a series of rocks harder than any we have experienced sets in. The river enters the granite!

We can see but a little way into the granite gorge, but it looks threatening.

After breakfast we enter on the waves. At the very introduction, it inspires awe. The canyon is narrower than we have ever before seen it; the water is swifter; there are but few broken rocks in the channel; but the walls are set, on either side, with pinnacles and crags; and sharp, angular buttresses, bristling with wind and wave-polished spires, extend far out into the river.

Ledges of rocks jut into the stream, their tops sometimes just below the surface, sometimes rising few or many feet above; and island ledges, and island pinnacles, and island towers break the swift course of the stream into chutes, and eddies, and whirlpools. We soon reach a place where a creek comes in from the left, and just below, the channel is choked with boulders, which have washed down this lateral canyon and formed a dam, over which there is a fall of thirty or forty feet; but on the boulders we can get foothold, and we make a portage.

Three more such dams are found. Over one we make a

portage; at the other two we find chutes, through which we can run.

As we proceed, the granite rises higher, until nearly a thousand feet of the lower part of the walls are composed of this rock.

About eleven o'clock we hear a great roar ahead, and approach it very cautiously. The sound grows louder and louder as we run, and at last we find ourselves above a long, broken fall, with ledges and pinnacles of rock obstructing the river. There is a descent of, perhaps, seventy-five or eighty feet in a third of a mile, and the rushing waters break into great waves on the rocks, and lash themselves into a mad, white foam. We can land just above, but there is no foothold on either side by which we can make a portage. It is nearly a thousand feet to the top of the granite, so it will be impossible to carry our boats around, though we can climb to the summit up a side gulch, and, passing along a mile or two, can descend to the river. This we find on examination; but such a portage would be impracticable for us, and we must run the rapid, or abandon the river. There is no hesitation. We step into our boats, push off and away we go, first on smooth but swift water, then we strike a glassy wave, and ride to its top, down again into the trough, up again on a higher wave, and down and up on waves higher and still higher, until we strike one just as it curls back, and a breaker rolls over our little boat. Still, on we speed, shooting past projecting rocks, till the little boat is caught in a whirlpool, and spun around several times. At last we pull out again into the stream, and now the other boats have passed us. The open compartment of the "Emma Dean" is filled with water, and every breaker rolls over us. Hurled back from a rock, now on this side, now on that, we are carried into an eddy, in which we struggle for a few minutes, and are then out again, the breakers still rolling over us. Our boat is unmanageable, but she cannot sink, and we drift down another hundred yards, through breakers; how, we scarcely know. We find the other boats have turned into an eddy at the foot of the fall, and are waiting to catch us as we come, for the men have seen that our boat is swamped. They push out as we come near, and pull us in against the wall. We bail our boat, and on we go again.

The walls, now, are more than a mile in height—a vertical distance difficult to appreciate. Stand on the south steps of the Treasury building, in Washington, and look down Pennsylvania Avenue to the Capitol Park, and measure this distance overhead, and imagine cliffs to extend to that altitude, and you will understand what I mean; or, stand at Canal Street, in New York, and look up Broadway to Grace Church, and you have about the distance; or, stand at Lake Street bridge in Chicago, and look down to the Central Depot, and you have it again.

A thousand feet of this is up through granite crags, then steep slopes and perpendicular cliffs rise, one above another, to the summit. The gorge is black and narrow below, red and gray and flaring above, with crags and angular projections on the walls, which, cut in many places by side canyons, seem to be a vast wilderness of rocks. Down in these grand, gloomy depths we glide, ever listening, for the mad waters keep up their roar; ever watching, ever peering ahead, for the narrow canyon is winding, and the river is closed in so that we can see but a few hundred yards, and what there may be below we know not; but we listen for falls, and watch for rocks, or stop now and then, in the bay of a recess, to admire the gigantic scenery. And ever, as we go, there is some new pinnacle or tower, some crag or peak, some distant view of the upper plateau, some strange-shaped rock, or some deep, narrow side canyon. Then we come to another broken fall, which appears more difficult than the one we ran this morning.

A small creek comes in on the right, and the first fall of the water is over boulders, which have been carried down by this lateral stream. We land at its mouth, and stop for an hour or two to examine the fall. It seems possible to let down with lines, at least a part of the way, from point to point, along the right-hand wall. So we make a portage over the first rocks, and find footing on some boulders below. Then we let down one of the boats to the end of her line, when she reaches a corner of the projecting rock, to which one of the men clings, and steadies her, while I examine an eddy below. I think we can pass the other boats down by us, and catch them in the eddy. This is soon done and the men in the boats in the eddy pull us to their side. On the shore of this little eddy there is about two feet of gravel beach above the water. Standing on this beach, some of the men take the line of the little boat and let it drift down against another projecting angle. Here is a little shelf, on which a man from my boat climbs, and a shorter line is passed to him, and he fastens the boat to the side of the cliff. Then the second one is let down, bringing the line of the third. When the second boat is tied up, the two men standing on the beach above spring into the last boat, which is pulled up along-side of ours. Then we let down the boats, for twenty-five or thirty yards, by walking along the shelf, landing them again in the mouth of a side canyon. Just below this there is another pile of boulders, over which we make another portage. From the foot of these rocks we can climb to another shelf, forty or fifty feet above the water.

On this bench, we camp for the night. We find a few sticks, which have lodged in the rocks. It is raining hard, and we have no shelter, but kindle a fire and have our supper. We sit on the rocks all night, wrapped in our ponchos, getting what sleep we can.

August 15. This morning we find we can let down for three or four hundred yards, and it is managed in this way: We pass along the wall, by climbing from projecting point to point, sometimes near the water's edge, at other places, fifty or sixty feet above, and hold the boat with a line, while two men remain aboard, and prevent her from being dashed against the rocks, and keep the line from getting caught on the wall. In two hours we have brought them all down, as far as it is possible, in this way. A few yards below, the river strikes with great violence against a projecting rock, and our boats are pulled up in a little bay above. We must now manage to pull out of this, and clear the point below. The little boat is held by the bow obliquely up the stream. We jump in, and pull out only a few strokes, and sweep clear of the dangerous rock. The other boats follow in the same manner, and the rapid is passed.

It is not easy to describe the labor of such navigation. We must prevent the waves from dashing the boats against the cliffs. Sometimes, where the river is swift, we must put a bight of rope about a rock, to prevent her being snatched from us by a wave; but where the plunge is too great, or the chute too swift, we must let her leap, and catch her below, or the undertow will drag her under the falling water, and she sinks. Where we wish to run her out a little way from shore, through a channel between rocks, we first throw in little sticks of driftwood, and watch their course, to see where we must steer, so that she will pass the channel in safety. And so we hold, and let go, and pull, and lift, and ward, among rocks, around rocks, and over rocks.

And now we go on through this solemn, mysterious way. The river is very deep, the canyon very narrow, and still obstructed, so that there is no steady flow of the stream; but the waters wheel, and roll, and boil, and we are scarcely able to determine where we can go. Now, the boat is carried to the right, perhaps close to the wall; again, she is shot into the stream, and perhaps is dragged over to the other side, where, caught in a whirlpool, she spins about. We can neither land nor run as we please. The boats are entirely unmanageable; no order in their running can be preserved; now one, now another, is ahead, each crew laboring for its own preservation. In such a place we come to another rapid. Two of the boats run it perforce. One succeeds in landing, but there is no foothold by which to make a portage, and she is pushed out again into the stream. The next minute a great reflex wave fills the open compartment; she is waterlogged, and drifts unmanageable. Breaker after breaker rolls over her, and one capsizes her. The men are thrown out; but they cling to the boat, and she drifts down some distance, alongside of us, and we are able to catch her. She is soon bailed out, and the men are aboard once more; but the oars are lost, so a pair from the "Emma Dean" is spared. Then for two miles we find smooth water.

Clouds are playing in the canyon today. Sometimes they roll down in great masses, filling the gorge with gloom; sometimes they hang above, from wall to wall, and cover the canyon with a roof of impending storm; and we can peer long distances up and down this canyon corridor, with its cloud roof overhead, its walls of black granite, and its river bright with the sheen of broken waters. Then, a gust of wind sweeps down a side gulch, and, making a rift in the clouds, reveals the blue heavens, and a stream of sunlight pours in. Then, the clouds drift away into the distance, and hang around crags, and peaks, and pinnacles, and towers, and walls, and cover them with a mantle, that lifts from time to time, and sets them all in sharp relief. Then, baby clouds creep out of side canyons, glide around points, and creep back again, into more distant gorges. Then, clouds, set in strata, across the canyon, with intervening vista views, to cliffs and rocks beyond. The clouds are children of the heavens, and when they play among the rocks, they lift them to the region above.

It rains! Rapidly little rills are formed above, and these soon grow into brooks, and the brooks grow into creeks, and tumble over the walls in innumerable cascades, adding their wild music to the roar of the river. When the rain ceases, the rills, brooks, and creeks run dry. The waters that fall, during a rain, on these steep rocks, are gathered at once into the river; they could scarcely be poured in more suddenly, if some vast spout ran from the clouds to the stream itself. When a storm bursts over the canyon, a side gulch is dangerous, for a sudden flood may come, and the inpouring waters will raise the river, so as to hide the rocks before your eyes.

Early in the afternoon, we discover a stream, entering from the north, a clear, beautiful creek, coming down through a gorgeous red canyon. We land, and camp on a sand beach, above its mouth, under a great, overspreading tree, with willow-shaped leaves. . . .

August 24. The canyon is wider today. The walls rise to a vertical height of nearly three thousand feet. In many places the river runs under a cliff, in great curves, forming amphitheaters, half domeshaped.

Though the river is rapid, we meet with no serious obstructions, and run twenty miles. It is curious how anxious we are to make up our reckoning every time we stop, now that our diet is confined to plenty of coffee, very little spoiled flour, and very few dried apples. It has come to be a race for a dinner. Still, we make such fine progress, all hands are in good cheer, but not a moment of daylight is lost.

August 25. We make twelve miles this morning, when we come to monuments of lava, standing in the river; low rocks, mostly, but some of them shafts more than a hundred feet high. Going on down, three or four miles, we find them increasing in number. Great quantities of cooled lava and many cinder cones are seen on either side; and then we come to an abrupt cataract. Just over the fall, on the right wall, a cinder cone, or extinct volcano, with a well-defined crater, stands on the very brink of the canyon. This, doubtless, is the one we saw two or three days ago. From this volcano vast floods of lava have been poured down into the river, and a stream of the molten rock has run up the canyon, three or four miles, and down, we know not how far. Just where it poured over the canyon wall is the fall. The whole north side, as far as we can see, is lined with the black basalt, and high up on the opposite wall are patches of the same material, resting on the benches, and filling old alcoves and caves, giving to the wall a spotted appearance.

The rocks are broken in two, along a line which here crosses the river, and the beds, which we have seen coming down the canyon for the last thirty miles, have dropped 800 feet, on the lower side of the line, forming what geologists call a fault. The volcanic cone stands directly over the fissure thus formed. On the side of the river opposite, mammoth springs burst out of this crevice, one or two hundred feet above the river, pouring in a stream quite equal in volume to the Colorado Chiquito.

This stream seems to be loaded with carbonate of lime, and the water, evaporating, leaves an incrustation on the rocks; and this process has been continued for a long time, for extensive deposits are noticed, in which are basins, with bubbling springs. The water is salty.

We have to make a portage here, which is completed in about three hours, and on we go.

We have no difficulty as we float along, and I am able to observe the wonderful phenomena connected with this flood of lava. The canyon was doubtless filled to a height of twelve or fifteen hundred feet, perhaps by more than one flood. This would dam the water back; and in cutting through this great lava bed, a new channel has been formed, sometimes on one side, sometimes on the other. The cooled lava, being of firmer texture than the rocks of which the walls are composed, remains in some places; in others a narrow channel has been cut, leaving a line of basalt on either side. It is possible that the lava cooled faster on the sides against the walls, and that the centre ran out; but of this we can only conjecture. There are other places, where almost the whole of the lava is gone, patches of it only being seen where it has caught on the walls. As we float down, we can see that it ran out into side canyons. In some places this basalt has a fine, columnar structure, often in concentric prisms, and masses of these concentric columns have coalesced. In some places, when the flow occurred, the canyon was probably at about the same depth as it is now, for we can see where the basalt has rolled out on the sands, and, what seems curious to me, the sands are not melted or metamorphosed to any appreciable extent. In places the bed of the river is of sandstone or limestone, in other places of lava, showing that it has all been cut out again where the sandstones and limestones appear; but there is a little yet left where the bed is of lava.

What a conflict of water and fire there must have been here! Just imagine a river of molten rock, running down into a river of melted snow. What a seething and boiling of the waters; what clouds of steam rolled into the heavens!

Thirty-five miles today. Hurrah!

August 26. The canyon walls are steadily becoming higher as we advance. They are still bold, and nearly vertical up to the terrace. We still see evidence of the eruption discovered yesterday, but the thickness of the basalt is decreasing, as we go down the stream; yet it has been reinforced at points by streams that have come down from volcanoes standing on the terrace above, but which we cannot see from the river below.

Since we left the Colorado Chiquito, we have seen no evidences that the tribe of Indians inhabiting the plateaus on either side ever come down to the river; but about eleven o'clock today we discover an Indian garden, at the foot of the wall on the right, just where a little stream, with a narrow flood plain, comes down through a side canyon. Along the valley, the Indians have planted corn, using the water which burst out in springs at the foot of the cliff, for irrigation. The corn is looking quite well, but is not sufficiently advanced to give us roasting ears; but there are some nice, green squashes. We carry ten or a dozen of these on board our boats, and hurriedly leave, not willing to be caught in the robbery, yet excusing ourselves by pleading our great want. We run down a short distance, to where we feel certain no Indians can follow; and what a kettle of squash sauce we make! True, we have no salt with which to season it, but it makes a fine addition to our unleavened bread and coffee. Never was fruit so sweet as these stolen squashes.

After dinner we push on again, making fine time, finding many rapids, but none so bad that we cannot run them with safety, and when we stop, just at dusk, and foot up our reckoning, we find we have run thirty-five miles again.

What a supper we make! unleavened bread, green squash sauce, and strong coffee. We have been for a few days on half rations, but we have no stint of roast squash.

A few days like this, and we are out of prison.

August 27. This morning the river takes a more southerly direction. The dip of the rocks is to the north, and we are rapidly running into lower formations. Unless our course changes, we shall very soon run again into the granite. This gives us some anxiety. Now and then the river turns to the west, and excites hopes that are soon destroyed by another turn to the south. About nine o'clock we come to the dreaded rock. It is with no little misgiving that we see the river enter these black, hard walls. At its very entrance we have to make a portage; then we have to let down with lines past some ugly rocks. Then we run a mile or two farther, and then the rapids below can be seen.

About eleven o'clock we come to a place in the river where it seems much worse than any we have yet met in all its course.

A little creek comes down from the left. We land first on the right, and clamber up over the granite pinnacles for a mile or two, but can see no way by which we can let down, and to run it would be sure destruction. After dinner we cross to examine it on the left. High above the river we can walk along on the top of the granite, which is broken off at the edge, and set with crags and pinnacles, so that it is very difficult to get a view of the river at all. In my eagerness to reach a point where I can see the roaring fall below, I go too far on the wall, and can neither advance nor retreat. I stand with one foot on a little projecting rock, and cling with my hand fixed in a little crevice. Finding I am caught here, suspended 400 feet above the river, into which I should fall if my footing fails, I call for help. The men come, and pass me a line, but I cannot let go of the rock long enough to take hold of it. Then they bring two or three of the largest oars. All this takes time which seems very precious to me; but at last they arrive. The blade of one of the oars is pushed into a little crevice in the rock beyond me, in such a manner that they can hold me pressed against the wall. Then another is fixed in such a way that I can step on it, and thus I am extricated.

Still another hour is spent in examining the river from this side, but no good view of it is obtained, so now we return to the side that was first examined, and the afternoon is spent in clambering among the crags and pinnacles, and carefully scanning the river again. We find that the lateral streams have washed boulders into the river, so as to form a dam, over which the water makes a broken fall of eighteen or twenty feet; then there is a rapid, beset with rocks, for two or three hundred yards, while, on the other side, points of the wall project into the river. Then there is a second fall below; how great, we cannot tell. Then there is a rapid, filled with huge rocks, for one or two hundred yards. At the bottom of it, from the right wall, a great rock projects quite halfway across the river. It has a sloping surface extending upstream, and the water, coming down with all the momentum gained in the falls and rapids above, rolls up this inclined plane many feet, and tumbles over to the left. I decide that it is possible to let down over the first fall, then run near the right cliff to a point just above the second, where we can pull out into a little chute, and, having run over that in safety, we must pull with all our power across the stream, to avoid the great rock below. On my return to the boat, I announce to the men that we are to run it in the morning. Then we cross the river, and go into camp for the night on some rocks, in the mouth of the little side canyon.

After supper Captain Howland asks to have a talk with me. We walk up the little creek a short distance, and I soon find that his object is to remonstrate against my determination to proceed. He thinks that we had better abandon the river here. Talking with him, I learn that his brother, William Dunn, and himself have determined to go no farther in the boats. So we return to camp. Nothing is said to the other men.

For the last two days, our course has not been plotted. I sit down and do this now, for the purpose of finding where we are by dead reckoning. It is a clear night, and I take out the sextant to make observation for latitude, and find that the astronomic determination agrees very nearly with that of the plot—quite as closely as might be expected, from a meridian observation on a planet. In a direct line, we must be about forty-five miles from the mouth of the Rio Virgen. If we can reach that point, we know that there are settlements up that river about twenty miles. This forty-five miles, in a direct line, will probably be eighty or ninety in the meandering line of the river. But then we know that there is comparatively open country for many miles above the mouth of the Virgen, which is our point of destination.

As soon as I determine all this, I spread my plot on the

sand, and wake Howland, who is sleeping down by the river, and show him where I suppose we are, and where several Mormon settlements are situated.

We have another short talk about the morrow, and he lies down again; but for me there is no sleep. All night long, I pace up and down a little path, on a few yards of sand beach, along by the river. Is it wise to go on? I go to the boats again, to look at our rations. I feel satisfied that we can get over the danger immediately before us; what there may be below I know not. From our outlook yesterday, on the cliffs, the canyon seemed to make another great bend to the south, and this, from our experience heretofore, means more and higher granite walls. I am not sure that we can climb out of the canyon here, and, when at the top of the wall, I know enough of the country to be certain that it is a desert of rock and sand, between this and the nearest Mormon town, which, on the most direct line, must be seventy-five miles away. True, the late rains have been favorable to us, should we go out, for the probabilities are that we shall find water still standing in holes, and, at one time, I almost conclude to leave the river. But for years I have been contemplating this trip. To leave the exploration unfinished, to say that there is a part of the canyon which I cannot explore, having already almost accomplished it, is more than I am willing to acknowledge, and I determine to go on.

I wake my brother, and tell him of Howland's determination, and he promises to stay with me; then I call up Hawkins, the cook, and he makes a like promise; then Sumner, and Bradley, and Hall, and they all agree to go on.

August 28. At last daylight comes, and we have breakfast, without a word being said about the future. The meal is as solemn as a funeral. After breakfast, I ask the three men if they still think it best to leave us. The elder Howland thinks it is, and

Dunn agrees with him. The younger Howland tries to persuade them to go on with the party, failing in which, he decides to go with his brother.

Then we cross the river. The small boat is very much disabled, and unseaworthy. With the loss of hands, consequent on the departure of the three men, we shall not be able to run all of the boats, so I decide to leave my "Emma Dean."

Two rifles and a shotgun are given to the men who are going out. I ask them to help themselves to the rations, and take what they think to be a fair share. This they refuse to do, saying they have no fear but that they can get something to eat; but Billy, the cook, has a pan of biscuits prepared for dinner, and these he leaves on a rock.

Before starting, we take our barometers, fossils, the minerals, and some ammunition from the boat, and leave them on the rocks. We are going over this place as light as possible. The three men help us lift our boats over a rock twenty-five or thirty feet high, and let them down again over the first fall, and now we are all ready to start. The last thing before leaving, I write a letter to my wife, and give it to Howland. Sumner gives him his watch, directing that it be sent to his sister, should he not be heard from again. The records of the expedition have been kept in duplicate. One set of these is given to Howland, and now we are ready. For the last time, they entreat us not to go on, and tell us that it is madness to set out in this place; that we can never get safely through it; and, further, that the river turns again to the south into the granite, and a few miles of such rapids and falls will exhaust our entire stock of rations, and then it will be too late to climb out. Some tears are shed; it is rather a solemn parting; each party thinks the other is taking the dangerous course.

My old boat left, I go on board of the "Maid of the Canon." The three men climb a crag, that overhangs the river, to watch us off. The "Maid of the Canon" pushes out. We glide rapidly along the foot of the wall, just grazing one great rock, then pull out a little into the chute of the second fall, and plunge over it. The open compartment is filled when we strike the first wave below, but we cut through it, and then the men pull with all their power toward the left wall, and swing clear of the dangerous rock below all right. We are scarcely a minute in running it, and find that, although it looked bad from above, we have passed many places that were worse.

The other boat follows without more difficulty. We land at the first practicable point below and fire our guns, as a signal to the men above that we have come over in safety. Here we remain a couple of hours, hoping that they will take the small boat and follow us. We are behind a curve in the canyon, and cannot see up to where we left them, and so we wait until their coming seems hopeless, and push on.

And now we have a succession of rapids and falls until noon, all of which we run in safety. Just after dinner we come to another bad place. A little stream comes in from the left, and below there is a fall, and still below another fall. Above, the river tumbles down, over and among the rocks, in whirlpools and great waves, and the waters are lashed into mad, white foam. We run along the left, above this, and soon see that we cannot get down on this side, but it seems possible to let down on the other. We pull upstream again, for two or three hundred yards, and cross. Now there is a bed of basalt on this northern side of the canyon, with a bold escarpment, that seems to be a hundred feet high. We can climb it, and walk along its summit to a point where we are just at the head of the fall. Here the basalt is broken down again, so it seems to us, and I direct the men to take a line to the top of the cliff, and let the boats down along the wall. One man remains in the boat, to keep her clear of the rocks, and prevent her line from being caught on the projecting angles. I climb the cliff, and pass along to a point just over the fall, and descend by broken rocks, and find that the break of the fall is above the break of the wall, so that we cannot land; and that still below the river is very bad, and that there is no possibility of a portage. Without waiting further to examine and determine what shall be done, I hasten back to the top of the cliff, to stop the boats from coming down. When I arrive, I find the men have let one of them down to the head of the fall. She is in swift water, and they are not able to pull her back; nor are they able to go on with the line, as it is not long enough to reach the higher part of the cliff, which is just before them; so they take a bight around a crag. I send two men back for the other line. The boat is in very swift water, and Bradley is standing in the open compartment, holding out his oar to prevent her from striking against the foot of the cliff. Now she shoots out into the stream, and up as far as the line will permit, and then, wheeling, drives headlong against the rock, then out and back again, now straining on the line, now striking against the rock. As soon as the second line is brought, we pass it down to him, but his attention is all taken up with his own situation, and he does not see that we are passing the line to him. I stand on a projecting rock, waving my hat to gain his attention, for my voice is drowned by the roaring of the falls. Just at this moment, I see him take his knife from its sheath, and step forward to cut the line. He has evidently decided that it is better to go over with the boat as it is, than to wait for her to be broken to pieces. As he leans over, the boat sheers again into the stream, the stem-post breaks away, and she is loose. With perfect composure Bradley seizes the great scull oar, places it in the stern rowlock, and pulls with all his power (and he is an athlete) to turn the bow of the boat downstream, for he wishes to go bow down, rather than to drift broadside on. One, two strokes he makes, and a third just as she goes over, and the boat is fairly turned, and she goes down almost beyond our sight, though we are more than a hundred feet above the river. Then she comes up again, on a great wave, and down and up, then around behind some great rocks, and is lost in the mad, white foam below. We stand frozen with fear, for we see no boat. Bradley is gone, so it seems. But now, away below, we see something coming out of the waves. It is evidently a boat. A moment more, and we see Bradley standing on deck, swinging his hat to show that he is all right. But he is in a whirlpool. We have the stem-post of his boat attached to the line. How badly she may be disabled we know not. I direct Sumner and Powell to pass along the cliff, and see if they can reach him from below. Rhodes, Hall, and myself run to the other boat, jump aboard, and put out, and away we go over the falls. A wave rolls over us, and our boat is unmanageable. Another great wave strikes us, the boat rolls over, and tumbles and tosses, I know not how. All I know is that Bradley is picking us up. We soon have all right again, and row to the cliff, and wait until Sumner and Powell can come. After a difficult climb they reach us. We run two or three miles farther, and turn again to the northwest, continuing until night, when we have run out of the granite once more.

August 29. We start very early this morning. The river still continues swift, but we have no serious difficulty, and at twelve o'clock emerge from the Grand Canyon of the Colorado.

We are in a valley now, and low mountains are seen in the distance, coming to the river below. We recognize this as the Grand Wash.

A few years ago, a party of Mormons set out from St. George, Utah, taking with them a boat, and came down to the mouth of the Grand Wash, where they divided, a portion of the party crossing the river to explore the San Francisco Mountains. Three men—Hamblin, Miller, and Crosby—taking the boat, went on down the river to Callville, landing a few miles below the mouth of the Rio Virgen. We have their manuscript journal with us, and so the stream is comparatively well known.

Tonight we camp on the left bank, in a mesquite thicket.

The relief from danger, and the joy of success, are great. When he who has been chained by wounds to a hospital cot, until his canvas tent seems like a dungeon cell, until the groans of those who lie about, tortured with probe and knife, are piled up, a weight of horror on his ears that he cannot throw off, cannot forget, and until the stench of festering wounds and anaesthetic drugs has filled the air with its loathsome burthen, at last goes out into the open field, what a world he sees! How beautiful the sky; how bright the sunshine; what "floods of delirious music" pour from the throats of birds; how sweet the fragrance of earth, and tree, and blossom! The first hour of convalescent freedom seems rich recompense for all—pain, gloom, terror.

Something like this are the feelings we experience tonight. Ever before us has been an unknown danger, heavier than immediate peril. Every waking hour passed in the Grand Canyon has been one of toil. We have watched with deep solicitude the steady disappearance of our scant supply of rations, and from time to time have seen the river snatch a portion of the little left, while we were ahungered. And danger and toil were endured in those gloomy depths, where ofttimes the clouds hid the sky by day, and but a narrow zone of stars could be seen at night. Only during the few hours of deep sleep, consequent on hard labor, has the roar of the waters been hushed. Now the danger is over; now the toil has ceased; now the gloom has disappeared; now the firmament is bounded only by the horizon; and what a vast expanse of constellations can be seen!

The river rolls by us in silent majesty; the quiet of the camp is sweet; our joy is almost ecstasy. We sit till long after midnight, talking of the Grand Canyon, talking of home.

George Catlin

BUFFALO COUNTRY

The buffalo calf, during the first six months is red, and has so much the appearance of a red calf in cultivated fields, that it could easily be mingled and mistaken amongst them. In the fall, when it changes its hair it takes a brown coat for the winter, which it always retains. In pursuing a large herd of buffaloes at the season when their calves are but a few weeks old, I have often been exceedingly amused with the curious manoeuvres of these shy little things. Amidst the thundering confusion of a throng of several hundreds or several thousands of these animals, there will be many of the calves that lose sight of their dams; and being left behind by the throng, and the swift passing hunters, they endeavour to secrete themselves, when they are exceedingly put to it on a level prairie, where nought can be seen but the short grass of six or eight inches in height, save an occasional bunch of wild sage, a few inches higher, to which the poor affrighted things will run, and dropping on their knees, will push their noses under it, and into the grass, where they will stand for hours, with their eyes shut, imagining themselves securely hid, whilst they are standing up quite straight upon their hind feet and can easily be seen at several miles distance. It is a familiar amusement for us accustomed to these scenes, to retreat back over the ground where we have just escorted the herd, and approach these little trembling things, which stubbornly maintain their positions, with their noses pushed under the grass, and their eyes strained upon us, as we dismount from our horses and are passing around them. From this fixed position they are sure not to move, until hands are laid upon them, and then for the skins of a novice, we can extend our sympathy; or if he can preserve the skin on his bones from the furious buttings of its head, we know how to congratulate him on his signal success and good luck. In these desperate struggles, for a moment, the little thing is conquered, and makes no further resistance. And I have often, in concurrence with a known custom of the country, held my hands over the eyes of the calf, and breathed a few strong breaths into its nostrils; after which I have, with my hunting companions, rode several miles into our encampment, with the little prisoner busily following the heels of my horse the whole way; as closely and as affectionately as its instinct would attach it to the company of its dam!

This is one of the most extraordinary things that I have met with in the habits of this wild country, and although I had often heard of it, and felt unable exactly to believe it, I am now willing to bear testimony to the fact, from the numerous instances which I have witnessed since I came into the country.

During the time that I resided at this post, in the spring of the year, on my way up the river, I assisted (in numerous hunts of the buffalo, with the Fur Company's men,) in bringing in, in the above manner, several of these little prisoners, which sometimes followed for five or six miles close to our horses' heels, and even into the Fur Company's Fort, and into the stables where our horses were led. In this way, before I left for the head waters of the Missouri, I think we had collected about a dozen, which Mr. Laidlaw was successfully raising with the aid of a good milch cow, and which were to be committed to the care of Mr. Chouteau to be transported by the return of the steamer, to his extensive plantation in the vicinity of St. Louis.*

It is truly a melancholy contemplation for the traveller in this country, to anticipate the period which is not far distant, when the last of these noble animals, at the hands of white and red men, will fall victims to their cruel and improvident rapacity; leaving these beautiful green fields, a vast and idle waste, unstocked and unpeopled for ages to come, until the bones of the one and the traditions of the other will have vanished, and left scarce an intelligible trace behind.

That the reader should not think me visionary in these contemplations, or romancing in making such assertions, I will hand him the following item of the extravagancies which are practiced in these regions, and rapidly leading to the results which I have just named.

When I first arrived at this place, on my way up the river, which was in the month of May, in 1832, and had taken up my lodgings in the Fur Company's Fort, Mr. Laidlaw, of whom I have before spoken, and also his chief clerk, Mr. Halsey, and

* The fate of these poor little prisoners, I was informed on my return to St. Louis a year afterwards, was a very disastrous one. The steamer having a distance of 1600 miles to perform, and lying a week or two on sand bars, in a country where milk could not be procured, they all perished but one, which is now flourishing in the extensive fields of this gentleman.

95

many of their men, as well as the chiefs of the Sioux, told me, that only a few days before I arrived, (when an immense herd of buffaloes had showed themselves on the opposite side of the river, almost blackening the plains for a great distance,) a party of five or six hundred Sioux Indians on horseback, forded the river about mid-day, and spending a few hours amongst them, recrossed the river at sun-down and came into the Fort with *fourteen hundred fresh buffalo tongues*, which were thrown down in a mass, and for which they required but a few gallons of whiskey, which was soon demolished, indulging them in a little, and harmless carouse.

This profligate waste of the lives of these noble and useful animals, when, from all that I could learn, not a skin or a pound of the meat (except the tongues), was brought in, fully supports me in the seemingly extravagant predictions that I have made as to their extinction, which I am certain is near at hand. In the above extravagant instance, at a season when their skins were without fur and not worth taking off, and their camp was so well stocked with fresh and dried meat, that they had no occasion for using the flesh, there is a fair exhibition of the improvident character of the savage, and also of his recklessness in catering for his appetite, so long as the present inducements are held out to him in his country, for its gratification.

In this singular country, where the poor Indians have no laws or regulations of society, making it a vice or an impropriety to drink to excess, they think it no harm to indulge in the delicious beverage, as long as they are able to buy whiskey to drink. They look to white men as wiser than themselves, and able to set them examples—they see none of these in their country but sellers of whiskey, who are constantly tendering it to them, and most of them setting the example by using it themselves; and they easily acquire a taste, that to be catered for, where whiskey is sold at sixteen dollars per gallon, soon impoverishes them, and must soon strip the skin from the last buffalo's back that lives in their country, to "be dressed by their squaws" and vended to the Traders for a pint of diluted alcohol. From the above remarks it will be seen, that not only the red men, but red men and white, have aimed destruction at the race of these animals; and with them, *beasts* have turned hunters of buffaloes in this country, slaying them, however, in less numbers, and for far more laudable purpose than that of selling their skins. The white wolves, of which I have spoken in a former epistle, follow the herds of buffaloes as I have said, from one season to another, glutting themselves on the carcasses of those that fall by the deadly shafts of their enemies, or linger with disease or old age to be dispatched by these sneaking cormorants, who are ready at all times kindly to relieve them from the pangs of a lingering death.

Whilst the herd is together, the wolves never attack them, as they instantly gather for combined resistance, which they effectually make. But when the herds are travelling, it often happens that an aged or wounded one, lingers at a distance behind, and when fairly out of sight of the herd, is set upon by these voracious hunters, which often gather to the number of fifty or more, and are sure at last to torture him to death, and use him up at a meal. The buffalo, however, is a huge and furious animal, and when his retreat is cut off, makes desperate and deadly resistance, contending to the last moment for the right of life—and oftentimes deals death by wholesale, to his canine assailants, which he is tossing into the air or stamping to death under his feet.

During my travels in these regions, I have several times come across such a gang of these animals surrounding an old or a wounded bull, where it would seem, from appearances, that they had been for several days in attendance, and at intervals desperately engaged in the effort to take his life. But a short time since, as one of my hunting companions and myself were returning to our encampment with our horses loaded with meat, we discovered at a distance, a huge bull, encircled with a gang of white wolves; we rode up as near as we could without driving them away, and being within pistol shot, we had a remarkably good view, where I sat for a few moments and made a sketch in my note-book; after which, we rode up and gave the signal for them to disperse, which they instantly did, withdrawing themselves to the distance of fifty or sixty rods, when we found, to our great surprise, that the animal had made desperate resistance, until his eyes were entirely eaten out of his head—the grizzle of his nose was mostly gone—his tongue was half eaten off, and the skin and flesh of his legs torn almost literally into strings. In this tattered and torn condition, the poor old veteran stood bracing up in the midst of his devourers, who had ceased hostilities for a few minutes, to enjoy a sort of parley, recovering strength and preparing to resume the attack in a few moments again. In this group, some were reclining, to gain breath, whilst others were sneaking about and licking their chaps in anxiety for a renewal of the attack; and others, less lucky, had been crushed to death by the feet or the horns of the bull. I rode nearer to the pitiable object as he stood bleeding and trembling before me, and said to him, "Now is your time, old fellow, and you had better be off." Though blind and nearly destroyed, there seemed evidently to be a recognition of a friend in me, as he straightened up, and, trembling with excitement, dashed off at full speed upon the prairie, in a straight line. We turned our horses and resumed our march, and when we had advanced a mile or more, we looked back, and on our left, where we saw again the ill-fated animal surrounded by his tormentors, to whose insatiable voracity he unquestionably soon fell a victim.

Thus much I wrote of the buffaloes, and of the accidents that befall them, as well as of the fate that awaits them; and before I closed my book, I strolled out one day to the shade of a plum-tree, where I laid in the grass on a favourite bluff, and wrote thus:—

"It is generally supposed, and familiarly said, that a man *'falls'* into a rêverie; but I seated myself in the shade a few minutes since, resolved to *force* myself into one; and for this purpose I laid open a small pocket-map of North America, and excluding my thoughts from every other object in the world, I soon succeeded in producing the desired illusion. This little chart, over which I bent, was seen in all its parts, as nothing but the green and vivid reality. I was lifted up upon an imaginary pair of wings, which easily raised and held me floating in the open air, from whence I could behold beneath me the Pacific and the Atlantic Oceans—the great cities of the East, and the mighty rivers. I could see the blue chain of the great lakes at the North—the Rocky Mountains, and beneath them and near their base, the vast, and almost boundless plains of grass, which were speckled with the bands of grazing buffaloes!

"The world turned gently around, and I examined its surface; continent after continent passed under my eye, and yet amidst them all, I saw not the vast and vivid green, that is spread like a carpet over the Western wilds of my own country. I saw not elsewhere in the world, the myriad herds of buffaloes—my eyes scanned in vain, for they were not. And when I turned again to the wilds of my native land, I beheld them all in motion! For the distance of several hundreds of miles from North to South, they were wheeling about in vast columns and herds—some were scattered, and ran with furious wildness—some lay dead, and others were pawing the earth for a hiding-place—some were sinking down and dying, gushing out their life's blood in deep-drawn sighs—and others were contending in furious battle for the life they possessed, and the ground that they stood upon. They had long since assembled from the thickets, and secret haunts of the deep forest, into the midst of the treeless and bushless plains, as the place for their safety. I could see in an hundred places, amid the wheeling bands, and on their skirts and flanks, the leaping wild horse darting among them. I saw not the arrows, nor heard the twang of the sinewy bows that sent them; but I saw their victims fall!—on other steeds that rushed along their sides, I saw the glistening lances, which seemed to lay across them; their blades were blazing in the sun, till dipped in blood, and then I lost them! In other parts (and there were many), the vivid flash of *fire-arms* was seen—*their* victims fell too, and over their dead bodies hung suspended in air, little clouds of whitened smoke, from under which the flying horsemen had darted forward to mingle again with, and deal death to, the trampling throng.

"So strange were men mixed (both red and white) with the countless herds that wheeled and eddyed about, that all below seemed one vast extended field of battle—whole armies, in some places, seemed to blacken the earth's surface;—in other parts, regiments, battalions, wings, platoons, rank and file, and '*Indian-file*'—all were in motion; and death and destruction seemed to be the watch-word amongst them. In their turmoil, they sent up great clouds of dust, and with them came the mingled din of groans and trampling hoofs, that seemed like the rumbling of a dreadful cataract, or the roaring of distant thunder. Alternate pity and admiration harrowed up in my bosom and my brain, many a hidden thought; and amongst them a few of the beautiful notes that were once sung, and exactly in point: '*Quadrupedante putrem sonitu quatit ungula campum.*' Even such was the din amidst the quadrupeds of these vast plains. And from the craggy cliffs of the Rocky Mountains also were seen descending into the valley, the myriad Tartars, who had not horses to ride, but before their well-drawn bows the fattest of the herds were falling. Hundreds and thousands were strewed upon the plains—they were flayed, and their reddened carcasses left; and about them bands of wolves, and dogs, and buzzards were seen devouring them. Contiguous, and in sight, were the distant and feeble smokes of wigwams and villages, where the skins were dragged, and dressed for white man's luxury! where they were all sold for *whiskey*, and the poor Indians laid drunk, and were crying. I cast my eyes into the towns and cities of the East, and there I beheld buffalo robes hanging at almost every door for traffic; and I saw also the curling smokes of a thousand *Stills*—and I said, 'Oh insatiable man, is thy avarice such! wouldst thou tear the skin from the back of the last animal of this noble race, *and rob thy fellow-man of his meat, and for it give him poison!*' "

Many are the rudenesses and wilds in Nature's works, which are destined to fall before the deadly axe and desolating hands of cultivating man; and so amongst her ranks of *living*, of beast and human, we often find noble stamps, or beautiful colours, to which our admiration clings; and even in the overwhelming march of civilized improvements and refinements do we love to cherish their existence, and lend our efforts to preserve them in their primitive rudeness. Such of Nature's works are always worthy of our preservation and protection; and the further we become separated (and the face of the country) from that pristine wildness and beauty, the more pleasure does the mind of enlightened man feel in recurring to those scenes, when he can have them preserved for his eyes and his mind to dwell upon.

Of such "rudenesses and wilds," Nature has no where presented more beautiful and lovely scenes, than those of the vast prairies of the West; and of *man* and *beast*, no nobler specimens than those who inhabit them—the *Indian* and the *buffalo*—joint and original tenants of the soil, and fugitives together from the approach of civilized man; they have fled to the great plains of the West, and there, under an equal doom, they have taken up their *last abode*, where their race will expire, and their bones will bleach together.

It may be that *power* is *right*, and *voracity* a *virtue*; and that these people, and these noble animals, are *righteously* doomed to an issue that *will* not be averted. It can be easily proved—we have a civilized science that can easily do it, or anything else that may be required to cover the iniquities of civilized man in catering for his unholy appetites. It can be proved that the weak and ignorant have no *rights*— that there can be no virtue in darkness—that God's gifts have no meaning or merit until they are appropriated by civilized man—by him brought into the light, and converted to his use and luxury. We have a mode of reasoning (I forget what it is called) by which all this can be proved, and even more. The *word* and the *system* are entirely of *civilized* origin; and latitude is admirably given to them in proportion to the increase of civilized wants, which often require a *judge* to overrule the laws of nature. I say that *we* can prove such things; but an *Indian* cannot. It is a mode of reasoning unknown to him in his nature's simplicity, but admirably adapted to subserve the interests of the enlightened world, who are always their own judges, when dealing with the savage; and who, in the present refined age, have many appetites that can only be lawfully indulged, by proving God's laws defective.

It is not enough in this polished and extravagant age, that we get from the Indian his lands, and the very clothes from his back, but the food from their mouths must be stopped, to add a new and useless article to the fashionable world's luxuries. The ranks must be thinned, and the race exterminated, of this noble animal, and the Indians of the great plains left without the means of supporting life, that white men may figure a few years longer, enveloped in buffalo robes—that they may spread them, for their pleasure and elegance, over the backs of their sleighs, and trail them ostentatiously amidst the busy throng, as things of beauty and elegance that had been made for them!

Reader! listen to the following calculations, and forget

them not. The buffaloes (the quadrupeds from whose backs your beautiful robes were taken, and whose myriads were once spread over the whole country, from the Rocky Mountains to the Atlantic Ocean) have recently fled before the appalling appearance of civilized man, and taken up their abode and pasturage amid the almost boundless prairies of the West. An instinctive dread of their deadly foes, who made an easy prey of them whilst grazing in the forest, has led them to seek the midst of the vast and treeless plains of grass, as the spot where they would be least exposed to the assaults of their enemies; and it is exclusively in those desolate fields of silence (yet of beauty) that they are to be found—and over these vast steppes, or prairies, have they fled, like the Indian, towards the "setting sun"; until their bands have been crowded together, and their limits confined to a narrow strip of country on this side of the Rocky Mountains.

This strip of country, which extends from the province of Mexico to lake Winnipeg on the North, is almost one entire plain of grass, which is, and ever must be, useless to cultivating man. It is here, and here chiefly, that the buffaloes dwell; and with, and hovering about them, live and flourish the tribes of Indians, whom God made for the enjoyment of that fair land and its luxuries.

It is a melancholy contemplation for one who has travelled as I have, through these realms, and seen this noble animal in all its pride and glory, to contemplate it so rapidly wasting from the world, drawing the irresistible conclusion too, which one must do, that its species is soon to be extinguished, and with it the peace and happiness (if not the actual existence) of the tribes of Indians who are joint tenants with them, in the occupancy of these vast and idle plains.

And what a splendid contemplation too, when one (who has travelled these realms, and can duly appreciate them) imagines them as they *might* in future be seen, (by some great protecting policy of government) preserved in their pristine beauty and wildness, in a *magnificent park*, where the world could see for ages to come, the native Indian in his classic attire, galloping his wild horse, with sinewy bow, and shield and lance, amid the fleeting herds of elks and buffaloes. What a beautiful and thrilling specimen for America to preserve and hold up to the view of her refined citizens and the world, in future ages! A *nation's Park*, containing man and beast, in all the wild and freshness of their nature's beauty!

I would ask no other monument to my memory, nor any other enrolment of my name amongst the famous dead, than the reputation of having been the founder of such an institution.

Such scenes might easily have been preserved, and still could be cherished on the great plains of the West, without detriment to the country or its borders; for the tracts of country on which the buffaloes have assembled, are uniformly sterile, and of no available use to cultivating man.

It is on these plains, which are stocked with buffaloes, that the finest specimens of the Indian race are to be seen. It is here, that the savage is decorated in the richest costume. It is here, and here only, that his wants are all satisfied, and even the *luxuries* of life are afforded him in abundance. And here also is he the proud and honourable man (before he has had teachers or laws), above the imported wants, which beget meanness and vice; stimulated by ideas of honour and virtue, in which the God of Nature has certainly not curtailed him.

There are, by a fair calculation, more than 300,000 Indians, who are now subsisted on the flesh of the buffaloes, and by those animals supplied with all the luxuries of life which they desire, as they know of none others. The great variety of uses to which they convert the body and other parts of that animal, are almost incredible to the person who has not actually dwelt amongst these people, and closely studied their modes and customs. Every part of their flesh is converted into food, in one shape or another, and on it they entirely subsist. The robes of the animals are worn by the Indians instead of blankets—their skins when tanned, are used as coverings for their lodges, and for their beds; undressed, they are used for constructing canoes—for saddles, for bridles—l'arrêts, lassos, and thongs. The horns are shaped into ladles and spoons—the brains are used for dressing the skins—their bones are used for saddle trees—for war clubs, and scrapers for graining the robes—and others are broken up for the marrow-fat which is contained in them. Their sinews are used for strings and backs to their bows—for thread to string their beads and sew their dresses. The feet of the animals are boiled, with their hoofs, for the glue they contain, for fastening their arrow points, and many other uses. The hair from the head and shoulders, which is long, is twisted and braided into halters, and the tail is used for a fly brush. In this wise do these people convert and use the various parts of this useful animal, and with all these luxuries of life about them, and their numerous games, they are happy (God bless them) in the ignorance of the disastrous fate that awaits them.

Yet this interesting community, with its sports, its wildnesses, its languages, and all its manners and customs, could be perpetuated, and also the buffaloes, whose numbers would increase and supply them with food for ages and centuries to come, if a system of non-intercourse could be established and preserved. But such is not to be the case—the buffalo's doom is sealed, and with their extinction must assuredly sink into real despair and starvation, the inhabitants of these vast plains, which afford for the Indians, no other possible means of subsistence; and they must at last fall a prey to wolves and buzzards, who will have no other bones to pick.

It seems hard and cruel, (does it not?) that we civilized people with all the luxuries and comforts of the world about us, should be drawing from the backs of these useful animals the skins for our luxury, leaving their carcasses to be devoured by the wolves—that we should draw from that country, some 150 or 200,000 of their robes annually, the greater part of which are taken from animals that are killed expressly for the robe, at a season when the meat is not cured and preserved, and for each of which skins the Indian has received but a pint of whiskey!

Such is the fact, and that number or near it, are annually destroyed, in addition to the number that is necessarily killed for the subsistence of 300,000 Indians, who live entirely upon them. It may be said, perhaps, that the Fur Trade of these great western realms, which is now limited chiefly to the purchase of buffalo robes, is of great and national importance, and should and must be encouraged. To such a suggestion I would reply, by merely enquiring, (independently of the poor Indians' disasters,) how much more advantageously would such a capital be employed, both for the weal of the country and for the owners, if it were invested in machines for the manufacture of *woollen robes*, of equal and superior value and beauty; thereby encouraging the growers of wool, and the industrious manufacturer, rather than cultivating a taste for the use of buffalo skins; which is just to be acquired, and then, from necessity, to be dispensed with, when a few years shall have destroyed the last of the animals producing them.

It may be answered, perhaps, that the necessaries of life are given in exchange for these robes; but what, I would ask, are the necessities in Indian life, where they have buffaloes in abundance to live on? The Indian's necessities are entirely artificial—are all created; and when the buffaloes shall have disappeared in his country, which will be within *eight* or *ten* years, I would ask, who is to supply him with the necessaries of life then? and I would ask, further, (and leave the question to be answered ten years hence), when the skin shall have been stripped from the back of the last animal, who is to resist the ravages of 300,000 starving savages; and in their trains, 1,500,000 wolves, whom direst necessity will have driven from their desolate and gameless plains, to seek for the means of subsistence along our exposed frontier? God has everywhere supplied man in a state of Nature, with the necessaries of life, and before we destroy the game of his country, or teach him new desires, he has no wants that are not satisfied.

Amongst the tribes who have been impoverished and repeatedly removed, the necessaries of life are extended with a better grace from the hands of civilized man; 90,000 of such have already been removed, and they draw from Government some 5 or 600,000 dollars annually in cash; *which money passes immediately into the hands of white men*, and for it the necessaries of life *may be* abundantly furnished. But who, I would ask, are to furnish the Indians who have been instructed in this unnatural mode—living upon *such* necessaries, and even luxuries of life, extended to them by the hands of white men, when those annuities are at an end, and the skin is stripped from the last of the animals which God gave them for their subsistence?

Reader, I will stop here, lest you might forget to answer these important queries—these are questions which I know will puzzle the world—and, perhaps it is not right that I should ask them.

Mark Twain

ROUGHING IT

NEW ACQUAINTANCES—THE CAYOTE—A DOG'S
EXPERIENCES—A DISGUSTED DOG—THE RELATIVES OF THE
CAYOTE—MEALS TAKEN AWAY FROM HOME

Another night of alternate tranquillity and turmoil. But morning came, by and by. It was another glad awakening to fresh
breezes, vast expanses of level greensward, bright sunlight, an
impressive solitude utterly without visible human beings or
human habitations, and an atmosphere of such amazing mag-
nifying properties that trees that seemed close at hand were
more than three miles away. We resumed undress uniform,

climbed a-top of the flying coach, dangled our legs over the side, shouted occasionally at our frantic mules, merely to see them lay their ears back and scamper faster, tied our hats on to keep our hair from blowing away, and leveled an outlook over the world-wide carpet about us for things new and strange to gaze at. Even at this day it thrills me through and through to think of the life, the gladness and the wild sense of freedom that used to make the blood dance in my veins on those fine overland mornings!

Along about an hour after breakfast we saw the first prairie-dog villages, the first antelope, and the first wolf. If I remember rightly, this latter was the regular *cayote* (pronounced ky-o-te) of the farther deserts. And if it *was*, he was not a pretty creature or respectable either, for I got well acquainted with his race afterward, and can speak with confidence. The cayote is a long, slim, sick and sorry-looking skeleton, with a gray wolf-skin stretched over it, a tolerably bushy tail that forever sags down with a despairing expression of forsakenness and misery, a furtive and evil eye, and a long, sharp face, with slightly lifted lip and exposed teeth. He has a general slinking expression all over. The cayote is a living, breathing allegory of Want. He is *always* hungry. He is always poor, out of luck and friendless. The meanest creatures despise him, and even the fleas would desert him for a velocipede. He is so spiritless and cowardly that even while his exposed teeth are pretending a threat, the rest of his face is apologizing for it. And he is *so* homely!—so scrawny, and ribby, and coarse-haired, and pitiful. When he sees you he lifts his lip and lets a flash of his teeth out, and then turns a little out of the course he was pursuing, depresses his head a bit, and strikes a long, soft-footed trot through the sage-brush, glancing over his shoulder at you, from time to time, till he is about out of easy pistol range, and then he stops and takes a deliberate survey of you; he will trot fifty yards and stop again—another fifty and stop again; and finally the gray of his gliding body blends with the gray of the sage-brush, and he disappears. All this is when you make no demonstration against him; but if you do, he develops a livelier interest in his journey, and instantly electrifies his heels and puts such a deal of real estate between himself and your weapon, that by the time you have raised the hammer you see that you need a mini rifle, and by the time you have got him in line you need a rifled cannon, and by the time you have "drawn a bead" on him you see well enough that nothing but an unusually long-winded streak of lightning could reach him where he is now. But if you start a swift-footed dog after him, you will enjoy it ever so much—especially if it is a dog that has a good opinion of himself, and has been brought up to think he knows something about speed. The cayote will go swinging gently off on that deceitful trot of his, and every little while he will smile a fraudful smile over his shoulder that will fill that dog entirely full of encouragement and worldly ambition, and make him lay his head still lower to the ground, and stretch his neck further to the front, and pant more fiercely, and stick his tail out straighter behind, and move his furious legs with a yet wilder frenzy, and leave a broader and broader, and higher and denser cloud of desert sand smoking behind, and marking his long wake across the level plain! And all this time the dog is only a short twenty feet behind the cayote, and to save the soul of him he cannot understand why it is that he cannot get perceptibly closer; and he begins to get aggravated, and it makes him madder and madder to see how gently the cayote glides along and never pants or sweats or ceases to smile; and he grows still more and more incensed to see how shamefully he has been taken in by an entire stranger, and what an ignoble swindle that long, calm, soft-footed trot is; and next he notices that he is getting fagged, and that the cayote actually has to slacken speed a little to keep from running away from him—and *then* that town-dog is mad in earnest, and he begins to strain and weep and swear, and paw the sand higher than ever, and reach for the cayote with concentrated and desperate energy. This "spurt" finds him six feet behind the gliding enemy, and two miles from his friends. And then, in the instant that a wild new hope is lighting up his face, the cayote turns and smiles blandly upon him once more, and with a something about it which seems to say: "Well, I shall have to tear myself away from you, bub—business is business, and it will not do for me to be fooling along this way all day"—and forthwith there is a rushing sound, and the sudden splitting of a long crack through the atmosphere, and behold that dog is solitary and alone in the midst of a vast solitude!

It makes his head swim. He stops, and looks all around; climbs the nearest sand-mound, and gazes into the distance; shakes his head reflectively, and then, without a word, he turns and jogs along back to his train, and takes up a humble position under the hindmost wagon, and feels unspeakably mean, and looks ashamed, and hangs his tail at half-mast for a week. And for as much as a year after that, whenever there is a great hue and cry after a cayote, that dog will merely glance in that direction without emotion, and apparently observe to himself, "I believe I do not wish any of the pie."

The cayote lives chiefly in the most desolate and forbidding deserts, along with the lizard, the jackass-rabbit and the raven, and gets an uncertain and precarious living, and earns it. He seems to subsist almost wholly on the carcasses of oxen, mules and horses that have dropped out of emigrant trains and died, and upon windfalls of carrion, and occasional legacies of offal bequeathed to him by white men who have been opulent enough to have something better to butcher than condemned army bacon. He will eat anything in the world that his first cousins, the desert-frequenting tribes of Indians will, and they will eat anything they can bite. It is a curious fact that these latter are the only creatures known to history who will eat nitro-glycerine and ask for more if they survive.

The cayote of the deserts beyond the Rocky Mountains has a peculiarly hard time of it, owing to the fact that his relations, the Indians, are just as apt to be the first to detect a seductive scent on the desert breeze, and follow the fragrance to the late ox it emanated from, as he is himself; and when this occurs he has to content himself with sitting off at a little distance watching those people strip off and dig out everything edible, and walk off with it. Then he and the waiting ravens explore the skeleton and polish the bones. It is considered that the cayote, and the obscene bird, and the Indian of the desert, testify their blood kinship with each other in that they live together in the waste places of the earth on terms of perfect confidence and friendship, while hating all other creatures and yearning to assist at their funerals. He does not mind going a hundred miles to breakfast, and a hundred and fifty to dinner, because he is sure to have three or four days between meals, and he can just as well be traveling and looking at the scenery as lying around doing nothing and adding to the burdens of his parents.

We soon learned to recognize the sharp, vicious bark of the cayote as it came across the murky plain at night to disturb our dreams among the mail-sacks; and remembering his forlorn aspect and his hard fortune, may shift to wish him the blessed novelty of a long day's good luck and a limitless larder the morrow.

John Muir

THE RANGE OF LIGHT

Arriving on the summit of this dividing crest, one of the most exciting pieces of pure wilderness was disclosed that I ever discovered in all my mountaineering. There, immediately in front, loomed the majestic mass of Mount Ritter, with a glacier swooping down its face nearly to my feet, then curving westward and pouring its frozen flood into a dark blue lake, whose shores were bound with precipices of crystalline snow; while a deep chasm drawn between the divide and the glacier separated the massive picture from everything else. I could see only the one sublime mountain, the one glacier, the one lake; the whole veiled with one blue shadow

—rock, ice, and water close together without a single leaf or sign of life. After gazing spellbound, I began instinctively to scrutinize every notch and gorge and weathered buttress of the mountain, with reference to making the ascent. The entire front above the glacier appeared as one tremendous precipice, slightly receding at the top, and bristling with spires and pinnacles set above one another in formidable array. Massive lichen-stained battlements stood forward here and there, hacked at the top with angular notches, and separated by frosty gullies and recesses that have been veiled in shadow ever since their creation; while to right and left, as far as I could see, were huge, crumbling buttresses, offering no hope to the climber. The head of the glacier sends up a few finger-like branches through narrow *couloirs*; but these seemed too steep and short to be available, especially as I had no ax with which to cut steps, and the numerous narrow-throated gullies down which stones and snow are avalanched seemed hopelessly steep, besides being interrupted by vertical cliffs; while the whole front was rendered still more terribly forbidding by the chill shadow and the gloomy blackness of the rocks.

Descending the divide in a hesitating mood, I picked my way across the yawning chasm at the foot, and climbed out upon the glacier. There were no meadows now to cheer with their brave colors, nor could I hear the dun-headed sparrows, whose cheery notes so often relieve the silence of our highest mountains. The only sounds were the gurgling of small rills down in the veins and crevasses of the glacier, and now and then the rattling report of falling stones, with the echoes they shot out into the crisp air.

I could not distinctly hope to reach the summit from this side, yet I moved on across the glacier as if driven by fate. Contending with myself, the season is too far spent, I said, and even should I be successful, I might be storm-bound on the mountain; and in the cloud-darkness, with the cliffs and crevasses covered with snow, how could I escape? No; I must wait till next summer. I would only approach the mountain now, and inspect it, creep about its flanks, learn what I could of its history, holding myself ready to flee on the approach of the first storm-cloud. But we little know until tried how much of the uncontrollable there is in us, urging across glaciers and torrents, and up dangerous heights, let the judgment forbid as it may.

I succeeded in gaining the foot of the cliff on the eastern extremity of the glacier, and there discovered the mouth of a narrow avalanche gully, through which I began to climb, intending to follow it as far as possible, and at least obtain some fine wild views for my pains. Its general course is oblique to the plane of the mountain-face, and the metamorphic slates of which the mountain is built are cut by cleavage planes in such a way that they weather off in angular blocks, giving rise to irregular steps that greatly facilitate climbing on the sheer places. I thus made my way into a wilderness of crumbling spires and battlements, built together in bewildering combinations, and glazed in many places with a thin coating of ice, which I had to hammer off with stones. The situation was becoming gradually more perilous, but, having passed several dangerous spots, I dared not think of descending; for, so steep was the entire ascent, one would inevitably fall to the glacier in case a single misstep were made. Knowing, therefore, the tried danger beneath, I became all the more anxious concerning the developments to be made above, and began to be conscious of a vague foreboding of what actually befell; not that I was given to fear, but rather because my instincts, usually so positive and true, seemed vitiated in some way, and were leading me astray. At length, after attaining an elevation of about 12,800 feet, I found myself at the foot of a sheer drop in the bed of the avalanche channel I was tracing, which seemed absolutely to bar further progress. It was only about forty-five or fifty feet high, and somewhat roughened by fissures and projections; but these seemed so slight and insecure, as footholds, that I tried hard to avoid the precipice altogether, by scaling the wall of the channel on either side. But, though less steep, the walls were smoother than the obstructing rock, and repeated efforts only showed that I must either go right ahead or turn back. The tried dangers beneath seemed even greater than that of the cliff in front; therefore, after scanning its face again and again, I began to scale it, picking my holds with intense caution. After gaining a point about halfway to the top, I was suddenly brought to a dead stop, with arms outspread, clinging close to the face of the rock, unable to move hand or foot either up or down. My doom appeared fixed. I *must* fall. There would be a moment of bewilderment, and then a lifeless rumble down the one general precipice to the glacier below.

When this final danger flashed upon me, I became nerve-shaken for the first time since setting foot on the mountains, and my mind seemed to fill with a stifling smoke. But this terrible eclipse lasted only a moment, when life blazed forth again with preternatural clearness. I seemed suddenly to become possessed of a new sense. The other self, bygone experiences, instinct, or Guardian Angel,—call it what you will,—came forward and assumed control. Then my trembling muscles became firm again, every rift and flaw in the rock was seen as through a microscope, and my limbs moved with positiveness and precision with which I seemed to have nothing at all to do. Had I been borne aloft upon wings, my deliverance could not have been more complete.

Above this memorable spot, the face of the mountain is still more savagely hacked and torn. It is a maze of yawning chasms and gullies, in the angles of which rise beetling crags and piles of detached boulders that seemed to have been gotten ready to be launched below. But the strange influx of strength I had received seemed inexhaustible. I found a way without effort, and soon stood upon the topmost crag in the blessed light.

How truly glorious the landscape circled around this noble summit!—giant mountains, valleys innumerable, glaciers and meadows, rivers and lakes, with the wide blue sky bent tenderly over them all. But in my first hour of freedom from that terrible shadow, the sunlight in which I was laving seemed all in all.

Looking southward along the axis of the range, the eye is first caught by a row of exceedingly sharp and slender spires, which rise openly to a height of about a thousand feet, above a series of short, residual glaciers that lean back against their bases; their fantastic sculpture and the unrelieved sharpness with which they spring out of the ice rendering them peculiarly wild and striking. These are "The Minarets." Beyond them you behold a sublime wilderness of mountains, their snowy summits towering together in crowded abundance, peak beyond peak, swelling higher, higher as they sweep on southward, until the culminating point of the range is reached on Mount Whitney, near the head of the Kern River, at an elevation of nearly 14,700 feet above the level of the sea.

Westward, the general flank of the range is seen flowing sublimely away from the sharp summits, in smooth undulations; a sea of huge gray granite waves dotted with lakes and meadows, and fluted with stupendous cañons that grow steadily deeper as they recede in the distance. Below this gray region lies the dark forest zone, broken here and there by upswelling ridges and domes; and yet beyond lies a yellow, hazy belt, marking the broad plain of the San Joaquin, bounded on its farther side by the blue mountains of the coast.

Turning now to the northward, there in the immediate foreground is the glorious Sierra Crown, with Cathedral Peak, a temple of marvelous architecture, a few degrees to the left of it; the gray, massive form of Mammoth Mountain to the right; while Mounts Ord, Gibbs, Dana, Conness, Tower Peak,

Castle Peak, Silver Mountain, and a host of noble companions, as yet nameless, make a sublime show along the axis of the range.

Eastward, the whole region seems a land of desolation covered with beautiful light. The torrid volcanic basin of Mono, with its one bare lake fourteen miles long; Owen's Valley and the broad lava table-land at its head, dotted with craters, and the massive Inyo Range, rivaling even the Sierra in height; these are spread, map-like, beneath you, with countless ranges beyond, passing and overlapping one another and fading on the glowing horizon.

At a distance of less than 3,000 feet below the summit of Mount Ritter you may find tributaries of the San Joaquin and Owen's rivers, bursting forth from the ice and snow of the glaciers that load its flanks; while a little to the north of here are found the highest affluents of the Tuolumne and Merced. Thus, the fountains of four of the principal rivers of California are within a radius of four or five miles.

Lakes are seen gleaming in all sorts of places, round, or oval, or square, like very mirrors; others narrow and sinuous, drawn close around the peaks like silver zones, the highest reflecting only rocks, snow, and the sky. But neither these nor the glaciers, nor the bits of brown meadow and moorland that occur here and there, are large enough to make any marked impression upon the mighty wilderness of mountains. The eye, rejoicing in its freedom, roves about the vast expanse, yet returns again and again to the fountain peaks. Perhaps some one of the multitude excites special attention, some gigantic castle with turret and battlement, or some Gothic cathedral more abundantly spired than Milan's. But, generally, when looking for the first time from an all-embracing standpoint like this, the inexperienced observer is oppressed by the incomprehensible grandeur, variety, and abundance of the mountains rising shoulder to shoulder beyond the reach of vision; and it is only after they have been studied one by one, long and lovingly, that their far-reaching harmonies become manifest. Then, penetrate the wilderness where you may, the main telling features, to which all the surrounding topography is subordinate, are quickly perceived, and the most complicated clusters of peaks stand revealed harmoniously correlated and fashioned like works of art—eloquent monuments of the ancient ice-rivers that brought them into relief from the general mass of the range. The cañons, too, some of them a mile deep, mazing wildly through the mighty host of mountains, however lawless and ungovernable at first sight they appear, are at length recognized as the necessary effects of causes which followed each other in harmonious sequence—Nature's poems carved on tables of stone—the simplest and most emphatic of her glacial compositions.

Could we have been here to observe during the glacial period, we should have overlooked a wrinkled ocean of ice as continuous as that now covering the landscapes of Greenland; filling every valley and cañon with only the tops of the fountain peaks rising darkly above the rock-encumbered ice-waves like islets in a stormy sea—those islets the only hints of the glorious landscapes now smiling in the sun. Standing here in the deep, brooding silence all the wilderness seems motionless, as if the work of creation were done. But in the midst of this outer steadfastness we know there is incessant motion and change. Ever and anon, avalanches are falling from yonder peaks. These cliff-bound glaciers, seemingly wedged and immovable, are flowing like water and grinding the rocks beneath them. The lakes are lapping their granite shores and wearing them away, and every one of these rills and young rivers is fretting the air into music, and carrying the mountains to the plains. Here are the roots of all the life of the valleys, and here more simply than elsewhere is the eternal flux of nature manifested. Ice changing to water, lakes to meadows, and mountains to plains.

And while we thus contemplate Nature's methods of landscape creation, and, reading the records she has carved on the rocks, reconstruct, however imperfectly, the landscapes of the past, we also learn that as these we now behold have succeeded those of the pre-glacial age, so they in turn are withering and vanishing to be succeeded by others yet unborn.

But in the midst of these fine lessons and landscapes, I had to remember that the sun was wheeling far to the west, while a new way down the mountain had to be discovered to some point on the timber line where I could have a fire; for I had not even burdened myself with a coat. I first scanned the western spurs, hoping some way might appear through which I might reach the northern glacier, and cross its snout; or pass around the lake into which it flows, and thus strike my morning track. This route was soon sufficiently unfolded to show that, if practicable at all, it would require so much time that reaching camp that night would be out of the question. I therefore scrambled back eastward, descending the southern slopes obliquely at the same time. Here the crags seemed less formidable, and the head of a glacier that flows northeast came in sight, which I determined to follow as far as possible, hoping thus to make my way to the foot of the peak on the east side, and thence across the intervening cañons and ridges to camp.

The inclination of the glacier is quite moderate at the head, and, as the sun had softened the *névé*, I made safe and rapid progress, running and sliding, and keeping up a sharp outlook for crevasses. About half a mile from the head, there is an ice-cascade, where the glacier pours over a sharp declivity and is shattered into massive blocks separated by deep, blue fissures. To thread my way through the slippery mazes of the crevassed portion seemed impossible, and I endeavored to avoid it by climbing off to the shoulder of the mountain. But the slopes rapidly steepened and at length fell away in sheer precipices, compelling a return to the ice. Fortunately, the day had been warm enough to loosen the ice-crystals so as to admit of hollows being dug in the rotten portions of the blocks, thus enabling me to pick my way with far less difficulty than I had anticipated. Continuing down over the snout, and along the left lateral moraine, was only a confident saunter, showing that the ascent of the mountain by way of this glacier is easy, provided one is armed with an ax to cut steps here and there.

The lower end of the glacier was beautifully waved and barred by the outcropping edges of the bedded ice-layers which represent the annual snowfalls, and to some extent the irreg-ularities of structure caused by the weathering of the walls of crevasses, and by separate snowfalls which have been followed by rain, hail, thawing and freezing, etc. Small rills were gliding and swirling over the melting surface with a smooth, oily ap-pearance, in channels of pure ice—their quick, compliant movements contrasting most impressively with the rigid, in-visible flow of the glacier itself, on whose back they all were riding.

Night drew near before I reached the eastern base of the mountain, and my camp lay many a rugged mile to the north; but ultimate success was assured. It was now only a matter of endurance and ordinary mountain-craft. The sunset was, if pos-sible, yet more beautiful than that of the day before. The Mono landscape seemed to be fairly saturated with warm, purple light. The peaks marshaled along the summit were in shadow, but through every notch and pass streamed vivid sunfire, soothing and irradiating their rough black angles, while companies of small, luminous clouds hovered above them like very angels of light.

Darkness came on, but I found my way by the trends of the cañons and the peaks projected against the sky. All ex-citement died with the light, and then I was weary. But the joyful sound of the waterfall across the lake was heard at last, and soon the stars were seen reflected in the lake itself. Taking my bearings from these, I discovered the little pine thicket in which my nest was, and then I had a rest such as only a tired mountaineer may enjoy. After lying loose and lost for awhile, I made a sunrise fire, went down to the lake, dashed water on my head, and dipped a cupful for tea. The revival brought about by bread and tea was as complete as the exhaustion from excessive enjoyment and toil. Then I crept beneath the pine-tassels to bed. The wind was frosty and the fire burned low, but my sleep was none the less sound, and the evening constellations had swept far to the west before I awoke. . . .

One of the most beautiful and exhilarating storms I ever enjoyed in the Sierra occurred in December, 1874, when I happened to be exploring one of the tributary valleys of the Yuba River. The sky and the ground and the trees had been thoroughly rain-washed and were dry again. The day was intensely pure, one of those incomparable bits of California winter, warm and balmy and full of white sparkling sunshine, redolent of all the purest influences of the spring, and at the same time enlivened with one of the most bracing windstorms conceivable. Instead of camping out, as I usually do, I then chanced to be stopping at the house of a friend. But when the storm began to sound, I lost no time in pushing out into the woods to enjoy it. For on such occasions Nature has always something rare to show us, and the danger to life and limb is hardly greater than one would experience crouching deprecatingly beneath a roof.

It was still early morning when I found myself fairly adrift. Delicious sunshine came pouring over the hills, lighting the tops of the pines, and setting free a steam of summery fragrance that contrasted strangely with the wild tones of the storm. The air was mottled with pine-tassels and bright green plumes, that went flashing past in the sunlight like birds pursued. But there was not the slightest dustiness, nothing less pure than leaves, and ripe pollen, and flecks of withered bracken and moss. I heard trees falling for hours at the rate of one every two or three minutes; some uprooted, partly on account of the loose, water-soaked condition of the ground; others broken straight across, where some weakness caused by fire had determined the spot. The gestures of the various trees made a delightful study. Young Sugar Pines, light and feathery as squirrel-tails, were bowing almost to the ground; while the grand old patriarchs, whose massive boles had been tried in a hundred storms, waved solemnly above them, their long, arching branches streaming fluently on the gale, and every needle thrilling and ringing and shedding off keen lances of light like a diamond. The Douglas Spruces, with long sprays drawn out in level tresses, and needles massed in a gray, shimmering glow, presented a most striking appearance as they stood in bold relief along the hilltops. The madroños in the dells, with their red bark and large glossy leaves tilted every way, reflected the sunshine in throbbing spangles like those one so often sees on the rippled surface of a glacier lake. But the Silver Pines were now the most impressively beautiful of all. Colossal spires 200 feet in height waved like supple goldenrods chanting and bowing low as if in worship, while the whole mass of their long, tremulous foliage was kindled into one continuous blaze of white sun-fire. The force of the gale was such that the most steadfast monarch of them all rocked down to its roots with a motion plainly perceptible when one leaned against it. Nature was holding high festival, and every fiber of the most rigid giants thrilled with glad excitement.

I drifted on through the midst of this passionate music and motion, across many a glen, from ridge to ridge; often halting in the lee of a rock for shelter, or to gaze and listen. Even when the grand anthem had swelled to its highest pitch, I could distinctly hear the varying tones of individual trees,—Spruce, and Fir, and Pine, and leafless Oak,—and even the infinitely gentle rustle of the withered grasses at my feet. Each was expressing itself in its own way,—singing its own song, and making its own peculiar gestures,—manifesting a richness of variety to be found in no other forest I have yet seen. The coniferous woods of Canada, and the Carolinas, and Florida, are made up of trees that resemble one another about as nearly as blades of grass, and grow close together in much the same way. Coniferous trees, in general, seldom possess individual character, such as is manifest among Oaks and Elms. But the California forests are made up of a greater number of distinct species than any other in the world. And in them we find, not only a marked differentiation into special groups, but also a marked individuality in almost every tree, giving rise to storm effects indescribably glorious.

Toward midday, after a long, tingling scramble through copses of hazel and ceanothus, I gained the summit of the highest ridge in the neighborhood; and then it occurred to me that it would be a fine thing to climb one of the trees to obtain a wider outlook and get my ear close to the Æolian music of its topmost needles. But under the circumstances the choice of a tree was a serious matter. One whose instep was not very strong seemed in danger of being blown down, or of being struck by others in case they should fall; another was branchless to a considerable height above the ground, and at the same time too large to be grasped with arms and legs in climbing; while others were not favorably situated for clear views. After cautiously casting about, I made choice of the tallest of a group of Douglas Spruces that were growing close together like a tuft of grass, no one of which seemed likely to fall unless all the rest fell with it. Though comparatively young, they were about 100 feet high, and their lithe, brushy tops were rocking and swirling in wild ecstasy. Being accustomed to climb trees in making botanical studies, I experienced no difficulty in reaching the top of this one, and never before did I enjoy so noble an exhilaration of motion. The slender tops fairly flapped and swished in the passionate torrent, bending and swirling backward and forward, round and round, tracing indescribable combinations of vertical and horizontal curves, while I clung with muscles firm braced, like a bobolink on a reed.

In its widest sweeps my tree-top described an arc of from twenty to thirty degrees, but I felt sure of its elastic temper, having seen others of the same species still more severely tried—bent almost to the ground indeed, in heavy snows—without breaking a fiber. I was therefore safe, and free to take the wind into my pulses and enjoy the excited forest from my superb outlook. The view from here must be extremely beautiful in any weather. Now my eye roved over the piny hills and dales as over fields of waving grain, and felt the light running in ripples and broad swelling undulations across the valleys from ridge to ridge, as the shining foliage was stirred by corresponding waves of air. Oftentimes these waves of reflected light would break up suddenly into a kind of beaten foam, and again, after chasing one another in regular order, they would seem to bend forward in concentric curves, and disappear on some hillside, like sea-waves on a shelving shore. The quantity of light reflected from the bent needles was so great as to make whole groves appear as if covered with snow, while the black shadows beneath the trees greatly enhanced the effect of the silvery splendor.

Excepting only the shadows there was nothing somber in all this wild sea of pines. On the contrary, notwithstanding this was the winter season, the colors were remarkably beautiful. The shafts of the pine and libocedrus were brown and purple, and most of the foliage was well tinged with yellow; the laurel groves, with the pale undersides of their leaves turned upward, made masses of gray; and then there was many a dash of chocolate color from clumps of manzanita, and jet of vivid crimson from the bark of the madroños, while the ground on the hill-sides, appearing here and there through openings between the groves, displayed masses of pale purple and brown.

The sounds of the storm corresponded gloriously with this wild exuberance of light and motion. The profound bass of the naked branches and boles booming like waterfalls; the quick, tense vibrations of the pine-needles, now rising to a shrill, whistling hiss, now falling to a silky murmur; the rustling of laurel groves in the dells, and the keen metallic click of leaf on leaf—all this was heard in easy analysis when the attention was calmly bent.

The varied gestures of the multitude were seen to fine advantage, so that one could recognize the different species at a distance of several miles by this means alone, as well as by their forms and colors, and the way they reflected the light. All seemed strong and comfortable, as if really enjoying the storm, while responding to its most enthusiastic greetings. We hear much nowadays concerning the universal struggle for existence, but no struggle in the common meaning of the word was manifest here; no recognition of danger by any tree; no deprecation; but rather an invincible gladness as remote from exultation as from fear.

I kept my lofty perch for hours, frequently closing my eyes to enjoy the music by itself or to feast quietly on the delicious fragrance that was streaming past. The fragrance of the woods was less marked than that produced during warm rain, when so many balsamic buds and leaves are steeped like tea; but, from the chafing of resiny branches against each other, and the incessant attrition of myriads of needles, the gale was spiced to a very tonic degree. And besides the fragrance from these local sources there were traces of scents brought from afar. For this wind came first from the sea, rubbing against its fresh, briny waves, then distilled through the redwoods, threading rich ferny gulches, and spreading itself in broad undulating

currents over many a flower-enameled ridge of the coast mountains, then across the golden plains, up the purple foot-hills,
and into these piny woods with the varied incense gathered by
the way.

Winds are advertisements of all they touch, however much
or little we may be able to read them; telling their wanderings
even by their scents alone. Mariners detect the flowery perfume
of land-winds far at sea, and sea-winds carry the fragrance of
dulse and tangle far inland, where it is quickly recognized,
though mingled with the scents of a thousand land-flowers. As
an illustration of this, I may tell here that I breathed sea-air on
the Firth of Forth, in Scotland, while a boy; then was taken to
Wisconsin, where I remained nineteen years; then, without in
all this time having breathed one breath of the sea, I walked
quietly, alone, from the middle of the Mississippi Valley to the
Gulf of Mexico, on a botanical excursion, and while in Florida,
far from the coast, my attention wholly bent on the splendid
tropical vegetation about me, I suddenly recognized a sea-
breeze, as it came sifting through the palmettos and blooming
vine-tangles, which at once awakened and set free a thousand
dormant associations, and made me a boy again in Scotland, as
if all the intervening years had been annihilated.

Most people like to look at mountain rivers, and bear them
in mind; but few care to look at the winds, though far more
beautiful and sublime, and though they become at times about
as visible as flowing water. When the north winds in winter are
making upward sweeps over the curving summits of the High
Sierra, the fact is sometimes published with flying snow-banners
a mile long. Those portions of the winds thus embodied can
scarce be wholly invisible, even to the darkest imagination.
And when we look around over an agitated forest, we may see
something of the wind that stirs it, by its effects upon the trees.
Yonder it descends in a rush of water-like ripples, and sweeps
over the bending pines from hill to hill. Nearer, we see de-

tached plumes and leaves, now speeding by on level currents, now whirling in eddies, or, escaping over the edges of the whirls, soaring aloft on grand, upswelling domes of air, or tossing on flame-like crests. Smooth, deep currents, cascades, falls, and swirling eddies, sing around every tree and leaf, and over all the varied topography of the region with telling changes of form, like mountain rivers conforming to the features of their channels.

After tracing the Sierra streams from their fountains to the plains, marking where they bloom white in falls, glide in crystal plumes, surge gray and foam-filled in boulder-choked gorges, and slip through the woods in long, tranquil reaches— after thus learning their language and forms in detail, we may at length hear them chanting all together in one grand anthem, and comprehend them all in clear inner vision, covering the range like lace. But even this spectacle is far less sublime and not a whit more substantial than what we may behold of these storm-streams of air in the mountain woods.

We all travel the milky way together, trees and men; but it never occurred to me until this storm-day, while swinging in the wind, that trees are travelers, in the ordinary sense. They make many journeys, not extensive ones, it is true; but our own little journeys, away and back again, are only little more than tree-wavings—many of them not so much.

When the storm began to abate, I dismounted and sauntered down through the calming woods. The storm-tones died away, and, turning toward the east, I beheld the countless hosts of the forests hushed and tranquil, towering above one another on the slopes of the hills like a devout audience. The setting sun filled them with amber light, and seemed to say, while they listened, "My peace I give unto you."

As I gazed on the impressive scene, all the so-called ruin of the storm was forgotten, and never before did these noble woods appear so fresh, so joyous, so immortal.

RAGGYLUG

The Story of a Cottontail Rabbit

Raggylug, or Rag, was the name of a young cottontail rabbit. It was given him from his torn and ragged ear, a life-mark that he got in his first adventure. He lived with his mother in Olifant's swamp, where I made their acquaintance and gathered, in a hundred different ways, the little bits of proof and scraps of truth that at length enabled me to write this history.

Those who do not know the animals well may think I have humanized them, but those who have lived so near them as to know somewhat of their ways and their minds will not think so.

Truly rabbits have no speech as we understand it, but they have a way of conveying ideas by a system of sounds, signs, scents, whisker-touches, movements, and example that answers the purpose of speech; and it must be remembered that though in telling this story I freely translate from rabbit into English, *I repeat nothing that they did not say.*

I

The rank swamp grass bent over and concealed the snug nest where Raggylug's mother had hidden him. She had partly covered him with some of the bedding, and, as always, her last warning was to "lay low and say nothing, whatever happens." Though tucked in bed, he was wide awake and his bright eyes were taking in that part of his little green world that was straight above. A bluejay and a red-squirrel, two notorious thieves, were loudly berating each other for stealing, and at one time Rag's home bush was the centre of their fight; a yellow warbler caught a blue butterfly but six inches from his nose, and a scarlet and black ladybug, serenely waving her knobbed feelers, took a long walk up one grassblade, down another, and across the nest and over Rag's face—and yet he never moved nor even winked.

After a while he heard a strange rustling of the leaves in the near thicket. It was an odd, continuous sound, and though it went this way and that way and came ever nearer, there was no patter of feet with it. Rag had lived his whole life in the Swamp (he was three weeks old) and yet had never heard anything like this. Of course his curiosity was greatly aroused. His mother had cautioned him to lay low, but that was understood to be in case of danger, and this strange sound without footfalls could not be anything to fear.

The low rasping went past close at hand, then to the right, then back, and seemed going away. Rag felt he knew what he was about; he wasn't a baby; it was his duty to learn what it was. He slowly raised his roly-poly body on his short fluffy legs, lifted his little round head above the covering of his nest and peeped out into the woods. The sound had ceased as soon as he moved. He saw nothing, so took one step forward to a clear view, and instantly found himself face to face with an enormous Black Serpent.

"Mammy," he screamed in mortal terror as the monster darted at him. With all the strength of his tiny limbs he tried to run. But in a flash the Snake had him by one ear and whipped around him with his coils to gloat over the helpless little baby bunny he had secured for dinner.

"Mam-my—Mam-my," gasped poor little Raggylug as the cruel monster began slowly choking him to death. Very soon the little one's cry would have ceased, but bounding through the woods straight as an arrow came Mammy. No longer a shy, helpless little Molly Cottontail, ready to fly from a shadow: the mother's love was strong in her. The cry of her baby had filled her with the courage of a hero, and—hop, she went over that horrible reptile. Whack, she struck down at him with her sharp hind claws as she passed, giving him such a stinging blow that he squirmed with pain and hissed with anger.

"M-a-m-m-y" came feebly from the little one. And Mammy came leaping again and again and struck harder and fiercer until the loathsome reptile let go the little one's ear and tried to bite the old one as she leaped over. But all he got was a mouthful of wool each time, and Molly's fierce blows began to tell, as long bloody rips were torn in the Black Snake's scaly armor.

Things were now looking bad for the Snake; and bracing himself for the next charge, he lost his tight hold on Baby Bunny, who at once wriggled out of the coils and away into the underbrush, breathless and terribly frightened, but unhurt save that his left ear was much torn by the teeth of that dreadful Serpent.

Molly now had gained all she wanted. She had no notion of fighting for glory or revenge. Away she went into the woods and the little one followed the shining beacon of her snow-white tail until she led him to a safe corner of the Swamp.

II

Old Olifant's Swamp was a rough, brambly tract of second-growth woods, with a marshy pond and a stream through the middle. A few ragged remnants of the old forest still stood in it and a few of the still older trunks were lying about as dead logs in the brushwood. The land about the pond was of that willow-grown sedgy kind that cats and horses avoid, but that cattle do not fear. The drier zones were overgrown with briars and young trees. The outermost belt of all, that next to the fields, was of thrifty, gummy-trunked young pines whose living needles in air and dead ones on earth offer so delicious an odor to the nostrils of the passer-by, and so deadly a breath to those seedlings that would compete with them for the worthless waste they grow on.

All around for a long way were smooth fields, and the only wild tracks that ever crossed these fields were those of a thoroughly bad and unscrupulous fox that lived only too near.

The chief indwellers of the swamp were Molly and Rag. Their nearest neighbors were far away, and their nearest kin were dead. This was their home, and here they lived together, and here Rag received the training that made his success in life.

Molly was a good little mother and gave him a careful bringing up. The first thing he learned was "to lay low and say nothing." His adventure with the snake taught him the wisdom of this. Rag never forgot that lesson; afterward he did as he was told, and it made the other things come more easily.

The second lesson he learned was "freeze." It grows out of the first, and Rag was taught it as soon as he could run.

"Freezing" is simply doing nothing, turning into a statue. As soon as he finds a foe near, no matter what he is doing, a well-trained Cottontail keeps just as he is and stops all movement, for the creatures of the woods are of the same color as the things in the woods and catch the eye only while moving. So when enemies chance together, the one who first sees the other can keep himself unseen by "freezing" and thus have all the advantage of choosing the time for attack or escape. Only those who live in the woods know the importance of this; every wild creature and every hunter must learn it; all learn to do it well, but not one of them can beat Molly Cottontail in the doing. Rag's mother taught him this trick by example. When the white cotton cushion that she always carried to sit on went bobbing away through the woods, of course Rag ran his hardest to keep up. But when Molly stopped and "froze," the natural wish to copy made him do the same.

But the best lesson of all that Rag learned from his mother was the secret of the Brierbrush. It is a very old secret now, and to make it plain you must first hear why the Brierbrush quarrelled with the beasts.

> *Long ago the Roses used to grow on bushes that had no thorns. But the Squirrels and Mice used to climb after them, the Cattle used to knock them off with their horns, the Possum would twitch them off with his long tail, and the Deer, with his sharp hoofs, would break them down. So the Brierbrush armed itself with spikes to protect its roses and declared eternal war on all creatures that climbed trees, or had horns, or hoofs, or long tails. This left the Brierbrush at peace with none but Molly Cottontail, who could not climb, was hornless, hoofless, and had scarcely any tail at all.*

> *In truth the Cottontail had never harmed a Brierrose, and having now so many enemies the Rose took the Rabbit into especial friendship, and when dangers are threatening poor Bunny he flies to the nearest Brierbrush, certain that it is ready with a million keen and poisoned daggers to defend him.*

So the secret that Rag learned from his mother was, "The Brierbrush is your best friend."

Much of the time that season was spent in learning the lay of the land, and the bramble and brier mazes. And Rag learned them so well that he could go all around the swamp by two different ways and never leave the friendly briers at any place for more than five hops.

It is not long since the foes of the Cottontails were disgusted to find that man had brought a new kind of bramble and planted it in long lines throughout the country. It was so strong that no creatures could break it down, and so sharp that the toughest skin was torn by it. Each year there was more of it and each year it became a more serious matter to the wild creatures. But Molly Cottontail had no fear of it. She was not brought up in the briers for nothing. Dogs and foxes, cattle and sheep, and even man himself might be torn by those fearful spikes: but Molly understands it and lives and thrives under it. And the further it spreads the more safe country there is for the Cottontail. And the name of this new and dreaded bramble is—*the barbed-wire fence.*

III

Molly had no other children to look after now, so Rag had all her care. He was unusually quick and bright as well as strong, and he had uncommonly good chances; so he got on remarkably well.

All the season she kept him busy learning the tricks of the trail, and what to eat and drink and what not to touch. Day by day she worked to train him; little by little she taught him, putting into his mind hundreds of ideas that her own life or early training had stored in hers, and so equipped him with the knowledge that makes life possible to their kind.

Close by her side in the clover-field or the thicket he would sit and copy her when she wobbled her nose "to keep her smeller clear," and pull the bite from her mouth or taste her lips to make sure he was getting the same kind of fodder. Still copying her, he learned to comb his ears with his claws and to dress his coat and to bite the burrs out of his vest and socks. He learned, too, that nothing but clear dewdrops from the briers were fit for a rabbit to drink, as water which has once touched the earth must surely bear some taint. Thus he began the study of woodcraft, the oldest of all sciences.

As soon as Rag was big enough to go out alone, his mother taught him the signal code. Rabbits telegraph each other by thumping on the ground with their hind feet. Along the ground sound carries far; a thump that at six feet from the earth is not heard at twenty yards will, near the ground, be heard at least one hundred yards. Rabbits have very keen hearing, and so might hear this same thump at two hundred yards, and that would reach from end to end of Olifant's Swamp. A single *thump* means "look out" or "freeze." A slow *thump thump* means "come." A fast *thump thump* means "danger;" and a very fast *thump thump thump* means "run for dear life."

At another time, when the weather was fine and the blue-jays were quarrelling among themselves, a sure sign that no dangerous foe was about, Rag began a new study. Molly, by flattening her ears, gave the sign to squat. Then she ran far away in the thicket and gave the thumping signal for "come." Rag set out at a run to the place but could not find Molly. He thumped, but got no reply. Setting carefully about his search he found her foot-scent and following this strange guide, that the beasts all know so well and man does not know at all, he worked out the trail and found her where she was hidden. Thus he got his first lesson in trailing, and thus it was that the games of hide and seek they played became the schooling for the serious chase of which there was so much in his after life.

Before that first season of schooling was over he had learnt all the principal tricks by which a rabbit lives and in not a few problems showed himself a veritable genius.

He was an adept at "tree," "dodge," and "squat," he could play "log-lump," with "wind" and "baulk" with "back-track" so well that he scarcely needed any other tricks. He had not yet tried it, but he knew just how to play "barb-wire," which is a new trick of the brilliant order; he had made a special study of "sand," which burns up all scent, and he was deeply versed in "change-off," "fence," and "double" as well as "hole-up," which is a trick requiring longer notice, and yet he never forgot that "lay-low" is the beginning of all wisdom and "brierbush" the only trick that is always safe.

He was taught the signs by which to know all his foes and then the way to baffle them. For hawks, owls, foxes, hounds, curs, minks, weasels, cats, skunks, coons, and men, each have a different plan of pursuit, and for each and all of these evils he was taught a remedy.

And for knowledge of the enemy's approach he learnt to depend first on himself and his mother, and then on the bluejay. "Never neglect the bluejay's warning," said Molly; "he is a mischief-maker, a marplot, and a thief all the time, but nothing escapes him. He wouldn't mind harming us, but he cannot, thanks to the briers, and his enemies are ours, so it is well to heed him. If the woodpecker cries a warning you can trust him, he is honest; but he is a fool beside the bluejay, and though the bluejay often tells lies for mischief you are safe to believe him when he brings ill news."

The barb-wire trick takes a deal of nerve and the best of legs. It was long before Rag ventured to play it, but as he came to his full powers it became one of his favorites.

"It's fine play for those who can do it," said Molly. "First you lead off your dog on a straightaway and warm him up a bit by nearly letting him catch you. Then keeping just one hop ahead, you lead him at a long slant full tilt into a breast-high barb-wire. I've seen many a dog and fox crippled, and one big hound killed outright this way. But I've also seen more than one rabbit lose his life in trying it."

Rag early learnt what some rabbits never learn at all, that "hole-up" is not such a fine ruse as it seems; it may be the certain safety of a wise rabbit, but soon or late is a sure death-trap to a fool. A young rabbit always thinks of it first, an old rabbit never tries it till all others fail. It means escape from a man or dog, a fox or a bird of prey, but it means sudden death if the foe is a ferret, mink, skunk, or weasel.

There were but two ground-holes in the Swamp. One on the Sunning Bank, which was a dry sheltered knoll in the South-end. It was open and sloping to the sun, and here on fine days the Cottontails took their sunbaths. They stretched out among the fragrant pine needles and winter-green in odd cat-like positions, and turned slowly over as though roasting and wishing all sides well done. And they blinked and panted, and squirmed as if in dreadful pain; yet this was one of the keenest enjoyments they knew.

Just over the brow of the knoll was a large pine stump. Its grotesque roots wriggled out above the yellow sand-bank like dragons, and under their protecting claws a sulky old woodchuck had digged a den long ago. He became more sour and ill-tempered as weeks went by, and one day waited to quarrel with Olifant's dog instead of going in so that Molly Cottontail was able to take possession of the den an hour later.

This, the pine-root hole, was afterward very coolly taken

by a self-sufficient young skunk who with less valor might have enjoyed greater longevity, for he imagined that even man with a gun would fly from him. Instead of keeping Molly from the den for good, therefore, his reign, like that of a certain Hebrew king, was over in seven days.

The other, the fern-hole, was in a fern thicket next the clover field. It was small and damp, and useless except as a last retreat. It also was the work of a woodchuck, a well-meaning friendly neighbor, but a hare-brained youngster whose skin in the form of a whip-lash was now developing higher horse-power in the Olifant working team.

"Simple justice," said the old man, "for that hide was raised on stolen feed that the team would a' turned into horse-power anyway."

The Cottontails were now sole owners of the holes, and did not go near them when they could help it, lest anything like a path should be made that might betray these last retreats to an enemy.

There was also the hollow hickory, which, though nearly fallen, was still green, and had the great advantage of being open at both ends. This had long been the residence of one Lotor, a solitary old coon whose ostensible calling was frog-hunting, and who, like the monks of old, was supposed to abstain from all flesh food. But it was shrewdly suspected that he needed but a chance to indulge in a diet of rabbit. When at last one dark night he was killed while raiding Olifant's hen-house, Molly, so far from feeling a pang of regret, took possession of his cosy nest with a sense of unbounded relief.

IV

Bright August sunlight was flooding the Swamp in the morning. Everything seemed soaking in the warm radiance. A little brown swamp-sparrow was teetering on a long rush in the pond. Beneath him there were open spaces of dirty water that brought down a few scraps of the blue sky, and worked it and the yellow duckweed into an exquisite mosaic, with a little wrong-side picture of the bird in the middle. On the bank behind was a great vigorous growth of golden green skunk-cabbage, that cast dense shadow over the brown swamp tussocks.

The eyes of the swamp-sparrow were not trained to take in the color glories, but he saw what we might have missed; that two of the numberless leafy brown bumps under the broad cabbage-leaves were furry living things, with noses that never ceased to move up and down whatever else was still.

It was Molly and Rag. They were stretched under the skunk-cabbage, not because they liked its rank smell, but because the winged ticks could not stand it at all and so left them in peace.

Rabbits have no set time for lessons, they are always learning; but what the lesson is depends on the present stress, and that must arrive before it is known. They went to this place for a quiet rest, but had not been long there when suddenly a warning note from the ever-watchful bluejay caused Molly's nose and ears to go up and her tail to tighten to her back. Away across the Swamp was Olifant's big black and white dog, coming straight toward them.

"Now," said Molly, "squat while I go and keep that fool out of mischief." Away she went to meet him and she fearlessly dashed across the dog's path.

"Bow-ow-ow," he fairly yelled as he bounded after Molly, but she kept just beyond his reach and led him where the million daggers struck fast and deep, till his tender ears were scratched raw, and guided him at last plump into a hidden barbed-wire fence, where he got such a gashing that he went homeward howling with pain. After making a short double, a loop and a baulk in case the dog should come back, Molly returned to find that Rag in his eagerness was standing bolt upright and craning his neck to see the sport.

This disobedience made her so angry that she struck him with her hind foot and knocked him over in the mud.

One day as they fed on the near clover field a red-tailed hawk came swooping after them. Molly kicked up her hind legs to make fun of him and skipped into the briers along one of their old pathways, where of course the hawk could not follow. It was the main path from the Creekside Thicket to the Stove-pipe brush-pile. Several creepers had grown across it, and Molly, keeping one eye on the hawk, set to work and cut the creepers off. Rag watched her, than ran on ahead, and cut some more that were across the path. "That's right," said Molly, "always keep the runways clear, you will need them often enough. Not wide, but clear. Cut everything like a creeper across them and some day you will find you have cut a snare."

"A what?" asked Rag, as he scratched his right ear with his left hind foot.

"A snare is something that looks like a creeper, but it doesn't grow and it's worse than all the hawks in the world," said Molly, glancing at the now far-away red-tail, "for there it hides night and day in the runway till the chance to catch you comes."

"I don't believe it could catch me," said Rag, with the pride of youth as he rose on his heels to rub his chin and whiskers high up on a smooth sapling. Rag did not know he was doing this, but his mother saw and knew it was a sign, like the changing of a boy's voice, that her little one was no longer a baby but would soon be a grown-up Cottontail.

V

There is magic in running water. Who does not know it and feel it? The railroad builder fearlessly throws his bank across the wide bog or lake, or the sea itself, but the tiniest rill of running water he treats with great respect, studies its wish and its way and gives it all it seems to ask. The thirst-parched traveller in the poisonous alkali deserts holds back in deadly fear from the sedgy ponds till he finds one down whose centre is a thin, clear line, and a faint flow, the sign of running, living water, and joyfully he drinks.

There is magic in running water, no evil spell can cross it. Tam O'Shanter proved its potency in time of sorest need. The wild-wood creature with its deadly foe following tireless on the trail scent, realizes its nearing doom and feels an awful spell. Its strength is spent, its every trick is tried in vain till the good Angel leads it to the water, the running, living water, and dashing in it follows the cooling stream, and then with force renewed takes to the woods again.

There is magic in running water. The hounds come to the very spot and halt and cast about; and halt and cast in vain. Their spell is broken by the merry stream, and the wild thing lives its life.

And this was one of the great secrets that Raggylug learned from his mother—"after the Brierrose, the Water is your friend."

One hot, muggy night in August, Molly led Rag through the woods. The cotton-white cushion she wore under her tail twinkled ahead and was his guiding lantern, though it went out as soon as she stopped and sat on it. After a few runs and stops to listen, they came to the edge of the pond. The hylas in the trees above them were singing *"sleep, sleep,"* and away out on a sunken log in the deep water, up to his chin in the cooling bath, a bloated bullfrog was singing the praises of a *"jug o' rum."*

"Follow me still," said Molly, in rabbit, and "flop" she went into the pond and struck out for the sunken log in the middle. Rag flinched but plunged with a little "ouch," gasping and wobbling his nose very fast but still copying his mother. The same movements as on land sent him through the water, and thus he found he could swim. On he went till he reached the sunken log and scrambled up by his dripping mother on the high dry end, with a rushy screen around them and the Water that tells no tales. After this in warm black nights when that old fox from Springfield came prowling through the Swamp, Rag would note the place of the bullfrog's voice, for in case of direst need it might be a guide to safety. And thenceforth the words of the song that the bullfrog sang were, *"Come, come, in danger come."*

This was the latest study that Rag took up with his mother—it was really a post-graduate course, for many little rabbits never learn it at all.

VI

No wild animal dies of old age. Its life has soon or late a tragic end. It is only a question of how long it can hold out against its foes. But Rag's life was proof that once a rabbit passes out of his youth he is likely to outlive his prime and be killed only in the last third of life, the downhill third we call old age.

The Cottontails had enemies on every side. Their daily life was a series of escapes. For dogs, foxes, cats, skunks, coons, weasels, minks, snakes, hawks, owls, and men, and even insects were all plotting to kill them. They had hundreds of adventures, and at least once a day they had to fly for their lives and save themselves by their legs and wits.

More than once that hateful fox from Springfield drove them to taking refuge under the wreck of a barbed-wire hogpen by the spring. But once there they could look calmly at him while he spiked his legs in vain attempts to reach them.

Once or twice Rag when hunted had played off the hound against a skunk that had seemed likely to be quite as dangerous as the dog.

Once he was caught alive by a hunter who had a hound and a ferret to help him. But Rag had the luck to escape next day, with a yet deeper distrust of ground holes. He was several times run into the water by the cat, and many times was chased by hawks and owls, but for each kind of danger there was a safeguard. His mother taught him the principal dodges, and he improved on them and made many new ones as he grew older. And the older and wiser he grew the less he trusted to his legs, and the more to his wits for safety.

Ranger was the name of a young hound in the neighborhood. To train him his master used to put him on the trail of one of the Cottontails. It was nearly always Rag that they ran, for the young buck enjoyed the runs as much as they did, the spice of danger in them being just enough for zest. He would say:

"Oh, mother! here comes the dog again, I must have a run to-day."

"You are too bold, Raggy, my son!" she might reply. "I fear you will run once too often."

"But, mother, it is such glorious fun to tease that fool dog, and it's all good training. I'll thump if I am too hard pressed, then you can come and change off while I get my second wind."

On he would come, and Ranger would take the trail and follow till Rag got tired of it. Then he either sent a thumping telegram for help, which brought Molly to take charge of the dog, or he got rid of the dog by some clever trick. A description of one of these shows how well Rag had learned the arts of the woods.

He knew that his scent lay best near the ground, and was strongest when he was warm. So if he could get off the ground, and be left in peace for half an hour to cool off, and for the trail to stale, he knew he would be safe. When, therefore, he tired of the chase, he made for the Creekside brier-patch, where he "wound"—that is, zigzagged—till he left a course so crooked that the dog was sure to be greatly delayed in working it out. He then went straight to D in the woods, passing one hop to windward of the high log E. Stopping at D, he followed his back trail to F, here he leaped aside and ran toward G. Then, returning on his trail to J, he waited till the hound passed on his trail at I. Rag then got back on his old trail at H, and followed it to E, where, with a scent-baulk or great leap aside, he reached the high log, and running to its higher end, he sat like a bump.

Ranger lost much time in the bramble maze, and the scent was very poor when he got it straightened out, and came to D. Here he began to circle to pick it up, and after losing much time, struck the trail which ended suddenly at G. Again he was at fault, and had to circle to find the trail. Wider and wider the circles, until at last, he passed right under the log Rag was on. But a cold scent, on a cold day, does not go downward much. Rag never budged nor winked, and the hound passed.

Again the dog came round. This time he crossed the low part of the log, and stopped to smell it. "Yes, clearly it was rabbity," but it was a stale scent now; still he mounted the log.

It was a trying moment for Rag, as the great hound came sniff-sniffing along the log. But his nerve did not forsake him; the wind was right; he had his mind made up to bolt as soon as Ranger came half way up. But he didn't come. A yellow cur would have seen the rabbit sitting there, but the hound did not, and the scent seemed stale, so he leaped off the log, and Rag had won.

VII

Rag had never seen any other rabbit than his mother. Indeed he had scarcely thought about there being any other. He was more and more away from her now, and yet he never felt lonely, for rabbits do not hanker for company. But one day in December, while he was among the red dogwood brush, cutting a new path to the great Creekside thicket, he saw all at once against the sky over the Sunning Bank the head and ears of a strange rabbit. The new-comer had the air of a well-pleased discoverer and soon came hopping Rag's way along one of *his* paths into *his* Swamp. A new feeling rushed over him, that boiling mixture of anger and hatred called jealousy.

The stranger stopped at one of Rag's rubbing-trees—that is, a tree against which he used to stand on his heels and rub his chin as far up as he could reach. He thought he did this simply because he liked it; but all buck-rabbits do so, and several ends are served. It makes the tree rabbity, so that other rabbits know that this swamp already belongs to a rabbit family and is not open for settlement. It also lets the next one know by the scent if the last caller was an acquaintance, and the height from the ground of the rubbing-places shows how tall the rabbit is.

Now to his disgust Rag noticed that the newcomer was a head taller than himself, and a big, stout buck at that. This was a wholly new experience and filled Rag with a wholly new feeling. The spirit of murder entered his heart; he chewed very hard with nothing in his mouth, and hopping forward onto a smooth piece of hard ground he struck slowly:

"*Thump—thump—thump*," which is a rabbit telegram for, "Get out of my swamp, or fight."

The new-comer made a big V with his ears, sat upright for a few seconds, then, dropping on his fore-feet, sent along the ground a louder, stronger, "*Thump—thump—thump*."

And so war was declared.

They came together by short runs side-wise, each one trying to get the wind of the other and watching for a chance advantage. The stranger was a big, heavy buck with plenty of muscle, but one or two trifles such as treading on a turnover and failing to close when Rag was on low ground showed that he had not much cunning and counted on winning his battles by his weight. On he came at last and Rag met him like a little fury. As they came together they leaped up and struck out with their hind feet. *Thud, thud* they came, and down went poor little Rag. In a moment the stranger was on him with his teeth and Rag was bitten, and lost several tufts of hair before he could get up. But he was swift of foot and got out of reach. Again he charged and again he was knocked down and bitten severely. He was no match for his foe, and it soon became a question of saving his own life.

Hurt as he was he sprang away, with the stranger in full chase, and bound to kill him as well as to oust him from the Swamp where he was born. Rag's legs were good and so was his wind. The stranger was big and so heavy that he soon gave up the chase, and it was well for poor Rag that he did, for he was getting stiff from his wounds as well as tired. From that day began a reign of terror for Rag. His training had been against owls, dogs, weasels, men, and so on, but what to do when chased by another rabbit, he did not know. All he knew was to lay low till he was found, then run.

Poor little Molly was completely terrorized; she could not help Rag and sought only to hide. But the big buck soon found her out. She tried to run from him, but she was not now so swift as Rag. The stranger made no attempt to kill her, but he made love to her, and because she hated him and tried to get away, he treated her shamefully. Day after day he worried her by following her about, and often, furious at her lasting hatred, he would knock her down and tear out mouthfuls of her soft fur till his rage cooled somewhat, when he would let her go for a while. But his fixed purpose was to kill Rag, whose escape seemed hopeless. There was no other swamp he could go to, and whenever he took a nap now he had to be ready at any moment to dash for his life. A dozen times a day the big stranger came creeping up to where he slept, but each time the watchful Rag awoke in time to escape. To escape yet not to escape. He saved his life indeed, but oh! what a miserable life it had become. How maddening to be thus helpless, to see his little mother daily beaten and torn, as well as to see all his favorite feeding-grounds, the cosy nooks, and the pathways he had made with so much labor, forced from him by this hateful brute. Unhappy Rag realized that to the victor belong the spoils, and he hated him more than ever he did fox or ferret.

How was it to end? He was wearing out with running and watching and bad food, and little Molly's strength and spirit were breaking down under the long persecution. The stranger was ready to go to all lengths to destroy poor Rag, and at last stooped to the worst crime known among rabbits. However much they may hate each other, all good rabbits forget their feuds when their common enemy appears. Yet one day when a great goshawk came swooping over the Swamp, the stranger, keeping well under cover himself, tried again and again to drive Rag into the open.

Once or twice the hawk nearly had him, but still the briers saved him, and it was only when the big buck himself came near being caught that he gave it up. And again Rag escaped, but was no better off. He made up his mind to leave, with his mother, if possible, next night and go into the world in quest of some new home when he heard old Thunder, the hound, sniffing and searching about the outskirts of the swamp, and he resolved on playing a desperate game. He deliberately crossed the hound's view, and the chase that then began was fast and furious. Thrice around the Swamp they went till Rag had made sure that his mother was hidden safely and that his hated foe was in his usual nest. Then right into that nest and plump over him he jumped, giving him a rap with one hind foot as he passed over his head.

"You miserable fool, I kill you yet," cried the stranger, and up he jumped only to find himself between Rag and the dog and heir to all the peril of the chase.

On came the hound baying hotly on the straight-away scent. The buck's weight and size were great advantages in a rabbit fight, but now they were fatal. He did not know many tricks. Just the simple ones like "double," "wind," and "hole-up," that every baby Bunny knows. But the chase was too close for doubling and winding, and he didn't know where the holes were.

It was a straight race. The brierrose, kind to all rabbits alike, did its best, but it was no use. The baying of the hound was fast and steady. The crashing of the brush and the yelping of the hound each time the briers tore his tender ears were borne to the two rabbits where they crouched in hiding. But suddenly these sounds stopped, there was a scuffle, then loud and terrible screaming.

Rag knew what it meant and it sent a shiver through him, but he soon forgot that when all was over and rejoiced to be once more the master of the dear old Swamp.

VIII

Old Olifant had doubtless a right to burn all those brush-piles in the east and south of the Swamp and to clear up the wreck of the old barbed-wire hog-pen just below the spring. But it was none the less hard on Rag and his mother. The first were their various residences and outposts, and the second their grand fastness and safe retreat.

They had so long held the Swamp and felt it to be their very own in every part and suburb,—including Olifant's grounds and buildings—that they would have resented the appearance of another rabbit even about the adjoining barnyard.

Their claim, that of long, successful occupancy, was exactly the same as that by which most nations hold their land, and it would be hard to find a better right.

During the time of the January thaw the Olifants had cut the rest of the large wood about the pond and curtailed the Cottontails' domain on all sides. But they still clung to the dwindling Swamp, for it was their home and they were loath to move to foreign parts. Their life of daily perils went on, but they were still fleet of foot, long of wind, and bright of wit. Of late they had been somewhat troubled by a mink that had wandered up-stream to their quiet nook. A little judicious guidance had transferred the uncomfortable visitor to Olifant's henhouse. But they were not yet quite sure that he had been properly looked after. So for the present they gave up using the ground-holes, which were, of course, dangerous blind-alleys, and stuck closer than ever to the briers and the brush-piles that were left.

That first snow had quite gone and the weather was bright and warm until now. Molly, feeling a touch of rheumatism, was somewhere in the lower thicket seeking a teaberry tonic. Rag was sitting in the weak sunlight on a bank in the east side. The smoke from the familiar gable chimney of Olifant's house came fitfully drifting a pale blue haze through the underwoods and showing as a dull brown against the brightness of the sky. The sun-gilt gable was cut off midway by the banks of brierbrush, that purple in shadow shone like rods of blazing crimson and gold in the light. Beyond the house the barn with its gable and roof, new gilt as the house, stood up like a Noah's ark.

The sounds that came from it, and yet more the delicious smell that mingled with the smoke, told Rag that the animals were being fed cabbage in the yard. Rag's mouth watered at the idea of the feast. He blinked and blinked as he snuffed its odorous promises, for he loved cabbage dearly. But then he had been to the barnyard the night before after a few paltry clover-tops, and no wise rabbit would go two nights running to the same place.

Therefore he did the wise thing. He moved across where he could not smell the cabbage and made his supper of a bundle of hay that had been blown from the stack. Later, when about to settle for the night, he was joined by Molly, who had taken her teaberry and then eaten her frugal meal of sweet birch near the Sunning Bank.

Meanwhile the sun had gone about his business elsewhere, taking all his gold and glory with him. Off in the east a big black shutter came pushing up and rising higher and higher; it spread over the whole sky, shut out all light and left the world a very gloomy place indeed. Then another mischief-maker, the wind, taking advantage of the sun's absence, came on the scene and set about brewing trouble. The weather turned colder and colder; it seemed worse than when the ground had been covered with snow.

"Isn't this terribly cold? How I wish we had our stove-pipe brush-pile," said Rag.

"A good night for the pine-root hole," replied Molly, "but we have not yet seen the pelt of that mink on the end of the barn, and it is not safe till we do."

The hollow hickory was gone—in fact at this very moment its trunk, lying in the wood-yard, was harboring the mink they feared. So the Cottontails hopped to the south side of the pond and, choosing a brush-pile, they crept under and snuggled down for the night, facing the wind but with their noses in different directions so as to go out different ways in case of alarm. The wind blew harder and colder as the hours went by, and about midnight a fine icy snow came ticking down on the dead leaves and hissing through the brush heap. It might seem a poor night for hunting, but that old fox from Springfield was out. He came pointing up the wind in the shelter of the Swamp and chanced in the lee of the brush-pile, where he scented the sleeping Cottontails. He halted for a moment, then came stealthily sneaking up toward the brush under which his nose told him the rabbits were crouching. The noise of the wind and the sleet enabled him to come quite close before Molly heard the faint crunch of a dry leaf under his paw. She touched Rag's whiskers, and both were fully awake just as the fox sprang on them; but they always slept with their legs ready for a jump. Molly darted out into the blinding storm. The fox missed his spring but followed like a racer, while Rag dashed off to one side.

There was only one road for Molly; that was straight up the wind, and bounding for her life she gained a little over the unfrozen mud that would not carry the fox, till she reached the margin of the pond. No chance to turn now, on she must go.

Splash! splash! through the weeds she went, then plunge into the deep water.

And plunge went the fox close behind. But it was too much for Reynard on such a night. He turned back, and Molly, seeing only one course, struggled through the reeds into the deep water and struck out for the other shore. But there was a strong headwind. The little waves, icy cold, broke over her head as she swam, and the water was full of snow that blocked her way like soft ice, or floating mud. The dark line of the other shore seemed far, far away, with perhaps the fox waiting for her there.

But she laid her ears flat to be out of the gale, and bravely put forth all her strength with wind and tide against her. After a long, weary swim in the cold water, she had nearly reached the farther reeds when a great mass of floating snow barred her road; then the wind on the bank made strange, fox-like sounds that robbed her of all force, and she was drifted far backward before she could get free from the floating bar.

Again she struck out, but slowly—oh so slowly now. And when at last she reached the lee of the tall reeds, her limbs were numbed, her strength spent, her brave little heart was sinking, and she cared no more whether the fox were there or not. Through the reeds she did indeed pass, but once in the weeds her course wavered and slowed, her feeble strokes no longer sent her landward, the ice forming around her, stopped her altogether. In a little while the cold, weak limbs ceased to move, the furry nose-tip of the little mother Cottontail wobbled no more, and the soft brown eyes were closed in death.

But there was no fox waiting to tear her with ravenous jaws. Rag had escaped the first onset of the foe, and as soon as he regained his wits he came running back to change-off and so help his mother. He met the old fox going round the pond to meet Molly and led him far and away, then dismissed him with a barbed-wire gash on his head, and came to the bank and sought about and trailed and thumped, but all his searching was in vain; he could not find his little mother. He never saw her again, and he never knew whither she went, for she slept her never-waking sleep in the ice-arms of her friend the Water that tells no tales.

Poor little Molly Cottontail! She was a true heroine, yet only one of unnumbered millions that without a thought of heroism have lived and done their best in their little world, and died. She fought a good fight in the battle of life. She was good stuff; the stuff that never dies. For flesh of her flesh and brain of her brain was Rag. She lives in him, and through him transmits a finer fibre to her race.

And Rag still lives in the Swamp. Old Olifant died that winter, and the unthrifty sons ceased to clear the Swamp or mend the wire fences. Within a single year it was a wilder place than ever; fresh trees and brambles grew, and falling wires made many Cottontail castles and last retreats that dogs and foxes dared not storm. And there to this day lives Rag. He is a big strong buck now and fears no rivals. He has a large family of his own, and a pretty brown wife that he got I know not where. There, no doubt, he and his children's children will flourish for many years to come, and there you may see them any sunny evening if you have learnt their signal code, and choosing a good spot on the ground, know just how and when to thump it.

John Burroughs

BIRCH BROWSINGS

The region of which I am about to speak lies in the southern part of the State of New York, and comprises parts of three counties,—Ulster, Sullivan, and Delaware. It is drained by tributaries of both the Hudson and Delaware, and, next to the Adirondack section, contains more wild land than any other tract in the State. The mountains which traverse it, and impart to it its severe northern climate, belong properly to the Catskill range. On some maps of the State they are called the Pine Mountains, though with obvious local impropriety, as pine, so far as I have observed, is nowhere found upon them. "Birch Mountains" would be a more characteristic name, as on their summits birch is the prevaling tree . . .

In 1868 a party of three of us set out for a brief trouting excursion to a body of water called Thomas's Lake, situated in the same chain of mountains. On this excursion, more particularly than on any other I have ever undertaken, I was taught how poor an Indian I should make, and what a ridiculous figure a party of men may cut in the woods when the way is uncertain and the mountains high.

We left our team at a farmhouse near the head of the Mill Brook, one June afternoon, and with knapsacks on our shoulders struck into the woods at the base of the mountain, hoping to cross the range that intervened between us and the lake by sunset. We engaged a good-natured but rather indolent young man, who happened to be stopping at the house, and who had carried a knapsack in the Union armies, to pilot us a couple of miles into the woods so as to guard against any mistakes at the outset. It seemed the easiest thing in the world to find the lake. The lay of the land was so simple, according to accounts, that I felt sure I could go to it in the dark. "Go up this little brook to its source on the side of the mountain," they said. "The valley that contains the lake heads directly on the other side." What could be easier? But on a little further inquiry, they said we should "bear well to the left" when we reached the top of the mountain. This opened the doors again; "bearing well to the left" was an uncertain performance in strange woods. We might bear so well to the left that it would bring us ill. But why bear to the left at all, if the lake was directly opposite? Well, not quite opposite; a little to the left. There were two or three other valleys that headed in near there. We could easily find the right one. But to make assurance doubly sure, we engaged a guide, as stated, to give us a good start, and go with us beyond the bearing-to-the-left point. He had been to the lake the winter before and knew the way. Our course, the first half hour, was along an obscure wood-road which had been used for drawing ash logs off the mountain in winter. There was some hemlock, but more maple and birch. The woods were dense and free from underbrush, the ascent gradual. Most of the way we kept the voice of the creek in our ear on the right. I approached it once, and found it swarming with trout. The water was as cold as one ever need wish. After a while the ascent grew steeper, the creek became a mere rill that issued from beneath loose, moss-covered rocks and stones, and with much labor and puffing we drew ourselves up the rugged declivity. Every mountain has its steepest point, which is usually near the summit, in keeping, I suppose, with the providence that makes the darkest hour just before day. It is steep, steeper, steepest, till you emerge on the smooth level or gently rounded space at the top, which the old ice-gods polished off so long ago.

We found this mountain had a hollow in its back where the ground was soft and swampy. Some gigantic ferns, which we passed through, came nearly to our shoulders. We passed also several patches of swamp honeysuckles, red with blossoms.

Our guide at length paused on a big rock where the land began to dip down the other way, and concluded that he had gone far enough, and that we would now have no difficulty in finding the lake. "It must lie right down there," he said, pointing with his hand. But it was plain that he was not quite sure in his own mind. He had several times wavered in his course, and had shown considerable embarrassment when bearing to the left across the summit. Still we thought little of it. We were full of confidence, and, bidding him adieu, plunged down the mountain-side, following a spring run that we had no doubt led to the lake.

In these woods, which had a southeastern exposure, I first began to notice the wood thrush. In coming up the other side I had not seen a feather of any kind, or heard a note. Now the golden *trillide-de* of the wood thrush sounded through the silent woods. While looking for a fish-pole about half way down the mountain, I saw a thrush's nest in a little sapling about ten feet from the ground.

After continuing our descent till our only guide, the spring run, became quite a trout brook, and its tiny murmur a loud brawl, we began to peer anxiously through the trees for a glimpse of the lake, or for some conformation of the land that would indicate its proximity. An object which we vaguely discerned in looking under the near trees and over the more distant ones proved, on further inspection, to be a patch of plowed ground. Presently we made out a burnt fallow near it. This was a wet blanket to our enthusiasm. No lake, no sport, no trout for supper that night. The rather indolent young man had either played us a trick, or, as seemed more likely, had missed the way. . . .

When we lay down, there was apparently not a mosquito in the woods; but the "no-see-ems," as Thoreau's Indian aptly named the midges, soon found us out, and after the fire had gone down annoyed us much. My hands and wrists suddenly began to smart and itch in a most unaccountable manner. My first thought was that they had been poisoned in some way. Then the smarting extended to my neck and face, even to my scalp, when I began to suspect what was the matter. So, wrapping myself up more thoroughly, and stowing my hands away as best I could, I tried to sleep, being some time behind my companions, who appeared not to mind the "no-see-ems." I was further annoyed by some little irregularity on my side of the couch. The chambermaid had not beaten it up well. One huge lump refused to be mollified, and each attempt to adapt it to some natural hollow in my own body brought only a moment's relief. But at last I got the better of this also and slept. Late in the night I woke up, just in time to hear a golden-crowned thrush sing in a tree near by. It sang as loud and cheerily as at midday, and I thought myself after all, quite in luck. Birds occasionally sing at night, just as the cock crows. I have heard the hairbird, and the note of the kingbird; and the ruffed grouse infrequently drums at night. . . .

As soon as it was fairly light we were up and ready to resume our march. A small bit of bread-and-butter and a swallow or two of whiskey was all we had for breakfast that morning. Our supply of each was very limited, and we were anxious to save a little of both, to relieve the diet of trout to which we looked forward.

At an early hour we reached the rock where we had parted with the guide, and looked around us into the dense, trackless woods with many misgivings. To strike out now on our own hook, where the way was so blind and after the experience we had just had, was a step not to be carelessly taken. The tops of these mountains are so broad, and a short distance in the woods seems so far, that one is by no means master of the situation after reaching the summit. And then there are so many spurs and offshoots and changes of direction, added to the impossibility of making any generalization by the aid of the eye, that before one is aware of it he is very wide of his mark. . . .

After looking in vain for the line of marked trees, we moved off to the left in a doubtful, hesitating manner, keeping on the highest ground and blazing the trees as we went. We were afraid to go down hill, lest we should descend too soon; our vantage-ground was high ground. A thick fog coming on, we were more bewildered than ever. Still we pressed forward, climbing up ledges and wading through ferns for about two hours, when we paused by a spring that issued from beneath an immense wall of rock that belted the highest part of the mountain. There was quite a broad plateau here, and the birch wood was very dense, and the trees of unusual size.

After resting and exchanging opinions, we all concluded

that it was best not to continue our search incumbered as we were; but we were not willing to abandon it altogether, and I proposed to my companions to leave them beside the spring with our traps, while I made one thorough and final effort to find the lake. If I succeeded and desired them to come forward, I was to fire my gun three times; if I failed and wished to return, I would fire it twice, they of course responding.

So, filling my canteen from the spring, I set out again, taking the spring run for my guide. Before I had followed it two hundred yards it sank into the ground at my feet. I had half a mind to be superstitious and to believe that we were under a spell, since our guides played us such tricks. However, I determined to put the matter to a further test, and struck out boldly to the left. This seemed to be the keyword,—to the left, to the left. The fog had now lifted, so that I could form a better idea of the lay of the land. Twice I looked down the steep sides of the mountain, sorely tempted to risk a plunge. Still I hesitated and kept along on the brink. As I stood on a rock deliberating, I heard a crackling of the brush, like the tread of some large game, on a plateau below me. Suspecting the truth of the case, I moved stealthily down, and found a herd of young cattle leisurely browsing. We had several times crossed their trail, and had seen that morning a level, grassy place on the top of the mountain, where they had passed the night. Instead of being frightened, as I had expected, they seemed greatly delighted, and gathered around me as if to inquire the tidings from the outer world,—perhaps the quotations of the cattle market. They came up to me, and eagerly licked my hand, clothes, and gun. Salt was what they were after, and they were ready to swallow anything that contained the smallest percentage of it. They were mostly yearlings and as sleek as moles. They had a very gamy look. We were afterwards told that, in the spring, the farmers round about turn into these woods their young cattle, which do not come out again till fall. They are then in good condition,—not fat, like grass-fed cattle, but trim and supple, like deer. Once a month the owner hunts them up and salts them. They have their beats, and seldom wander beyond well-defined limits. It was interesting to see them feed. They browsed on the low limbs and bushes, and on the various plants, munching at everything without any apparent discrimination.

They attempted to follow me, but I escaped them by clambering down some steep rocks. I now found myself gradually edging down the side of the mountain, keeping around it in a spiral manner, and scanning the woods and the shape of the ground for some encouraging hint or sign. Finally the woods became more open, and the descent less rapid. The trees were remarkably straight and uniform in size. Black birches, the first I had seen, were very numerous. I felt encouraged. Listening attentively, I caught, from a breeze just lifting the drooping leaves, a sound that I willingly believed was made by a bullfrog. On this hint, I tore down through the woods at my highest speed. Then I paused and listened again. This time there was no mistaking it; it was the sound of frogs. Much elated, I rushed on. By and by I could hear them as I ran. *Pthrung, pthrung,* croaked the old ones; *pug, pug,* shrilly joined in the smaller fry.

Then I caught, through the lower trees, a gleam of blue, which I first thought was distant sky. A second look and I knew it to be water, and in a moment more I stepped from the woods and stood upon the shore of the lake. I exulted silently. There it was at last, sparkling in the morning sun, and as beautiful as a dream. It was so good to come upon such open space and such bright hues, after wandering in the dim, dense woods! The eye is as delighted as an escaped bird, and darts gleefully from point to point.

The lake was a long oval, scarcely more than a mile in circumference, with evenly wooded shores, which rose gradually on all sides. After contemplating the scene for a moment,

I stepped back into the woods, and, loading my gun as heavily as I dared, discharged it three times. The reports seemed to fill all the mountains with sound. The frogs quickly hushed, and I listened for the response. But no response came. Then I tried again and again, but without evoking an answer. One of my companions, however, who had climbed to the top of the high rocks in the rear of the spring, thought he heard faintly one report. It seemed an immense distance below him, and far around under the mountain. I knew I had come a long way, and hardly expected to be able to communicate with my companions in the manner agreed upon. I therefore started back, choosing my course without any reference to the circuitous route by which I had come, and loading heavily and firing at intervals. I must have aroused many long-dormant echoes from a Rip Van Winkle sleep. As my powder got low, I fired and hallooed alternately, till I came near splitting both my throat and gun. Finally, after I had begun to have a very ugly feeling of alarm and disappointment, and to cast about vaguely for some course to pursue in the emergency that seemed near at hand,—namely, the loss of my companions now I had found the lake,—a favoring breeze brought me the last echo of a response. I rejoined with spirit, and hastened with all speed in the direction whence the sound had come, but, after repeated trials, failed to elicit another answering sound. This filled me with apprehension again. I feared that my friends had been misled by the reverberations, and I pictured them to myself hastening in the opposite direction. Paying little attention to my course, but paying dearly for my carelessness afterward, I rushed forward to undeceive them. But they had not been deceived, and in a few moments an answering shout revealed them near at hand. I heard their tramp, the bushes parted, and we three met again.

In answer to their eager inquiries, I assured them that I had seen the lake, that it was at the foot of the mountain, and that we could not miss it if we kept straight down from where we then were.

My clothes were soaked with perspiration, but I shouldered my knapsack with alacrity, and we began the descent. I noticed that the woods were much thicker, and had quite a different look from those I had passed through, but thought nothing of it, as I expected to strike the lake near its head, whereas I had before come out at its foot. We had not gone far when we crossed a line of marked trees, which my companions were disposed to follow. It intersected our course nearly at right angles, and kept along and up the side of the mountain. My impression was that it led up from the lake, and that by keeping our own course we should reach the lake sooner than if we followed this line.

About half way down the mountain, we could see through the interstices the opposite slope. I encouraged my comrades by telling them that the lake was between us and that, and not more than half a mile distant. We soon reached the bottom, where we found a small stream and quite an extensive alder swamp, evidently the ancient bed of a lake. I explained to my half-vexed and half-incredulous companions that we were probably above the lake, and that this stream must lead to it. "Follow it," they said; "we will wait here till we hear from you."

So I went on, more than ever disposed to believe that we were under a spell, and that the lake had slipped from my grasp after all. Seeing no favorable sign as I went forward, I laid down my accoutrements, and climbed a decayed beech that leaned out over the swamp and promised a good view from the top. As I stretched myself up to look around from the highest attainable branch, there was suddenly a loud crack at the root. With a celerity that would at least have done credit to a bear, I regained the ground, having caught but a momentary glimpse of the country, but enough to convince me no lake was near. Leaving all incumbrances here but my gun, I still pressed on, loath to be thus baffled. After floundering through another alder swamp for nearly half a mile, I flattered myself that I was close on to the lake. I caught sight of a low spur of the mountain sweeping around like a half-extended arm, and I fondly imagined that within its clasp was the object of my search. But I found only more alder swamp. After this region was cleared, the creek began to descend the mountain very rapidly. Its banks became high and narrow, and it went whirling away with a sound that seemed to my ears like a burst of ironical laughter. I turned back with a feeling of mingled disgust, shame, and vexation. In fact I was almost sick, and when I reached my companions, after an absence of nearly two hours, hungry, fatigued, and disheartened, I would have sold my interest in Thomas's Lake at a very low figure. For the first time, I heartily wished myself well out of the woods. Thomas might keep his lake, and the enchanters guard his possession! I doubted if he had ever found it the second time, or if any one else ever had.

My companions, who were quite fresh, and who had not felt the strain of baffled purpose as I had, assumed a more encouraging tone. After I had rested awhile, and partaken sparingly of the bread and whiskey, which in such an emergency is a great improvement on bread and water, I agreed to their proposition that we should make another attempt. As if to reassure us, a robin sounded his cheery call near by, and the winter wren, the first I had heard in these woods, set his music-box going, which fairly ran over with fine, gushing, lyrical sounds. There can be no doubt but this bird is one of our finest songsters. If it would only thrive and sing well when caged, like the canary, how far it would surpass that bird! It has all the vivacity and versatility of the canary, without any of its shrillness. Its song is indeed a little cascade of melody.

We again retraced our steps, rolling the stone, as it were, back up the mountain, determined to commit ourselves to the line of marked trees. These we finally reached, and, after exploring the country to the right, saw that bearing to the left was still the order. The trail led up over a gentle rise of ground, and in less than twenty minutes we were in the woods I had passed through when I found the lake. The error I had made was then plain; we had come off the mountain a few paces too far to the right, and so had passed down on the wrong side of the ridge, into what we afterwards learned was the valley of Alder Creek.

We now made good time, and before many minutes I again saw the mimic sky glance through the trees. As we approached the lake a solitary woodchuck, the first wild animal we had seen since entering the woods, sat crouched upon the root of a tree a few feet from the water, apparently completely nonplussed by the unexpected appearance of danger on the land side. All retreat was cut off, and he looked his fate in the face without flinching. I slaughtered him just as a savage would have done, and from the same motive,—I wanted his carcass to eat. . . .

The birds were unusually plentiful and noisy about the head of this lake; robins, blue jays, and woodpeckers greeted me with their familiar notes. The blue jays found an owl or some wild animal a short distance above me, and, as is their custom on such occasions, proclaimed it at the top of their voices, and kept on till the darkness began to gather in the woods.

I also heard here, as I had at two or three other points in the course of the day, the peculiar, resonant hammering of some species of woodpecker upon the hard, dry limbs. It was unlike any sound of the kind I had ever before heard, and, repeated at intervals through the silent woods, was a very marked and characteristic feature. Its peculiarity was the or-dered succession of the raps, which gave it the character of a premeditated performance. There were first three strokes fol-lowing each other rapidly, then two much louder ones with longer intervals between them. I heard the drumming here, and the next day at sunset at Furlow Lake, the source of Dry

Brook, and in no instance was the order varied. There was melody in it, such as a woodpecker knows how to evoke from a smooth dry branch. It suggested something quite as pleasing as the liveliest bird-song, and was if anything more woodsy and wild. As the yellow-bellied woodpecker was the most abundant species in these woods, I attributed it to him. It is the one sound that still links itself with those scenes in my mind.

At sunset the grouse began to drum in all parts of the woods about the lake. I could hear five at one time, *thump, thump, thump, thump, thr-r-r-r-r-rr*. It was a homely, welcome sound. As I returned to camp at twilight, along the shore of the lake, the frogs also were in full chorus. The older ones ripped out their responses to each other with terrific force and volume. I know of no other animal capable of giving forth so much sound, in proportion to its size, as a frog. Some of these seemed to bellow as loud as a two-year-old bull. They were of immense size, and very abundant. No frog-eater had ever been there. Near the shore we felled a tree which reached far out in the lake. Upon the trunk and branches the frogs had soon collected in large numbers, and gamboled and splashed about the half-submerged top, like a parcel of schoolboys, making nearly as much noise.

After dark, as I was frying the fish, a panful of the largest trout was accidentally capsized in the fire. With rueful countenances we contemplated the irreparable loss our commissariat had sustained by this mishap; but remembering there was virtue in ashes, we poked the half-consumed fish from the bed of coals and ate them, and they were good.

We lodged that night on a brush-heap and slept soundly. The green, yielding beech-twigs, covered with a buffalo robe, were equal to a hair mattress. The heat and smoke from a large fire kindled in the afternoon had banished every "no-see-em" from the locality, and in the morning the sun was above the mountain before we awoke.

I immediately started again for the inlet, and went far up the stream toward its source. A fair string of trout for breakfast was my reward. The cattle with the bell were at the head of the valley, where they had passed the night. Most of them were two-year-old steers. They came up to me and begged for salt, and scared the fish by their importunities.

We finished our bread that morning, and ate every fish we could catch, and about ten o'clock prepared to leave the lake. The weather had been admirable, and the lake was a gem, and I would gladly have spent a week in the neighborhood; but the question of supplies was a serious one, and would brook no delay. . . .

We were now close to the settlement, and began to hear human sounds. One rod more, and we were out of the woods. It took us a moment to comprehend the scene. Things looked very strange at first; but quickly they began to change and to put on familiar features. Some magic scene-shifting seemed to take place before my eyes, till, instead of the unknown settlement which I at first seemed to look upon, there stood the farmhouse at which we had stopped two days before, and at the same moment we heard the stamping of our team in the barn. We sat down and laughed heartily over our good luck. Our desperate venture had resulted better than we had dared to hope, and had shamed our wisest plans. At the house our arrival had been anticipated about this time, and dinner was being put upon the table.

It was then five o'clock, so that we had been in the woods just forty-eight hours; but if time is only phenomenal, as the philosophers say, and life only in feeling, as the poets aver, we were some months, if not years, older at that moment than we had been two days before. Yet younger, too,—though this be a paradox,—for the birches had infused into us some of their own suppleness and strength.

PART TWO

CREATURES GREAT AND SMALL:
The Twentieth-Century Naturalists

Mary Austin

THE LAND OF
LITTLE RAIN

East away from the Sierras,
south from Panamint and Amargosa, east and south many an
uncounted mile, is the Country of Lost Borders.

Ute, Paiute, Mojave, and Shoshone inhabit its frontiers,
and as far into the heart of it as a man dare go. Not the law,
but the land sets the limit. Desert is the name it wears upon
the maps, but the Indian's is the better word. Desert is a loose
term to indicate land that supports no man; whether the land
can be bitted and broken to that purpose is not proven. Void
of life it never is, however dry the air and villainous the soil.

This is the nature of that country. There are hills, rounded, blunt, burned, squeezed up out of chaos, chrome and vermilion painted, aspiring to the snow-line. Between the hills lie high level-looking plains full of intolerable sun glare, or narrow valleys drowned in a blue haze. The hill surface is streaked with ash drift and black, unweathered lava flows. After rains water accumulates in the hollows of small closed valleys, and, evaporating, leaves hard dry levels of pure desertness that get the local name of dry lakes. Where the mountains are steep and the rains heavy, the pool is never quite dry, but dark and bitter, rimmed about with the efflorescence of alkaline deposits. A thin crust of it lies along the marsh over the vegetating area, which has neither beauty nor freshness. In the broad wastes open to the wind the sand drifts in hummocks about the stubby shrubs, and between them the soil shows saline traces. The sculpture of the hills here is more wind than water work, though the quick storms do sometimes scar them past many a year's redeeming. In all the Western desert edges there are essays in miniature at the famed, terrible Grand Canyon, to which, if you keep on long enough in this country, you will come at last.

Since this is a hill country one expects to find springs, but not to depend upon them; for when found they are often brackish and unwholesome, or maddening, slow dribbles in a thirsty soil. Here you find the hot sink of Death Valley, or high rolling districts where the air has always a tang of frost. Here are the long heavy winds and breathless calms on the tilted mesas where dust devils dance, whirling up into a wide, pale sky. Here you have no rain when all the earth cries for it, or quick downpours called cloudbursts for violence. A land of lost rivers, with little in it to love; yet a land that once visited must be come back to inevitably. If it were not so there would be little told of it.

This is the country of three seasons. From June on to November it lies hot, still, and unbearable, sick with violent unrelieving storms; then on until April, chill, quiescent, drinking its scant rain and scanter snows; from April to the hot season again, blossoming, radiant, and seductive. These months are only approximate; later or earlier the rain-laden wind may drift up the water gate of the Colorado from the Gulf, and the land sets its seasons by the rain.

The desert floras shame us with their cheerful adaptations to the seasonal limitations. Their whole duty is to flower and fruit, and they do it hardly, or with tropical luxuriance, as the rain admits. It is recorded in the report of the Death Valley expedition that after a year of abundant rains, on the Colorado desert was found a specimen of Amaranthus ten feet high. A year later the same species in the same place matured in the drought at four inches. One hopes the land may breed like qualities in her human offspring, not tritely to "try," but to do. Seldom does the desert herb attain the full stature of the type. Extreme aridity and extreme altitude have the same dwarfing effect, so that we find in the high Sierras and in Death Valley related species in miniature that reach a comely growth in mean temperatures. Very fertile are the desert plants in expedients to prevent evaporation, turning their foliage edgewise toward the sun, growing silky hairs, exuding viscid gum. The wind, which has a long sweep, harries and helps them. It rolls up dunes about the stocky stems, encompassing and protective, and above the dunes, which may be, as with the mesquite, three times as high as a man, the blossoming twigs flourish and bear fruit.

There are many areas in the desert where drinkable water lies within a few feet of the surface, indicated by the mesquite and the bunch grass (*Sporobolus airoides*). It is this nearness of unimagined help that makes the tragedy of desert deaths. It is related that the final breakdown of that hapless party that gave Death Valley its forbidding name occurred in a locality where shallow wells would have saved them. But how were they to know that? Properly equipped it is possible to go safely across that ghastly sink, yet every year it takes its toll of death, and yet men find there sun-dried mummies, of whom no trace or recollection is preserved. To underestimate one's thirst, to pass a given landmark to the right or left, to find a dry spring where one looked for running water—there is no help for any of these things.

Along springs and sunken watercourses one is surprised to find such water-loving plants as grow widely in moist ground; but the true desert breeds its own kind, each in its particular habitat. The angle of the slope, the frontage of a hill, the structure of the soil determines the plant. South-looking hills are nearly bare, and the lower tree line higher here by a thousand feet. Canyons running east and west will have one wall naked and one clothed. Around dry lakes and marshes the herbage preserves a set and orderly arrangement. Most species have well-defined areas of growth, the best index the voiceless land can give the traveler of his whereabouts.

If you have any doubt about it, know that the desert begins with the creosote. This immortal shrub spreads down into Death Valley and up to the lower timber line, odorous and medicinal, as you might guess from the name, wandlike, with shining fretted foliage. Its vivid green is grateful to the eye in a wilderness of gray and greenish white shrubs. In the spring it exudes a resinous gum which the Indians of those parts know how to use with pulverized rock for cementing arrow points to shafts. Trust Indians not to miss any virtues of the plant world!

Nothing the desert produces expresses it better than the unhappy growth of the tree yucca. Tormented, thin forests of it stalk drearily in the high mesas, particularly in that triangular slip that fans out eastward from the meeting of the Sierras and coastwise hills where the first swings across the southern end of the San Joaquin Valley. The yucca bristles with bayonet-pointed leaves, dull green, growing shaggy with age, tipped with panicles of fetid, greenish bloom. After death, which is slow, the ghostly hollow network of its woody skeleton, with hardly power to rot, makes the moonlight fearful. Before the yucca has come to flower, while yet its bloom is a creamy cone-shaped bud of the size of a small cabbage, full of sugary sap, the Indians twist it deftly out of its fence of daggers and roast it for their own delectation. So it is that in those parts where man inhabits one sees young plants of *Yucca arborensis* infrequently. Other yuccas, cacti, low herbs, a thousand sorts, one finds journeying east from the coastwise hills. There is neither poverty of soil nor species to account for the sparseness of desert growth, but simply that each plant requires more room. So much earth must be pre-empted to extract so much moisture. The real struggle for existence, the real brain of the plant, is underground; above there is room for a rounded perfect growth. In Death Valley, reputed the very core of desolation, are nearly two hundred identified species.

Above the lower tree line, which is also the snow line, mapped out abruptly by the sun, one finds spreading growth of piñon, juniper, branched nearly to the ground, lilac and sage, and scattering white pines.

There is no special preponderance of self-fertilized or wind-fertilized plants, but everywhere the demand for and evidence of insect life. Now where there are seeds and insects there will be birds and small mammals, and where these are will come the slinking, sharp-toothed kind that prey on them. Go as far as you dare in the heart of a lonely land, you cannot go so far that life and death are not before you. Painted lizards slip in and out of rock crevices, and pant on the white hot sands. Birds, hummingbirds even, nest in the cactus scrub; woodpeckers befriend the demoniac yuccas; out of the stark, treeless waste rings the music of the night-singing mockingbird. If it be summer and the sun well down, there will be a burrowing owl to call. Strange, furry, tricksy things dart across the open places, or sit motionless in the conning towers of the creosote. The poet may have "named all the birds without a gun," but not the fairy-footed, ground-inhabiting, furtive, small folk of the rainless regions. They are too many and too swift; how many you would not believe without seeing the footprint tracings in the sand. They are nearly all night workers, finding the days too hot and white. In mid-desert where there are no cattle, there are no birds of carrion, but if you go far in that direction the chances are that you will find yourself shadowed by their tilted wings. Nothing so large as a man can move unspied upon in that country, and they know well how the land deals with strangers. There are hints to be had here of the way in which a land forces new habits on its dwellers. The quick increase of sun at the end of spring sometimes overtakes birds in their nesting and effects a reversal of the ordinary manner of incubation. It becomes necessary to keep eggs cool rather than warm. One hot, stifling spring in the Little Antelope I had occasion to pass and repass frequently the nest of a pair of meadowlarks, located unhappily in the shelter of a very slender weed. I never caught them sitting except near night, but at midday they stood, or drooped above it, half fainting with pitifully parted bills, between their treasure and the sun. Sometimes both of them together with wings spread and half lifted continued a spot of shade in a temperature that constrained me at last in a fellow feeling to spare them a bit of canvas for permanent shelter. There was a fence in that country shutting in a cattle range, and along its fifteen miles of posts one could be sure of finding a bird or two in every strip of shadow; sometimes the sparrow and the hawk, with wings trailed and beaks parted, drooping in the white trace of noon.

If one is inclined to wonder at first how so many dwellers came to be in the loneliest land that ever came out of God's hands, what they do there and why stay, one does not wonder so much after having lived there. None other than this long brown land lays such a hold on the affections. The rainbow hills, the tender bluish mists, the luminous radiance of the spring, have the lotus charm. They trick the sense of time, so that once inhabiting there you always mean to go away without quite realizing that you have not done it. Men who have lived there, miners and cattle-men, will tell you this, not so fluently, but emphatically, cursing the land and going back to it. For one thing there is the divinest, cleanest air to be breathed anywhere in God's world. Some day the world will understand that, and the little oases on the windy tops of hills will harbor for healing its ailing, house-weary broods. There is promise there of great wealth in ores and earths, which is no wealth by reason of being so far removed from water and workable conditions; but men are bewitched by it and tempted to try the impossible.

You should hear Salty Williams tell how he used to drive eighteen- and twenty-mule teams from the borax marsh to Mojave, ninety miles, with the trail wagon full of water barrels. Hot days the mules would go so mad for drink that the clank of the water bucket set them into an uproar of hideous, maimed noises, and a tangle of harness chains, while Salty would sit on the high seat with the sun glare heavy in his eyes, dealing out curses of pacification in a level, uninterested voice until the clamor fell off from sheer exhaustion. There was a line of shallow graves along that road; they used to count on dropping a man or two of every new gang of coolies brought out in the hot season. But when he lost his swamper, smitten without warning at the noon halt, Salty quit his job; he said it was "too durn hot." The swamper he buried by the way with stones upon him to keep the coyotes from digging him up, and seven years later I read the penciled lines on the pine headboard, still bright and unweathered.

But before that, driving up on the Mojave stage, I met Salty again crossing Indian Wells, his face from the high seat, tanned and ruddy as a harvest moon, looming through the golden dust above his eighteen mules. The land had called him.

The palpable sense of mystery in the desert air breeds fables, chiefly of lost treasure. Somewhere within its stark borders, if one believes report, is a hill strewn with nuggets; one seamed with virgin silver; an old clayey water bed where Indians scooped up earth to make cooking pots and shaped them reeking with grains of pure gold. Old miners drifting about the desert edges, weathered into the semblance of the tawny hills, will tell you tales like these convincingly. After a little sojourn in that land you will believe them on their own account. It is a question whether it is not better to be bitten by the little horned snake of the desert that goes sidewise and strikes without coiling, than by the tradition of a lost mine.

And yet—and yet—is it not perhaps to satisfy expectation that one falls into the tragic key in writing of desertness? The more you wish of it the more you get, and in the meantime lose much of pleasantness. In that country which begins at the foot of the east slope of the Sierras and spreads out by less and less lofty hill ranges toward the Great Basin, it is possible to live with great zest, to have red blood and delicate joys, to pass and repass about one's daily performance an area that would make an Atlantic seaboard state, and that with no peril, and according to our way of thought, no particular difficulty. At any rate, it was not people who went into the desert merely to write it up who invented the fabled Ilassayampa, of whose waters, if any drink, they can no more see fact as naked fact, but all radiant with the color of romance. I, who must have drunk of it in my twice seven years' wanderings, am assured that it is worth while.

For all the toll the desert takes of a man it gives compensations, deep breaths, deep sleep, and the communion of the stars. It comes upon one with new force in the pauses of the night that the Chaldeans were a desert-bred people. It is hard to escape the sense of mastery as the stars move in the wide clear heavens to risings and settings unobscured. They look large and near and palpitant; as if they moved on some stately service not needful to declare. Wheeling to their stations in the sky, they make the poor world-fret of no account. Of no account you who lie out there watching, nor the lean coyote that stands off in the scrub from you and howls and howls.

Loren Eiseley

THE BIRD AND THE MACHINE

I suppose their little bones
have years ago been lost among the stones and winds of those
high glacial pastures. I suppose their feathers blew eventually
into the piles of tumbleweed beneath the straggling cattle
fences and rotted there in the mountain snows, along with dead
steers and all the other things that drift to an end in the corners
of the wire. I do not quite know why I should be thinking of
birds over the *New York Times* at breakfast, particularly the
birds of my youth half a continent away. It is a funny thing
what the brain will do with memories and how it will treasure
them and finally bring them into odd juxtapositions with other

things, as though it wanted to make a design, or get some meaning out of them, whether you want it or not, or even see it.

It used to seem marvelous to me, but I read now that there are machines that can do these things in a small way, machines that can crawl about like animals, and that it may not be long now until they do more things—maybe even make themselves—I saw that piece in the *Times* just now. And then they will, maybe—well, who knows—but you read about it more and more with no one making any protest, and already they can add better than we and reach up and hear things through the dark and finger the guns over the night sky.

This is the new world that I read about at breakfast. This is the world that confronts me in my biological books and journals, until there are times when I sit quietly in my chair and try to hear the little purr of the cogs in my head and the tubes flaring and dying as the messages go through them and the circuits snap shut or open. This is the great age, make no mistake about it; the robot has been born somewhat appropriately along with the atom bomb, and the brain they say now is just another type of more complicated feedback system. The engineers have its basic principles worked out; it's mechanical, you know; nothing to get superstitious about; and man can always improve on nature once he gets the idea. Well, he's got it all right and that's why, I guess, that I sit here in my chair, with the article crunched in my hand, remembering those two birds and that blue mountain sunlight. There is another magazine article on my desk that reads "Machines Are Getting Smarter Every Day." I don't deny it, but I'll still stick with the birds. It's life I believe in, not machines.

Maybe you don't believe there is any difference. A skeleton is all joints and pulleys, I'll admit. And when man was in his simpler stages of machine building in the eighteenth century, he quickly saw the resemblances. "What," wrote Hobbes,

179

"is the heart but a spring, and the nerves but so many strings, and the joints but so many wheels, giving motion to the whole body?" Tinkering about in their shops it was inevitable in the end that men would see the world as a huge machine "subdivided into an infinite number of lesser machines."

The idea took on with a vengeance. Little automatons toured the country—dolls controlled by clockwork. Clocks described as little worlds were taken on tours by their designers. They were made up of moving figures, shifting scenes and other remarkable devices. The life of the cell was unknown. Man, whether he was conceived as possessing a soul or not, moved and jerked about like these tiny puppets. A human being thought of himself in terms of his own tools and implements. He had been fashioned like the puppets he produced and was only a more clever model made by a greater designer.

Then in the nineteenth century, the cell was discovered, and the single machine in its turn was found to be the product of millions of infinitesimal machines—the cells. Now, finally, the cell itself dissolves away into an abstract chemical machine—and that into some intangible, inexpressible flow of energy. The secret seems to lurk all about, the wheels get smaller and smaller, and they turn more rapidly, but when you try to seize it the life is gone—and so, by popular definition, some would say that life was never there in the first place. The wheels and the cogs are the secret and we can make them better in time—machines that will run faster and more accurately than real mice to real cheese.

I have no doubt it can be done, though a mouse harvesting seeds on an autumn thistle is to me a fine sight and more complicated, I think, in his multiform activity, than a machine "mouse" running a maze. Also, I like to think of the possible shape of the future brooding in mice, just as it brooded once in a rather ordinary mousy insectivore who became a man. It leaves a nice fine indeterminate sense of wonder that even an electronic brain hasn't got, because you know perfectly well that if the electronic brain changes, it will be because of something man has done to it. But what man will do to himself he doesn't really know. A certain scale of time and a ghostly intangible thing called change are ticking in him. Powers and potentialities like the oak in the seed, or a red and awful ruin. Either way, it's impressive; and the mouse has it, too. Or those birds, I'll never forget those birds—yet before I measured their significance, I learned the lesson of time first of all. I was young then and left alone in a great desert—part of an expedition that had scattered its men over several hundred miles in order to carry on research more effectively. I learned there that time is a series of planes existing superficially in the same universe. The tempo is a human illusion, a subjective clock ticking in our own kind of protoplasm.

As the long months passed, I began to live on the slower planes and to observe more readily what passed for life there. I sauntered, I passed more and more slowly up and down the canyons in the dry baking heat of midsummer. I slumbered for long hours in the shade of huge brown boulders that had gathered in tilted companies out on the flats. I had forgotten the world of men and the world had forgotten me. Now and then I found a skull in the canyons, and these justified my remaining there. I took a serene cold interest in these discoveries. I had come, like many a naturalist before me, to view life with a wary and subdued attention. I had grown to take pleasure in the divested bone.

I sat once on a high ridge that fell away before me into a waste of sand dunes. I sat through hours of a long afternoon. Finally, as I glanced beside my boot an indistinct configuration caught my eye. It was a coiled rattlesnake, a big one. How long he had sat with me I do not know. I had not frightened him. We were both locked in the sleep-walking tempo of the earlier world, baking in the same high air and sunshine. Perhaps he had been there when I came. He slept on as I left, his coils, so ill discerned by me, dissolving once more among the stones and gravel from which I had barely made him out.

Another time I got on a higher ridge, among some tough little wind-warped pines half covered over with sand in a basin-like depression that caught everything carried by the air up to those heights. There were a few thin bones of birds, some cracked shells of indeterminable age, and the knotty fingers of pine roots bulged out of shape from their long and agonizing grasp upon the crevices of the rock. I lay under the pines in the sparse shade and went to sleep once more.

It grew cold finally, for autumn was in the air by then, and the few things that lived thereabouts were sinking down into an even chillier scale of time. In the moments between sleeping and waking I saw the roots about me and slowly, slowly, a foot in what seemed many centuries, I moved my sleep-stiffened hands over the scaling bark and lifted my numbed face after the vanishing sun. I was a great awkward thing of knots and aching limbs, trapped up there in some long, patient endurance that involved the necessity of putting living fingers into rock and by slow, aching expansion bursting those rocks asunder. I suppose, so thin and slow was the time of my pulse by then, that I might have stayed on to drift still deeper into the lower cadences of the frost, or the crystalline life that glisters in pebbles, or shines in a snowflake, or dreams in the meteoric iron between the worlds.

It was a dim descent, but time was present in it. Somewhere far down in that scale the notion struck me that one might come the other way. Not many months thereafter I joined some colleagues heading higher into a remote windy tableland where huge bones were reputed to protrude like boulders from the turf. I had drowsed with reptiles and moved with the century-long pulse of trees; now, lethargically, I was climbing back up some invisible ladder of quickening hours. There had been talk of birds in connection with my duties. Birds are intense, fast-living creatures—reptiles, I suppose one might say, that have escaped out of the heavy sleep of time, transformed fairy creatures dancing over sunlit meadows. It is a youthful fancy, no doubt, but because of something that happened up there among the escarpments of that range, it remains with me a life-long impression. I can never bear to see a bird imprisoned.

We came into that valley through the trailing mists of a spring night. It was a place that looked as though it might never have known the foot of man, but our scouts had been ahead of us and we knew all about the abandoned cabin of stone that lay far up on one hillside. It had been built in the land rush of the last century and then lost to the cattlemen again as the marginal soils failed to take to the plow.

There were spots like this all over that country. Lost graves marked by unlettered stones and old corroding rim-fire cartridge cases lying where somebody had made a stand among the boulders that rimmed the valley. They are all that remain of the range wars; the men are under the stones now. I could see our cavalcade winding in and out through the mist below us: torches, the reflection of the truck lights on our collecting tins, and the far-off bumping of a loose dinosaur thigh bone in the bottom of a trailer. I stood on a rock a moment looking down and thinking what it cost in money and equipment to capture the past.

We had, in addition, instructions to lay hands on the present. The word had come through to get them alive—birds, reptiles, anything. A zoo somewhere abroad needed restocking. It was one of those reciprocal matters in which science involves itself. Maybe our museum needed a stray ostrich egg and this was the payoff. Anyhow, my job was to help capture some birds and that was why I was there before the trucks.

The cabin had not been occupied for years. We intended

to clean it out and live in it, but there were holes in the roof and the birds had come in and were roosting in the rafters. You could depend on it in a place like this where everything blew away, and even a bird needed some place out of the weather and away from coyotes. A cabin going back to nature in a wild place draws them till they come in, listening at the eaves, I imagine, pecking softly among the shingles till they find a hole and then suddenly the place is theirs and man is forgotten.

Sometimes of late years I find myself thinking the most beautiful sight in the world might be the birds taking over New York after the last man has run away to the hills. I will never live to see it, of course, but I know just how it will sound because I've lived up high and I know the sort of watch birds keep on us. I've listened to sparrows tapping tentatively on the outside of air conditioners when they thought no one was listening, and I know how other birds test the vibrations that come up to them through the television aerials.

"Is he gone?" they ask, and the vibrations come up from below, "Not yet, not yet."

Well, to come back, I got the door open softly and I had the spotlight all ready to turn on and blind whatever birds there were so they couldn't see to get out through the roof. I had a short piece of ladder to put against the far wall where there was a shelf on which I expected to make the biggest haul. I had all the information I needed just like any skilled assassin. I pushed the door open, the hinges squeaking only a little. A bird or two stirred—I could hear them—but nothing flew and there was a faint starlight through the holes in the roof.

I padded across the floor, got the ladder up and the light ready, and slithered up the ladder till my head and arms were over the shelf. Everything was dark as pitch except for the starlight at the little place back of the shelf near the eaves. With the light to blind them, they'd never make it. I had them.

I reached my arm carefully over in order to be ready to seize whatever was there and I put the flash on the edge of the shelf where it would stand by itself when I turned it on. That way I'd be able to use both hands.

Everything worked perfectly except for one detail—I didn't know what kind of birds were there. I never thought about it at all, and it wouldn't have mattered if I had. My orders were to get something interesting. I snapped on the flash and sure enough there was a great beating and feathers flying, but instead of my having them, they, or rather he, had me. He had my hand, that is, and for a small hawk not much bigger than my fist he was doing all right. I heard him give one short metallic cry when the light went on and my hand descended on the bird beside him; after that he was busy with his claws and his beak was sunk in my thumb. In the struggle I knocked the lamp over on the shelf, and his mate got her sight back and whisked neatly through the hole in the roof and off among the stars outside. It all happened in fifteen seconds and you might think I would have fallen down the ladder, but no, I had a professional assassin's reputation to keep up, and the bird, of course, made the mistake of thinking the hand was the enemy and not the eyes behind it. He chewed my thumb up pretty effectively and lacerated my hand with his claws, but in the end I got him, having two hands to work with.

He was a sparrow hawk and a fine young male in the prime of life. I was sorry not to catch the pair of them, but as I dripped blood and folded his wings carefully, holding him by the back so that he couldn't strike again, I had to admit the two of them might have been more than I could have handled under the circumstances. The little fellow had saved his mate by diverting me, and that was that. He was born to it, and made no outcry now, resting in my hand hopelessly, but peering toward me in the shadows behind the lamp with a fierce, almost indifferent glance. He neither gave nor expected mercy and something out of the high air passed from him to me, stirring a faint embarrassment.

I quit looking into that eye and managed to get my huge carcass with its fist full of prey back down the ladder. I put the bird in a box too small to allow him to injure himself by struggle and walked out to welcome the arriving trucks. It had been a long day, and camp still to make in the darkness. In the morning that bird would be just another episode. He would go back with the bones in the truck to a small cage in a city where he would spend the rest of his life. And a good thing, too. I sucked my aching thumb and spat out some blood. An assassin has to get used to these things. I had a professional reputation to keep up.

In the morning, with the change that comes on suddenly in that high country, the mist that had hovered below us in the valley was gone. The sky was a deep blue, and one could see for miles over the high outcroppings of stone. I was up early and brought the box in which the little hawk was imprisoned out onto the grass where I was building a cage. A wind as cool as a mountain spring ran over the grass and stirred my hair. It was a fine day to be alive. I looked up and all around and at the hole in the cabin roof out of which the other little hawk had fled. There was no sign of her anywhere that I could see.

"Probably in the next county by now," I thought cynically, but before beginning work I decided I'd have a look at my last night's capture.

Secretively, I looked again all around the camp and up and down and opened the box. I got him right out in my hand with his wings folded properly and I was careful not to startle him. He lay limp in my grasp and I could feel his heart pound under the feathers but he only looked beyond me and up.

I saw him look that last look away beyond me into a sky so full of light that I could not follow his gaze. The little breeze flowed over me again, and nearby a mountain aspen shook all its tiny leaves. I suppose I must have had an idea then of what I was going to do, but I never let it come up into consciousness. I just reached over and laid the hawk on the grass.

He lay there a long minute without hope, unmoving, his eyes still fixed on that blue vault above him. It must have been that he was already so far away in heart that he never felt the release from my hand. He never even stood. He just lay with his breast against the grass.

In the next second after that long minute he was gone. Like a flicker of light, he had vanished with my eyes full on him, but without actually seeing even a premonitory wing beat. He was gone straight into that towering emptiness of light and crystal that my eyes could scarcely bear to penetrate. For another long moment there was silence. I could not see him. The light was too intense. Then from far up somewhere a cry came ringing down.

I was young then and had seen little of the world, but when I heard that cry my heart turned over. It was not the cry of the hawk I had captured; for, by shifting my position against the sun, I was now seeing further up. Straight out of the sun's eye, where she must have been soaring restlessly above us for untold hours, hurtled his mate. And from far up, ringing from peak to peak of the summits over us, came a cry of such unutterable and ecstatic joy that it sounds down across the years and tingles among the cups on my quiet breakfast table.

I saw them both now. He was rising fast to meet her. They met in a great soaring gyre that turned to a whirling circle and a dance of wings. Once more, just once, their two voices, joined in a harsh wild medley of question and response, struck and echoed against the pinnacles of the valley. Then they were gone forever somewhere into those upper regions beyond the eyes of men.

* * *

I am older now, and sleep less, and have seen most of what there is to see and am not very much impressed any more, I suppose, by anything. "What Next in the Attributes of Machines?" my morning headline runs. "It Might Be the Power to Reproduce Themselves."

I lay the paper down and across my mind a phrase floats insinuatingly: "It does not seem that there is anything in the construction, constituents, or behavior of the human being which it is essentially impossible for science to duplicate and synthesize. On the other hand . . ."

All over the city the cogs in the hard, bright mechanisms have begun to turn. Figures move through computers, names are spelled out, a thoughtful machine selects the fingerprints of a wanted criminal from an array of thousands. In the laboratory an electronic mouse runs swiftly through a maze toward the cheese it can neither taste nor enjoy. On the second run it does better than a living mouse.

"On the other hand . . ." Ah, my mind takes up, on the other hand the machine does not bleed, ache, hang for hours in the empty sky in a torment of hope to learn the fate of another machine, nor does it cry out with joy nor dance in the air with the fierce passion of a bird. Far off, over a distance greater than space, that remote cry from the heart of heaven makes a faint buzzing among my breakfast dishes and passes on and away.

Roger Caras

THE ENDLESS
MIGRATIONS

To the far west of the North
American continent the northern fur seals come each spring.
By May the large bulls, the beach masters, have their territories
staked. Until recently this occurred only on St. George and St.
Paul islands, in the Pribilofs, off Alaska, on Copper and Bering
islands, in the Commanders east of Asia's Kamchatka Peninsula,
and southwest of there on Robben Island, off Sakhalin. Now
a new population of these generally northern animals breeds
on San Miguel Island, off southern California.

The beach masters are followed, a week or two after their
squabbling and social adjustments have been made, by the

189

pregnant females. These come ashore, bear the pups of last year's breeding and then breed again almost immediately. The big bulls collect huge harems and drive all other males off. It is organic, organized mayhem, the compelling factor for which is the need for another generation. That governs all.

But now the summer had passed, and the Pribilof herd was beginning to break apart. Females with their pups, now able to swim and dive well enough to survive in the sea, and the huge males had been feeding over an ever increasing area throughout the summer. Each day they traveled farther to sea, and each day the pups grew stronger and more independent. Many had died from diseases or from being rolled on or crawled across by the huge males, but enough had survived, and the water and the beach were athrob with life. Heads bobbed and vanished by the thousands in the swell. A tension was building as the time for migration was close. Each day the sun came later and stayed for a shorter period of time. The fogs were incessant and the rain and often sleet fell for hours at a time. The seals, with their fur and blubber and the protection of the sea where they spent so much of their time were not bothered by the weather, but soon freezes would begin and the area so close to land would become dangerous. Ice in surf can be a punishing force, a crusher, slicer of young lives.

Small groups of animals began to stay at sea, and the beaches collected fewer returnees each evening. Again the ubiquitous killer whales, sleek giants, slipped through the sea and took their toll. A large bull orca can cut a fur seal in half with a single bite, and that happened often. Females were easier to take than giant bulls, and pups were the easiest of all. They could be clipped in half like a grape or a fish, and many vanished from their mothers' sides as the sea boiled and then erupted for a moment before becoming quiet again.

The first freezes began. Each morning the rocks on the

beach would be covered, and a colloidal skin of first ice was forming in the bays and inlets. The males moved off first. They would move about five hundred miles out to sea and hold there until the following spring. There was plenty of food in the unfrozen sea, and unfrozen sea water is 28 degrees or warmer. That is not a temperature that can trouble a northern fur seal. The females with their pups would move more than three times as far from the breeding grounds as the males. They would seek shorelines, Japan's and parts of California, and push south as far as seventeen or even eighteen hundred miles. Again there would be more than enough food and warmer waters that would be easier on the pups. Although they would start back in the spring earlier than the males, the males would get to the pupping beaches first and be waiting for the births to be over so remating could occur again immediately. But all the fur seals would stand to sea during the worst of the winter.

The Pacific is a vast world, and even with shorelines for much of their trip, the females required navigational aids. The sun's course was reliable, the coastline was on the left during the trip south, to the east and therefore on the seal's right on the trip north. It gave reliable clues in both the wash of the sea rolls and the taste of the fresh water flowing in from the continent. But what of the males? They wintered at sea, and yet come each spring, male and female adults and pups of the previous years would unerringly strike the beaches where each had been born. Some seals miss the mark, go astray, and are consumed by the fog-shrouded seas, but the count belongs to fog bank and sea storm. By land clue, sky clue, and the clues of the sea, the seals find their way and repeat each year the incredible migrations without which their kind could not exist.

The sea is often a place of giants. Animals too large to negotiate well on land are swift and even graceful in the liquid medium for which evolution has prepared them. Among these animals the walrus stands as a prime example. The males of this giant seal may weigh close to three thousand pounds. The females —as with all seals they are smaller than the males—weigh just under a ton, usually about eighteen hundred pounds when at full maturity and in prime condition. Walruses are bottom feeders, and, although they will occasionally feed on a seal carcass, they grub with their great ivory tusks and scrape out shellfish, which they crush with their platelike teeth. They swallow just the meat, not the shell. It takes a great many clams to feed a ton or a ton and a half of walrus.

Walrus are in large part passive migrators because they prefer shallow water. They seldom go much deeper than three hundred feet for their harvest of clams. They prefer even less depth than that and rest on ice floes as long as they drift across shallow seas. When the floes are carried by wind and current toward deeper places the walrus abandon them and head for those closer to shore. In the Atlantic the small western Greenland population drifts all the way to Hudson Bay and Baffin Island, wintering near open-water leads between windblown sheets and citadel towers of ice. They dive, feed, and then rest, and are carried by the wind or the current or both. Typically they feed in the morning and rest the balance of the day on land if it is near, on rocks, or on ice. Since the drift is constant, the walrus is migrating even while it is sleeping.

The larger Pacific herd drifts through the Bering Sea when the ice is at its peak of power—February and March—but as the summer breaks that grip and pries apart the shore ice and the land, the walrus float north into the Bering Sea and feed there. Gestation lasts a year and the mothers care for their young well into their second or even third year, so cows are often separated from the bulls. In the summer cows with calves reach the northern-most shores of Canada and Alaska, but when winter starts down upon them again they head toward Asia and different shores for the different season.

Although the migration of the walrus is a wide and grand affair it is far more passive than most marine animals could tolerate. Diving, grubbing, and feeding, the walrus want only shallow seas, restful floes, and a reasonable wind to drive them in the circles they wish to travel. When the winds and currents are contrary, the walrus will push its vast bulk into the heaviest seas with scores of its kind, but it would rather drift, adding its thread to the fabric of migration that embraces the whole planet.

As we contemplate the movements of each or any species, we must conclude that migration has evolved, just as the animals themselves have evolved, from earlier patterns. Fundamentally, migration is flight from adversity. The energy used (except, perhaps, in the case of the walrus) is extraordinary, and as a species has evolved to be truly migratory it has often built its life around that single fact of existence. Breeding cycle, allowable nesting time, the degree of precocity at birth, the food tolerances developed, the ability to avoid enemies and seek prey—all of these elements in an animal's being have had to be built around the fact that the animal moves from one place to another to avoid adversity in a logical biological response, and seeks, equally logically, the maximum opportunity for survival.

The finest of the threads in the fabric of migration is pulled and tugged through the loom by the least of feathered animals, the one-eighth-ounce hummingbirds. On this vast planet the smallest of birds, they are found only in the New World. The pugnacious, rapacious little sprites gather wherever there is good feeding. They prefer red flowers, but they will take the nectar they can find and insects as well. Their metabolism is atomic and they must feed much of the time. Each a jewel, each a small electric explosion of color and iridescence, the birds must feed a furnace and will gain in return incomparable powers of flight.

The male ruby-throated hummingbird, easternmost member of his family in North America, was three inches long and weighed but a fraction of an ounce, no more than that. Metallic green above, he had a glowing ruby throat that marked both his sex and his maturity. Remarkably, he beat his wings seventy times every second. It appeared as if he were suspended in a veil of shimmering gauze. His body was distinct, but his wings were only a blur. He could hover seemingly endlessly, he could fly sideways, backward, and straight up and straight down without noticeably changing his body attitude. His needlelike bill darted in and out of flowers, sucking in nectar and occasionally the smallest of insects that hid deep in the crevasse of a flower's heart.

It was autumn; it was growing cooler each day and the hours of sunlight were shortening. Like billions of other birds around the globe, the ruby-throated hummingbird had received the signal "Store fat." Other birds might store it by the ounce or even, in a few cases, the pound, but the ruby-throated hummingbird lays it on in weights equal to those of postage stamps. Still, it was that minute store of fat that would enable the sprite to accomplish an epic journey. The bird—actually a tropical bird that had managed to work its way north rather than a northern species that wintered in strange lands to the south— would seek Mexico, where virtually all North American hummingbirds, even the widespread ruby-throateds from as far away as Alaska, go when winter comes. Deep into Mexico they go, to areas least touched by the weather. They consume energy in vast quantities relative to their size and must have food, and food for them means budding flowers, nectar, nature's sugar.

From immediately after sunup until just before dawn the little mote sought out fall flowers and sucked in nectar. He consumed as well a slightly higher weight in insects than during the humid, rich summer months, when nectar was less difficult to find, when, in fact, it was everywhere in endless quantities. He fed and he fed, and the minute layer of fat built and built as it must if he was to live. The monumental journey lay ahead, to be revealed by his genes.

One morning the ruby-throated awoke with first light, moved off its perch deep in the heart of a bush, and spun in place. Like a bullet he was gone. He reached fifty miles an hour almost immediately and vanished from the field at the edge of which he had spent his huddled night. The field was behind him, then the town, and then the next, and with his wings ablur he was across the state and heading southwest. He would reach the coast, the northern rim of the Gulf of Mexico, and then face five hundred miles of open water. At no point in that five hundred miles could he set down for so much as an instant.

On the way to the Gulf's margin he fed. He would drop into a field, hover before a flower, drive his needle-sharp bill forward into its innermost parts, and suck, then be gone. He seldom alighted during the day, but he did settle deep in thick vegetation at night. With each first light he would be gone again—a small memory that left no imprint. He may have weighed no more than a goose's feather, but everything the migrating goose had packed into it was tucked into this incredibly small unit of life as well. Mexico, an ancient memory, beckoned in so powerful a voice, his whole body had become attuned to response. He would take the states of the United States one at a time, he would take the vast Gulf of Mexico, and arrive home at last, having come before the wind, before the storm, before the weather that would surely have ended his life if it had ever reached him for so much as an hour. He was a tropical bird en route home. He might breed in the north—indeed, he had been hatched on Long Island—but he was a tropical jewel on loan and was being called home, where his colors would blaze until it was safe to venture north on loan again.

He finally crossed the boundary and was at sea. There was a different taste to the air and a new feel to the wind. There were also new problems in navigation, for all features disappeared the moment he crossed the line between land and water. Rivers carrying silt down from the continent to the north colored the water a uniform clay. Only far beyond would it be blue again. Small, choppy gusts of wind eddied and brought salty mist and then just salt into the air through which the hummingbird darted southward. But he was still bulletlike, fired and aimed toward his southern genetic home.

The little bird's course was toward the southwest so that he would come ashore on Quintana Roo Territory, on the easternmost outreach of the Yucatán Peninsula, beyond an ancient barrier reef. He would drop briefly on the Isle of Cozumel, feed, then push due west for less than an hour, and finally come ashore where the Mayan Indians so long ago built the cliff-top city of Tulum.

Then, over Mexican territory at last, he would turn south, feeding at the edge of jungles and over marshes that followed one after another. In the forest itself, not far from the sea, giant pyramids and temples stood draped in green, almost totally overgrown with vines and brush, and many of these species bore flowers. The energy lost over the featureless Gulf was quickly replaced in swamps and jungles where ancient civilizations once flourished.

The hummingbird's ability to hover depended on a network of muscles that constituted nearly a quarter of the little animal's total weight. In relationship to the whole animal, though, the hummingbird's wing muscles were huge. His wings acted as propellers when he fed from a flower, for he had to hover, fly without moving. He hung from his wings much the way a helicopter does from its blades. The stroke was downward and forward, then upward and backward, with the bird's body inclined forward at a steep angle. The breath driven downward by the wings at maximum effort was about the same as the draft created by languidly waving a palm leaf on a hot, still, and humid day. The hummingbird's ability to fly sideways as well as backward, the nature of its figure-eight wing stroke, make it resemble a large flying insect more than a tiny bird.

The little ruby-throated bird had a kind of computer built into its flight mechanism which told it how much energy to expend for each maneuver and how much fat to burn. In lateral movements the wingbeat varied from forty-two to seventy or more beats every second; hovering under different circumstances also told it how many beats to permit and how powerful each beat would have to be.

When the hummingbird stopped to feed, it did so because of its own sense of how much fat it had consumed and how much more was critical for its continued flight. It was judging fuel reserves as critically as it was energy expenditure, managing a fine-tuned network of nerves and their genetically coded demands.

The journey, even at fifty miles an hour during the sunlit hours, required all of the bird's strength and its navigational skills as well. It did not use ground signals, but it had the sense of its diagonal flight across the path of the sun, and that was tuned perfectly to the time of the year. Other hummingbirds, which had wintered and mated and built nests of cobwebs in other parts of the species' range, flew at different angles to the sun and were prepared as well for what was required of them in direction adjustments. As to how birds driven off course by winds on either journey—to the north in the spring, to the south in the fall—could ever compensate for that angle across the sun's track is a mystery far from solved. These wondrous creatures inherit not only information about geography and celestial maps but their ability to adapt the data in an infinite number of ways.

The speed of its flight created problems for the bird, and for these other compensations were necessary. His wings moved so fast in creating their blur that friction was intense, and that friction meant energy loss. Food had to be consumed and fat stored for the sole purpose of overcoming friction, and it was a considerable demand. The wing oscillations, those parts of the cycles that were more insectlike than birdlike, created eddies, which, again, had to be accounted for. Nothing the bird did or could do was without its price, and the price was fat, fat ready to be burned and then replaced. The hummingbird, as it sought the Mexico of invisible memory, was a balancing mechanism. It was a flashing jewel, a speck of colored light, a sunspot in a blurred gauze veil of flight-sustaining motion. More than that, however, it was a balancer of calories, a mechanical wonder that wandered over salt water and land of many altitudes, dipping in to feed, then vanishing like a fairy light, remote yet common, seeable yet unexplainable, one of the miracles of nature.

As the hummingbirds dispersed into swamps and meadows of the Yucatán new threats appeared that were unknown at the northern ends of their cycles. Coming in to feed from a flower whose widespread petals promised nectar, the rubythroat was met with the open jaws of a lunging arboreal snake poised near the flower, awaiting its meal. It knew the hummingbirds were coming. It missed, and the ruby-throated darted away so fast that the snake wasn't sure which direction it had taken. But the snake drew back and prepared to wait again, for the bright red-brown rufous hummingbird with its flame-red throat was coming too. The purple-throated Lucifer hummingbird would follow in days, as would the black-chinned, with the blue-violet band on its throat. There would be a few calliope hummers, with purple-red rays against a white field on their throats and others too.

The birds would come singly, but come they would, to spend the winter darting from flower to flower in the bewildering world of the tropics. Some would push farther into Central America, but all would leave the storms and uncertainties of the United States and Canada behind. Where the orchids clustered on tiger-striped stalks ten feet high the hummers would avoid snakes and even flights of aggressive bees and at night, as they rested, blankets of marauding ants. Safety from weather and starvation came at a high price, and the hummingbirds, as they gathered from the lands to the north, had only their speed and ability in maneuvering to compensate for each new threat, plus memory triggered by new hours of sunlight, new temperature and humidity, and perhaps the nature of available food. The hummingbird near the Guatemala border is a different bird from the one outside of Boston or San Francisco. It has to be to survive.

The hummingbird, no less than the monarch butterfly or the tern or the goose, had to read its environment, because all actions of all migrating animals are in some way timed and tuned to the world that immediately surrounds them. There are beacons, there must be forces that say abandon this and seek the other for now it is time. Photoperiodism—the ability to respond to the photo-period, or hours of light received daily—is known to be important for many species. Not all, but many. Without question, a shortening day spells one kind of behavior, a lengthening day demands another. Hours of sunlight not only mean signals, they create conditions that can be survived, conditions that can kill, and those best avoided although sublethal. Sunlight means warmth, both the periods when it is available and the angle at which it is received.

Lunar periods too are detected and used by some birds

and sea creatures as well. Very often invertebrates in the sea whose reproduction means vertical migration for at least a part of their body, the part that contains their reproductive cells, use the moon, follow its blue cast to the surface of the sea to swim and release the stuff of further generations. Climate, however, is probably the greatest clue. Changes occur that are not only unmistakable but provide or eliminate feeding opportunities, cause changes within bodies and fur growth without— rain and fog, snow and ice, temperature, humidity, all climatic, all changeable, all loaded with opportunities offered or denied.

Of all these changes temperature may be the most important. Temperature as an environmental factor is immediate in the air, or reflective in the water, but when temperatures fall and rise, scores of species of animals respond. Migration may be a few feet into the earth, a few feet horizontally or upward in the water or over thousands of miles on land and in the air. An animal like the hummingbird waits for the signals, for its responses to them are as much a part of the creature as its tissues and its structure. No animal that has evolved to migrate can ignore the signals when they are received and survive to pass the information along to the next generation, and from there to all future generations to come. Every migratory journey is a test, a trying, and upon completion, a reaffirmation of perfection.

Evolution was and will forever be at the root of migration as well as the need to migrate. One riddle will remain unanswerable: When did it start? How far along on its road to becoming a Canada goose was the Canada goose when first it began fleeing before the storms of the north? Did the ice ages play a role? How close to being a robin was the robin, and the caribou to its present form? We shall never know, for when we find bones they cannot tell us if they are from an animal in permanent residence or only recently on the scene at the time of its death.

No bone, no rock, no trunk of crystal tree will ever tell us when a migratory species reached that advanced state of survival technique. Did Pteranodon, the flying dinosaur, the largest creature ever to fly, with its wingspread of twenty-seven feet, ever migrate before a prehistoric storm or cycle of weather? No rock will tell us. Did, indeed, Archaeopteryx, the world's first bird, still retaining many reptilian features, migrate, and, if so, was it a few miles or a few hundred? It probably didn't, but no mineral shale will split apart and reveal the answer. We can know migration only as an accomplished thing. It is here. By quantum measures, it has been here longer than humans have; its history is sealed in the archive that may never be opened, in a code that can probably never be cracked. The answer was not written in stone, so it may never be found.

The weather that now triggers migratory behavior engulfed as well species now gone forever. It was perhaps because they could not adapt and abandon one latitude for another that they died out. Migration is a survival tactic, not so much for individuals as for species—the kind, the mass, the total product of evolution. Some animals can winter any storm, some any sea, some can hibernate, and others, such as bears, can at least imitate hibernation by sleeping long and deep when the weather is unkind, and unable to serve their needs. There are many answers to the tilt of the earth, but migration is the one we see most clearly and know best because it happens en masse, it thunders around us or fills our skies.

Although it has been a product of evolution, although it is built into birds as surely as their hollow bones, migratory behavior is strangely adaptable. Such birds as the blackbirds have been pushing farther north in recent years because of climatic changes. They summer farther north now than a few generations ago and may travel farther yet in the generations ahead. Population pressures may play a part in these subtle but inexorable changes, but other factors, having to do with temperature and rainfall, may count as well. It is difficult to sort out unless the time given and the data are long and plentiful.

The eight-inch-long cedar waxwing had fed well the summer months through, but early rains had reduced the berry crop and made the late pickings lean. Midway between the robin and the sparrow in size, the crested, sleek brown bird with the yellow slash across the tip of its tail began fussing as the fall wore on and fed less than usual for there was so much less to eat. It trilled its *zeeeeee* call endlessly, rose often from perches, flew in circles, and settled again only briefly. Something was about to happen. Its kind had bred that summer from the Gulf of St. Lawrence west to central Manitoba and at least south to Georgia. Its young were scattered now, able to see to their own needs, and the adults had only one responsibility left—to survive. In some areas the weather held well and the berry crop was good. The waxwings there were less peripatetic, less anxious. They fed and rested and fed some more. They were laying on fat that would help them through the inevitable period of lean pickings. They would not migrate, for it was certain that the berry crop that fall was good enough to see them through to spring. In the areas where the rains and the early freezes had come there was no such certainty, and the cedar waxwings *zeeeeeed* and fussed and seemed in almost constant motion.

Then it happened. The cedar waxwing, wherever it lives during the summer months, is capable of being an eruptive migrant. The birds summer almost anywhere within their range and winter over as broad a front and in as great a depth of latitudes. One morning the cedar waxwings in New England were gone. At first light they had lifted from their perches and turned south and slightly west. The fat they bore was too thin. They could sense that fact and know in the way of instinctive knowledge that they could not get through the winter in the north with the preparations they would be able to make. Their fat layer was too thin, as well, to see them through marathon flights across continents. They pushed only as far as necessary to find a good berry crop, and there they fed for weeks. They pushed on again for a few hundred miles and again found the feeding suited to their needs. By the time winter was well upon the land they would be in southern Kentucky and would winter there.

Interestingly, although eruptive migrators such as the cedar waxwings may show up almost anywhere, winter discontinuously and spasmodically, they would find their way home again. They would, perhaps, remember the angle of the sun and reverse that knowledge when the time came again to breed. Although they might go years between migrations, at least as individuals, and fly to different places when they did move for a season, they could always find their way back again. It is a special gift of the eruptive ones, but it works as well as any of the more rigid systems that guide birds from an ancestral terminal to its equivalent at the other end of their journey.

The continent of North America's fabric of migration has been separated from those of Europe and Asia for more than seventy million years. The Atlantic Ocean has been broadening all that time, and the Panamanian land bridge has been a fragile link to the south at best, time and again vanishing beneath a changing sea. There has been some small exchange of birds and mammals with continental Asia right up to almost modern times by way of the Bering land bridge, and there is still an exchange of tropical forms from Central and South America up through Mexico, the southern finger of North America. But the continental mass of North America, the Nearctic Realm, has had to develop, evolve its own species, and they each in turn have had to evolve their own means of facing the storm or escaping from its clutches.

The Nearctic Realm, the northern portion of the Western

Hemisphere, has offered virtually every possible habitat type, and so there have always been answers, however many species had to seek them. Those species that move on north-south lines in their migrations have had to learn to adapt to each habitat as they moved through it. There is tundra in the far north and below that, barely farther south, boreal forest. Below the boreal has been the temperate forest, then the chaparral. Lakes and rivers have always sliced through all of these, providing still different habitats, different plant species, different prey and nesting opportunities. There is semidesert in the Nearctic Realm and desert too, steppe vegetation, wooded savanna, montane forest both old and young, and even tropical rain forest. In Florida there is the incomparable Everglades system, a fifty-mile-wide river, six feet deep, flowing toward the southwest. And from each of these habitat types, from each of these systems of survival life has come forth, sprayed outward, followed genetic codes, and woven our fabric until the land and all of its waters and the sky have been threaded through an endless pattern of movement.

From farthest north come the Arctic loons, feeding on animal and plant food alike. They are equipped to take advantage of a wide variety of opportunities. They fly as far south as Baja California in the west. An eastern counterpart, the common loon, comes from Labrador and Newfoundland and retreats down the Atlantic coast before the winter, often resting for months in the Gulf of Mexico.

The Arctic also surrenders in the fall the golden plover, whose epic round-trip journey covers fifteen thousand miles, allowing it to winter on the east central coast of South America, a strange habitat for an Arctic bird. It is, of course, overflown by the Arctic tern, which travels to the two ends of the earth, alternating polar regions on seasons' cues. To this species goes the prize for greatest migrator of all. No bird travels farther, and it is doubtful that even in the sea any animal has a greater migratory trail.

In the spring the beautiful, the almost incomparably graceful black-and-white tern arrived on the shore of the inlet well above 60 degrees north latitude. It moved up a stream but found the water still frozen and moved back to the inlet off the great bay and farther along its shore. Still tired and hungry from its monumental flight of nine thousand miles, it fed often, swooping low and scooping up small fish, infant fish such as char and salmon. It swallowed them on the wing or occasionally dropped to a rock to peck and pull apart the muscle bundles of its prize. Usually, though, it took only what was small enough to swallow in a single gulp.

Eventually the tern found the stream it needed. It was fast enough and shallow enough to promise easy fishing throughout the brief summer. A mate appeared, and eggs were soon seen in a small depression along the rocky shore. When the young had hatched they demanded even more attention than the eggs had done, for now there were chicks calling for food, not just warmth. The parents fished all day, every day, to prepare themselves for their journey and to ready their chicks as well. It was only a matter of weeks before the chicks would fly. That was essential to the terns' timetable. When the winter came, as it would early to such high latitudes, the chicks would have to be ready to fly, as their parents would, almost halfway around the world. Those that were unable to join the exodus would soon be engulfed by the seasons and perish. Nothing could help them.

And a little over a month after the chicks first flew they were all away. Down from the Arctic they came, across tundra to the forested zones, one after the other, into the temperate bands that encompass the earth. They crossed the tropics, flew across the equator, and, tossed by storms, riding unpredictable sea winds, they pressed south through tropic zones again, along forested coastlines, and then into high latitudes once more. Forty south came and went, then the fury of the fifties, and just short of sixty degrees south they found a fog-shrouded island with a shallow bay and inland streams fed by the never ending rain and melting snow. They would winter there, for although it was never really summer so far south, it was gentler than the northern end of the range, which was by then raging in a fury of wind and ice. By the time they reached their bleak summer island in the south, the place where their chicks had hatched was under seven feet of snow and the stream they fished was frozen down to the rocks below and buried as well.

There were many places along the way that were gentler than this bleak southern outpost, places where crinkling ice sheets did not welcome each morning's sun, sheets that often held to almost midday before becoming a colloidal mass and sinking out of sight. There were softer streams with easier winds slipping along their surfaces and many places with just as many fish. Yet the Arctic tern, in answer to population pressure and competition, had been elected by the checks and balances of fate to go farther, to seek a harshness other species could not tolerate. Ironically, the winter retreat of the Arctic tern is far harsher than the world many species abandon to seek the warmer tropics or southern temperate zones. The tern's summer is another species' winter.

And as the terns of the high north settled into their high southern retreat, the hummingbirds that had in fact abandoned easier places pushed deeper into the jungles and forests of Central and South America. Migrating species are seen at each end of their journey, in each of their two principal habitats, as half animals only. The ruby-throated hummingbird that had spent part of its year on Long Island and was now sparkling from vine to vine in pursuit of nectar from totally different species of plants was one animal in the north and almost another in the south. In the south it would not experience the urge to mate. It would not ravage cobwebs, ignoring spiders in their threat displays in order to build the softest of nests. It would select from a totally different botanical world and be alert to different dangers. The twining tree snakes and massive spiders did not exist on Long Island, but in the tropical forest the hummer had to be alert to them every time it selected a promising flower whose nectar it needed to support its incredible demands for near-instant energy. That part didn't change, that was constant. The tiny bird still burned fuel like a fire storm.

As the hummer had come south, it was as if there were a television set automatically changing channels. The picture changed from day to day, often from hour to hour, as did the opportunities and the challenges. One moment there would be flowers, shrubs, trees, places to rest and feed so that the journey could continue. Then a coast would pass below and there would be nothing of any use to the bird at all except a following wind. For hours, open water, dangerous eddies of wind, and still no food. The hummer began dipping into its fat reserves, and just when exhaustion threatened, just as the fuel was about to run out, another coast would appear, and the tiny mote of sun and feathers would drop low, seek a blossom, and feed on the wing. At times it would land on a bush and rest, but often it simply fed and rose and shot out of sight like a dart from a bow, heading farther south yet.

The route the hummer took had been evolved genetically so the coast would always appear before it was too late. Hummers that tried other routes, that impressed upon their genes course memories that did not include the coast and the flowers, in time perished. They did not return to the north, they did not breed and did not pass their memories along, for they were deadly ones. Their latitudes had been the same, but the coasts and the flowers did not comply.

The hummingbird from Long Island spotted a bell-like

flower hanging from a vine in a softly shaded glade of endless
tones of green. The flower was red and glistening with damp-
ness. The mite flitted first to one side of the prize and then to
another, and then dropped several inches so that it could approach from below.

Pushing upward, it tilted on the axis of its wings and thrust
its long spike of a beak up into the heart of the flower. It did
not see the angry bee until it was too late to avoid it. The bee
came out of the flower buzzing its rage and stung the bird once
on the side of the head with the spike at the rear end of its
abdomen. The hummingbird sickened immediately, for, although it was twice the size of the bee, it had been assaulted
chemically. Wobbling, struggling to stay on a balanced course,
it dropped away to a brushy mid-trunk stem on a hardwood
tropical forest tree. It caught its perch as it was about to slip
by and held on, fighting to steady itself. The venom the bee
had injected had reached the bird's nervous system and the
hummer's vision began to blur, and then, even as it still hung
on, it went blind. The bird tipped forward and back fighting
the sickness in the only way it could, by remaining on its perch.
If it let go it would fall to the cushion below. The fall would
not kill it, but the ants and beetles that scurried about down
there would soon tear it apart.

Blind as it was, the hummer could not see the snake as it
slipped across from a bush nearby. The snake did not allow its
weight to rest on the same shoot that supported the hummer,
for it too had instincts and knew in that way that its weight
would tilt the branch and signal its approach. It struck from its
mid-air position and took the blind hummer off its perch, barely
disturbing the leaves that surrounded but did not adequately
protect it. The hummer would not breed again on Long Island.
After its epic journey, after braving the seas and coastal winds,
after crossing through scores of gradations of habitat types, it
had been killed by a bee and a snake, species found only at

one end of the journey, species belonging to only one half the little bird's life. It had never encountered either species before, but there are no excuses when death is the reward for failure.

America's unique animal, the pronghorn, does not migrate in the usual sense of the word, but such words are not ever really definitive. The pronghorn, referred to erroneously as an antelope, is a strange animal. In all hoofed animals that grow fighting tools on their head there is a clear distinction. Antlers distinguish the deer family and are shed annually. Horns are never found in deer, but occur in bovines, goats, sheep, antelope, and their like. Horns are not shed—except in one case. The American pronghorn grows horns, not antlers, but sheds them. In fact, the pronghorn is not closely related to any other animal in the world. The family, Antilocapridae, arose in North America, and once there were other members. But the sharp edge of time and competition has worn that family down to this last lone species.

The male pronghorn sniffed the wind and looked out across the flat prairie in North Dakota toward other pronghorns grazing almost a mile away. The white rump patches on the other animals were flat, and the lone animal knew there was no danger coming from that direction. At the first sign of intrusion the first animal to spot it would flash its rump patch like a semaphore by flexing muscles that would cause the white hairs on its rump to stand erect. The warning would be picked up and repeated across the vast flatland until every pronghorn within miles had the message. The menace, whatever it might be, real or imagined, would find the land empty. With incredible eyesight and enormous speed, the pronghorn is not an easy victim for any destructive intent.

But the pronghorn cannot flash its rump signals with purpose against inexorable forces, and when the storms that are destined to bury their food under feet of snow start their march from the north, the pronghorn needs more than eyesight and speed to survive. It needs an ordering of behavior. No winds we know of and no temperature extremes are likely to alter pronghorn behavior. From well above 100 degrees in the summer to 70 degrees below zero in the winter the pronghorn easily handles an incredible range of 170–180 degrees in the cycle of the year. Wind-chill factors do not govern it; food or its lack do.

Sensing the storm, the pronghorn began trudging slowly toward the place where the other animals were pawing at the light covering of snow that had fallen late in the afternoon a few days before. Other stragglers were coming in from the southwest and east as well. The animals were bunching up, and then their movement started. They began moving toward higher ground no more than twenty miles to their south. They crossed a frozen shallow river and pulled themselves up the ice-covered bank with easy bounds. As the slope of the earth inclined more sharply the wind increased. But the earlier accumulations of snow began to thin out. While other hoofed animals in other parts of the continent made their way down from the high slopes to the shelter between the hills, the pronghorn moved the other way. Pronghorn do not like to work for their food, to dig through inches or feet of snow. Impervious to the weather as it plays on their fat, padded bodies, they depend on that weather to keep the ground as clear as possible in a land of harsh winters and frequently heavy snowfall.

As the storm struck, the snow began to fall. The flakes at first were large and moist—accumulation, really, of flakes coalescing in midair. Then, as evening came on, the flakes grew smaller, were driven harder, and seemed to have a sharper edge. Within the first hour the flat lower ground was under several inches of snow, and even the stream the pronghorn had crossed vanished in the sameness of the blanket that was thickening upon the earth. The pronghorn, now higher up, stood with their backs to the wind, and however hard the driving storm might try, it would not dislodge them. The pronghorn had made a short migration, but it had been very different from those accomplished by the other hoofed animals of the continent. They had moved up to the most exposed places they could find. All across the west where the pronghorn still exist and down toward Mexico where a few still remain, the pronghorn would sense the storm's coming and seek the highest ground, areas that would be swept the cleanest by the driving wind. It is their pattern alone, but no less than any other it is a pattern of survival.

As the hummingbirds from the northern reaches of their ranges crossed the Gulf of Mexico and the last of the monarch butterflies sought their valley in Michoacán, Mexico, as the gray whales sought their shallow bays in the Gulf of California and new flights of geese found their feeding areas in eastern Maryland, pronghorn, scattered across the vast open lands of the west in the United States and Canada, climbed as steadily and unopposed as the storm itself toward the highest ground. The same pressures were on every species: Wait it out, make it through to another year, and pass your genetic package along once more. And the worthy generally would live. The best usually do survive.

The twenty-one-inch-long parasitic jaeger, a form of predatory gull, or at least a close relative of the gull that has become over time a hawk of the sea, had spent the summer feeding along Arctic coasts, robbing and attacking gulls and terns, acting the pirate's role in every way. Now it was time to abandon the Arctic, for there would be no young to take, no nests to raid, no rookeries to worry in low, sweeping passes. It would head to sea and more southerly latitudes.

But the jaeger would not head for new lands, new shores; it would go to sea and remain there throughout the winter months hunting the surface waters. The next year only breeding males over two years old would return to the breeding grounds of the continental tundra and Arctic islands. If the nonbreeding males of one year were to return as well they would compete with the essential adults for food. That is a poor husbanding of resources, so nature would keep them at sea for another full year before allowing them access to the land where they were hatched. Aboard ships, scientists equipped with powerful glasses would still not be able to tell which of the three jaegers—pomarine, parasitic, or long-tailed—they were seeing. It is all but an impossible distinction at sea—resolvable, really, only on an examining table, with dead birds side by side. Incredibly swift, dark, falconlike, the jaegers—the parasitic the most common of the three—poured out of the Arctic before the weather snapped the top of the world shut in a vise of steel-hard ice and spread out as marauders across the sea. Where they go, how long they stay there, what directs their movements are all elements shaped over thousands of generations by population density, feeding opportunities, and the weather patterns that govern all.

Below the latitudes of the Arctic tundra lies the belt of boreal forest that in Asia is known as the taiga. Dominated by spruce, larch, and fir species, the forests are home to many birds and small, fast, nervous hunting mammals. Some species, such as the spruce grouse, stay the year in these frigid forests and manage to survive on conifer needles. They do not migrate at all no matter how harsh the season. It is an austere diet, but there is energy enough in the slender, shiny, waxy leaves to hold the birds until spring.

But there are the birds that do not stay and they include the pretty warblers of the family Parulidae. Wood warblers, butterflies of the bird world, are smartly colored birds generally smaller than sparrows. Flitting, flashing, moving like sparks more than like flesh, these tiny birds summer in the high boreal forests down through most of Canada to the northern tier of the United States and then, in the fall, vanish to the tropics. It is a monumental flight for so small a bird and often entails travels of thousands of miles over both land and water. They fly high and fast, skipping across the sky like stones across a still pond. Propelled, projected southward by the winter behind them, the wood warblers, a confusing array of yellow, orange, white, black, gray, brown, olive, and heaven alone knows how many other shades, taunt bird watchers along the way and then are gone. Very often, their world stretches from one continent to another, and different populations winter and summer in different locales, but the pattern and the purpose are the same. Adversity is left behind; promise, always promise, lies ahead. That is the nature of the fabric. The warblers alone create a crazy-quilt-patterned cloth, but as it is pulled taut by the millions of birds reaching their southern terminuses, that cloth becomes merely a thin layer in the grander fabric that encompasses an entire hemisphere.

Although migration is a great weeding process and the overall effect is for the best, the most nearly complete specimens do survive. There is chance, and its play upon individual animals of any quality can be devastating and as illogical as the roll of dice.

The five-inch black-and-white warbler had mated that year near where it had hatched two years earlier in northeastern British Columbia. It had left early in the fall because it had so far to travel before it settled for the winter deep in South America. It drifted east and then south through western North Dakota. Its hair-lined nest on the stump of an old tree, abandoned, was already under snow when the bird crossed over the international boundary into the United States.

Feeding mostly on insects, it moved a score of miles one day and almost none the next. It was becoming less casual as it encountered other birds of its own and other species, but the storm was not immediate in its threat. The bird moved and it fed and called its piping *weeee-zeeeee* to the world for no apparent reason except well-being. It was not establishing territory or attracting a mate. It moved down across the flatlands of mid-America and came at last to Texas. There it would encounter millions of small perching birds, passerines like itself, and in enormous sheets of life they would cross the Gulf, reaching out toward South America. Just as each bird had a particular place it had come from, it had a special place to go to in the schizophrenic life-style of the migrator—from well up in Canada on one continent to well down into Venezuela on another. One bird, two worlds, opposite sides of a great circle—that was the order of it.

But chance came into play in central Texas, up on the Edwards Plateau. There, almost without purpose, a boy playing truant wandered in the bushy, rocky terrain. He did not like his teacher, he was angry, and so he became the instrument of chance. The black-and-white warbler landed on a bush to seek insects. It needed a substantial fat reserve before it reached the coast and the Gulf. It had been husbanding what it had and understood in its special way that it needed more. The angry boy whose day was without form took aim. The noise of the .22 caliber rifle echoed for moments only and was gone, but the bird was dead. The boy walked over to it, toed it in far less than an interested way, and moved on into the shapelessness of his afternoon. Two species had interacted briefly, and chance, foolish chance, had had its way.

The wooded area was set in North Carolina. It covered no more than twelve acres but it was a rookery, one of many in the American south and southeast that contain over a million birds each. This one is believed to contain almost twenty million—twenty million birds in twelve acres of overburdened trees. On high, lone trees and telephone poles all around, hawks of several species sat glutted, patiently waiting to be hungry again.

There are several reasons why this wood had become so heavily laden with life each winter. The mechanical agriculture used in the fields that lay in all directions is wasteful. Seed is spilled, edible stubble left standing—all food for northern birds on southern retreat. The other wooded areas that used to cover the landscape have been cut down or at least cut so far back they no longer offer roosting opportunities for birdlife. Those few stands of trees large enough to be found within a radius of fifty miles have been killed by similar flocks of birds roosting in previous years. Soon the trees of Scotch Plains will be gone too, and the birds will move again.

The latitude is right as well, and so the birds come. During the day they spread out fifty, sixty, even seventy miles and pick over the fields abandoned by man for the winter months. Then about four o'clock they begin their return. First a single bird is seen heading for the trees, then two, then four, then eighty, then a thousand, then ten thousand. Within an hour the twenty million birds are back. Beneath the trees the ground is a foot deep in dung, and dead and dying birds lie there as well. The stench can be detected a quarter of a mile away. The mortality from disease and injury and exhaustion among so many birds is awesome. Those alive above jockey and shoulder each other all night. The purring, squealing disturbance is endless. Periodically another bird falls to earth and is buried in the dung. Red-winged blackbirds, doves, warblers, vireos, cowbirds, robins, grackles, starlings—a score of species all bunch together. There is no organization to be discerned, just the noise and the dead and the living, the dying and the aggressive and the passive, and those that will die the next day in the talons of a hawk sleeping nearby, perhaps half listening in anticipation of the easy hunt ahead.

The temperature dropped, but the stench had not lessened, or the noise. It was freezing, below that, and the birds pushed closer together. Some branches snapped and dumped their loads. Some of those dumped were too weak to make it back into the trees again or they could not fight hard enough to gain space. They died on the ground, many still gasping at first light but soon to be dead. The hardier would live, for the rookery is a sorting place, and the next year's breeders are singled out for at least a chance to fly home to where they could nest.

And then, in the east, the first flush, then the first fingers, then the first real light. The birds became restive, and it was morning. The birds that had taken over an hour to arrive the evening before, although they had arrived by the thousands, suddenly lifted off, and the sky went black. In thirteen minutes twenty million birds would be gone. Clouds that seemed more like an animated film than reality puffed away, like stack smoke in the distance, and then it was still except for the pitiful twitching in the dung and the leaves, signals of mortality from the last of the birds to die. They would have been unfit for the return flight come spring and therefore unfit to produce the next generation. Nature was taking the easy way out. Without concern for the individual but with unending love for the species, she would allow the weak to die here, in North Carolina, here in a strange land. The nest and eggs and life were far away whichever way you looked, backward or forward in time. Life is a deed performed and a promise given and only the best can traverse those two points.

Jack Rudloe

ON THE EDGE
OF THE ABYSS

Thomas Lee Mills, skipper of the shrimp trawler *Norma Yvonne*, puffed his cigarette in the darkness and turned on the Fathometer. The whirling needle noisily traced out a long, fuzzy black line over the moving sheet of graph paper. "We're at seventy fathoms [a fathom equals six feet, thus four hundred twenty feet] right now," he said quietly. "It won't be long before the bottom starts dropping off fast. I'd say that by three or four in the morning, we'll have a good two hundred fathoms of water beneath us."

We had been at sea for eight hours, moving due south off the Florida panhandle out in the Gulf of Mexico on an expedition to find and bring back some strange, rare deep-water creatures for the New York Aquarium.

"But I'll tell you," the big skipper continued, "the way these seas have been building, I ain't so sure we're gonna be able to work when we do get there. Now it used to be that you could work this royal red shrimp territory out there in between the northerlies. But crazy-acting as the weather's been, I don't know. Last year they'd blow through about once a week; then it would clear up and even get pretty out here. But now these fronts come down one right behind the other, and I'm afraid that's what's happening."

"A big boat like this ought to be able to take it," I said hopefully, leaning back in the comfortable pilothouse chair, watching the sweep of the illuminated radar beam. We were alone; not a blip or mark of any kind showed up anywhere on the screen.

The skipper laughed sardonically. "Oh, hell yes, this boat can take it. She weighs fifty tons; she don't ride the waves, she flattens 'em. But *you* ain't gonna stand up to it. When she hits those twenty-five-foot seas and goes to flamming she'll beat your guts out. And when you got them big seas breaking over the bow and flooding down the decks, it don't take but a second for a man to get washed overboard. Damned if I'm gonna get drowned out here trying to drag up a mess of monster sea roaches!"

Thomas Lee didn't think much of our expedition. He had agreed to run the *Norma Yvonne* for Aquila Seafoods in Bon Secour, Alabama, only because it was January, the coldest and most wretched month of the year, and there were no shrimp. His own sixty-eight-foot wooden trawler was tied to the dock, along with nearly all the other shrimp boats. He needed to make some money.

Finding a vessel large enough and well enough equipped to trawl the submarine De Soto Canyon off Pensacola, Florida, in two hundred fathoms where *Bathynomus giganteus* lived had proved an ordeal. Never before had these big, grotesque, jointed-legged creatures been placed alive on public display. Occasionally the deep-water shrimp fishermen would haul up one in their nets and bring it in dead as a curiosity.

Months earlier we had gone from dock to dock, fishing village to fishing village, asking whether any shrimpers were fishing for royal reds. But all we got was an emphatic "No! I doubt you'll find anyone still messing with them royal reds anymore."

In the 1950s the Bureau of Commercial Fisheries first discovered the royal red shrimps. They were doing some exploratory trawling in two hundred fathoms and brought up a deckload of big red succulent royal reds, *Hymenopenaeus robustus*, previously known from a handful of pickled specimens gathering dust on the shelves of the U.S. National Museum.

The crew cooked up the shrimp, and overnight they changed from a scientific curiosity to a gourmet's delight. Adventurous shrimpers, spurred on by the discovery, had rigged up their boats to work the deep water. But fishing out there, farther than any shrimp boat had ever gone before, was brutal on equipment. The heavy seas would snatch the rigs off the bottom and tangle them. As cables were wound in, standard winches used on trawlers would often burn up from the strain, leaving the crew to haul in more than a mile of steel cable by hand, sometimes in gales. Sharks attacked the nets, and when the violent squalls struck, there was no shore to run to for safety. After a few years of trying, most skippers stopped replacing their miles of rusting cables and went back to working inshore for the traditional pink, brown, and white shrimp.

But we located the *Norma Yvonne*, one of the few boats that had been rigged for fishing royal reds. With giant hydraulic winches, she was built to last. "But I'll tell you something,"

Thomas Lee commented. "It damn sure don't pay to fool with them red shrimp unless there ain't *nothing* in shallow water. Long as I can catch three or four boxes of pink shrimp in twenty or thirty fathoms, that's where I'm gonna work! I ain't got near the expense nor the risk in getting them." Then he winked at me. "And the trash fish ain't nearly as boogerish-looking as this deep-water stuff!"

As we headed farther and farther out, we studied the National Marine Fisheries Service's printout of when and where *Bathynomus* had been captured by research vessels over the past twenty-five years. Nixon Griffis sat quietly in the galley, puffing his cigarette. A few months earlier we had had another lunch together. After describing his last trip to Africa in search of the hammerheaded bat, he had asked me, "Jack, where can we catch a monster? We need something so unusual that it will really draw crowds to the New York Aquarium. The giant toadfish was great, but I want something spectacular, something that will boost membership in the New York Zoological Society."

The only monster I could think of was the giant sea roach, *Bathynomus giganteus*, that lived deep along the edge of the continental shelf off Florida. With a length of up to two feet, it was the world's largest isopod, a flat-bodied crustacean with ten pairs of sharp-hooked legs and segmented body. Most species of isopods live in the ocean, although many are found in fresh water and a few live on land. The inconspicuous and ubiquitous pill bugs, or "roly-polies," that hide under rocks in a garden are isopods. Imagine a giant pill bug, with its jointed body, hooked claws, huge triangular eyes, and a mouth filled with cutting plates, and you've got a real monster. *Bathynomus giganteus* is also a living fossil. Common in the seas some 60 million years ago, it survives today only in a few scattered places around the world's oceans. Beyond the Gulf of Mexico it is found in the Sea of Japan and the Bay of Bengal.

Anne and another marine biologist at Florida State University, Joe Halusky, sat across from Nixon mapping the earlier *Bathynomus* coordinates.

"Latitude twenty-nine degrees, four minutes North," Anne called out, "longitude eighty-eight degrees, forty minutes West."

Joe grinned happily. "I can't believe this." His pencil made another round circle next to the tight bunch of others. "It's just too good to be true. Look at all these little rascals: they're sitting right at the edge of the continental shelf, just at that drop-off between two and three hundred fathoms."

"How about it, Skipper?" Anne asked. "Can we drop it down right here?" She pointed to the middle of the dots.

Thomas Lee grunted, "I don't know about that! I got a chart of my own in the wheelhouse, and it's got all the bad bottoms marked off. There's coral reefs out here in two and three hundred fathoms that will tear a net all to pieces. The government boat don't mind tearing up nets or losing a whole rig: they got plenty of money to replace it. All it takes is one bad hang out here and a man can lose eleven thousand dollars in nets and cable before you can say 'Don't do it!' "

A moment later he returned with his chart drawn with big red squares. "We'll put over here at the 2950 loran line, what we shrimpers call the 'edge of the earth.' I believe we caught some of them sea roaches there before." Thomas Lee flipped through our scientific reprints scattered around on the table. He contemplated a line drawing of *Bathynomus*. "I don't pay no attention to the trash that comes up when I'm out here. I'm interested in only one thing—shrimp. But you don't forget an ugly-looking critter like that."

As the evening wore on, the seas began building and the ninety-foot steel-hulled trawler pitched and rolled. Dishes flew out of the cupboard, and the two crewmen, Frankie Nelson and Claude Underwood, hurried about battening everything down.

The sea would have been better in June or July, but we hoped that we could run out to the drop-off in between the weather fronts, work a few days, and hurry back before the next storm came down on us. We had deliberately chosen to go in the dead of winter when the surface water was the coldest. If we had any hopes of taking our specimens back alive, thermal shock had to be minimized. The surface water was down to nearly seventy degrees now, and the bottom was only forty-five degrees F. Bringing the trawls up from those depths would still kill off many species, but hot summer surface temperatures of eighty-five to ninety-five degrees F. were a guarantee that everything that came up would die.

We had gambled on the weather, and now according to the weather radio our bluff was apparently going to be called. A storm was coming through. If it stalled over land, we could work. If it didn't, the expedition would be over.

We turned in, and I lay there most of the night listening to the engines revving up and down, fighting the ever-rising waves. I dozed off, but the surge breaking over the bow got so strong that it sprayed in through a porthole above my head, making the night drag on even more uncomfortably. That old queasy, familiar feeling in my guts returned with each wave. All this blue water business was fascinating and adventurous, but at heart (and stomach) I was a sea*shore* naturalist, one who likes to walk along warm Florida beaches at low tide with buckets in hand picking up things.

At three in the morning, Frankie was shaking me. "Captain wants to talk to you."

The skipper's mood was somber. "We're at one hundred fathoms," he said solemnly, pointing to the depth device, "and from here on out it starts dropping off pretty quick. In two hours we'll be over two hundred fathoms." The Fathometer's needle was etching a sloping line going down, down, down to the bottom of the page.

"But, Jack, the way these seas have been building, I ain't so sure I'm going to put the rigs overboard. Now it's possible that they'll lay down at daybreak; they usually do. Then we'll make at least one deep-water drag, but after that I ain't promising nothing. If that wind goes to switching around, I don't want to be where it will drown us."

I returned to my bunk. All that planning, and all that work and money, and now it would probably come to nothing. How foolish I was to believe that we could actually go out there, and drag up a big sea roach at the snap of a finger. The odds were too great against us. I popped another pink antiseasickness pill and tried to get some sleep. There were thousands of square miles of deep ocean out there, rolling on for an eternity. And we, with our two little trawls, dragging a tiny part of it.

Thomas Lee's concerns were real. He had seen big steel-hulled shrimp boats return from the royal red grounds at the edge of the Gulf Loop current with all the pipe rigging torn off, the windows shattered, and the electronics smashed. They were the lucky ones. There were wooden boats that went out and never came back.

At dawn we assembled, yawning, on the deck, gloomily contemplating the red skies and scruffy angry little clouds that hung overhead. They were moving before the wind, traveling ever southward before the cold front. The seas were a vast panorama of gray water frothed with whitecaps and huge rolling waves.

The deckhands, Claude and Frankie, were busy working the rigging, wearing their heavy yellow slickers as the boat pitched from side to side. "All right, we're setting out at one hundred ninety fathoms," Thomas Lee shouted, raising his voice above the winds. "That's the best we can do right now. We'll be dragging on out to two hundred fathoms."

Then with a jerk on the lever he started the winch turning, and the ten-foot-long, five-hundred-pound, iron-clad otter doors were snatched out of their brackets with a loud crash and dangled from the outriggers. Frankie threw the nets overboard, and when the winch brake was released, the two trawls splashed into the sea. The green webbing sank down behind them, and the net started its long journey to the bottom.

The giant drums spun rapidly, spewing out fathom after fathom of strong steel cable. Thomas Lee hurried to the pilothouse to push down the throttle, and the *Norma Yvonne* churned forward, her great spinning cable spools becoming thinner and thinner. When he was sure there was enough cable out, he slowed, and the crew locked the winches. The big steel hull strained ahead, her outriggers stretched wide like the wings of a giant bird as she pulled her heavy trawls over the soft muddy bottom.

Now it was our turn to get busy. There was much to be done before we were ready for the nets to come up. As I gushed seawater into the large waiting Styrofoam containers from the deck hose, Joe Halusky made his way up and down the long steel ladder that reached into the giant ice hold, bringing up buckets of ice. Anne scooped the crushed ice into plastic bags and tied them tightly. The ice bags were to be used to chill the seawater temperature down to forty-five degrees F. without allowing any melted fresh water to dilute the seawater. That would be deadly to any living specimens.

Nixon checked our packing manifest. We strung electrical lines and duct tape around the deck and inspected air pumps to see that they were working. If any creature came up alive, the water would have to be aerated.

The deckhand Frankie Nelson watched our activity. His voice was skeptical. "I don't see what y'all rushing so hard about; it'll be three hours before we're ready to take up. 'Sides, you ain't gonna get much to stay alive nohow. Most all the fish I see come up from this deep water got their eyeballs burst out of their heads. This ain't like regular shrimping where everything on deck is a jumpin' and a floppin'."

Three hours dragged on, and we waited with anticipation.

By midmorning we entered the Gulf Loop current. Suddenly the waters turned a deep transparent blue and the air was almost balmy. Off in the distance, a blue marlin, an enormous thing with a long pointed bill, leaped high out of the water and landed with a great splash. All around us the waves rose like blue hills crested with patches of golden brown sargassum weed.

It was wonderful being in the Gulf Stream again. We took off our heavy jackets and walked around the deck in sweaters, and for a while the cold winter weather almost disappeared. But high overhead, clouds continued to move ominously across the sky, pushed by the impending arctic front. Finally, Thomas Lee started reeling in the nets. It was thirty minutes or more before we could see the two otter doors on the starboard side ascending rapidly toward the surface, coming together and closing the mouth of the net. The heavy wooden doors broke from the sea with a splash and dangled from the outrigger, dripping into that blue, blue sea. Then the other set of doors rose and hung from the other outrigger.

Frankie grabbed a long bamboo pole and deftly hooked the lazy line that connected the doors to the end of the trawl bag. The fifty-foot-long starboard net came up first. The young crewman wrapped it around the cathead and started bringing in the webbing.

With creaking groans the rope hoisted the green-webbed bag higher and higher until it dangled from the boom above the deck, showering water, gorged with life. I tried to peer through the covering of brightly colored, red, yellow, and white chafing gear, but all I could see were blotches of pale color, of red scales and white bodies and shrimp antennae.

The skipper went forward and snatched the release ropes that kept the trawl bag closed. Suddenly a few fish started spilling on the deck, then an avalanche of little orange shrimp, along with every imaginable creature. "My Lord, will you look at the shrimp!" declared Joe loudly.

The captain and crew didn't seem impressed. "There ain't no shrimp, very few that's royal reds. Most of them's peewee *Megalops*, and there ain't much of a sale for them! They're too damn little."

I surveyed the pile dumbfounded. With more than fifteen years of collecting in the shallows of that very same Gulf of Mexico, I expected that I would know some of the creatures piled on the deck before me. But I couldn't classify a single one. There were long flat eels with wicked-looking teeth, huge horny-plated prehistoric gooseneck barnacles in twelve-inch-long clusters. The deck bounced with big leathery-skinned white sea anemones, and scattered everywhere were moon snails whose shells glowed with brilliant iridescent and opalescent colors. But they were the exceptions. The deep-water creatures lacked the shallows' diversity of color, most of them were either jet black, pale orange, or pasty white.

We were dragging at the very edge of the Gulf's continental shelf where the bottom begins its sharp plunge to the abyssal plains. The animals that live on this frontier are more closely related to the bizarre, spindly, spiny forms of life that dwell on the ocean floor three miles deep than they are to the familiar shallow-water species. But what they lack in color, they make up for in diversity of form, texture, and shape.

Nearly all the fish, the speckled stargazers, the red armored sea robins, and the toothy brown goosefishes, were dead on deck, their mouths sprung open, their tongues protruding, their dead eyes popping grotesquely out of their heads. All of these casualties had swim bladders, gas-filled sacs used to regulate their buoyancy. At two hundred fathoms they existed under a pressure of 550 pounds per square inch. As the net rapidly ascended, the pressure decreased, and the gases rapidly expanded, blowing up the swim bladders like balloons and tearing the fish apart. Some of the crustaceans and fish that lacked the swim bladders were still alive, although others had gone into shock from the abrupt temperature change.

Our eyes peered over the near-lifeless pile for any kind of movement, looking especially for the big, segmented, armored *Bathynomus giganteus*. But it was nowhere to be found.

Before the second net was dumped on deck, Anne cried excitedly, "Hey! Look at this! It's a *Chimaera*, and it's alive!"

As she pulled it out of the pile, Nixon jumped up with excitement. "That's a rattail fish; I've only seen pictures of them in books. We've got to get that back alive. That will make the expedition even if we never catch a *Bathynomus*."

In Greek mythology the Chimera was a fire-breathing she-monster. This creature had a soft, bulbous nose; its skin was soft, almost mushy like the squids on deck; and its elongated coal-black body tapered to a point. A wicked spine protruded from its back, and the network of lateral lines was prominently marked over its soft, clammy, scaleless body. But most impressive of all were its huge bright-green luminous eyes. They were like two big crystal balls, so big you could look into them and see the world.

But when Anne hurried her specimen over to the waiting Styrofoam box and gently eased it in, it began to swim beautifully. Its fan-shaped pectoral fins spread out like an angelfish's, and it hovered gracefully in the water. Suddenly it became beautiful, more beautiful than any fish I had ever seen.

When the second net was opened, more creatures piled out and there were more *Chimaera*. There were also intricately striped, tan, chain-dogfish sharks with black markings and slanted green eyes. They were alive and healthy. Sharks and their kin, which include rays, skates, sawfishes, and *Chimaera*, have no swim bladders and can withstand the rapid decrease in pressure. But unhappily many of the cigar-shaped bioluminescent green dogfish sharks, sexually mature at nine inches long and lacking a dorsal fin, had perished from the abrupt temperature shock.

Before long, our boxes were filling up with every imaginable creature, small spindly orange lobsters, hermit crabs living inside red sea anemones, delicate creatures with long flexing legs. We worked rapidly, raking through the catch. Time is critical on a shrimp trawler, especially with a thousand pounds of gasping creatures dumped on deck. If you don't get the animals into life-giving water within a few minutes, most perish.

Even with the two big piles of life, it didn't take long for the crew to cull it off. In thirty minutes, they had four baskets heaped high with the fluffy bright royal red shrimp.

The crew shoveled all the other creatures overboard, including three or four hundred pounds of the small orange shrimp, *Peniopsis megalops.*

"Yeah, I know it's a waste," the skipper said, "but until they get some way to machine-process them, they ain't worth fooling with. They're just too small and too fragile."

Like many of the Gulf species, *Peniopsis megalops* was an untapped resource. Yet Thomas Lee spoke of seeing Russian and Japanese trawlers working out in the deep water, saving almost everything that came aboard. Even with the two-hundred-mile limit, foreign exploitation would probably continue. According to international treaty, if we didn't utilize the tasty little shrimp and other unexploited species, they were fair game for other countries that would.

Thomas Lee looked over the three and a half baskets of royal reds. "We ain't gonna get rich this way." But he looked pleased.

"Hey, Jack," Nixon called, "come look at the *Chimaera.* I'm afraid they're dying."

The fish were now belly up, or lying flat on the bottom. I blasted pure oxygen into the water, hoping to revive them, but I knew it was hopeless. Their delicate bodies had undergone too much of a shock. Moments later I sadly lifted them out of the tank and put them into a plastic bag to preserve later.

"Skipper, we need to make another tow," I said. "We didn't get the sea roach, and we lost our other little monsters."

Joe had been in the wheelhouse, studying the loran readings and his charts. He was excited. "Look, the way I figure it, we were dragging just a little bit too shallow to catch *Bathynomus*. We were at one hundred ninety fathoms most of the time, and they're just a few fathoms deeper. I'll bet if we put over in two hundred and fifty fathoms we'll catch one."

"That's about an hour's running," said the skipper, rubbing his unshaven chin. "Hell, we'll try it. The seas ain't picking up any worse, and they always say that if you're catching those little red *Megalops* shrimp, step off into deeper water, and you'll hit the royal reds. Last year about this time we hauled the nets up and caught six thousand pounds!" Shrimping fever was getting to him.

Now that the nets were out of water, the *Norma Yvonne* churned ahead, straight into the big rolling swells. Her twin diesels strained, as the boat began to beat loudly. It was misery on deck: we watched the creatures slopping back and forth in the Styrofoam boxes. Nixon puffed his cigarette, looking ashen and pale. Before long I felt that all-too-familiar weakness.

But we had to keep on working. The water had to be changed periodically, replaced with new prechilled seawater, when the old became slimy from the mucous secretions of the sea creatures. Anne wasn't much help. "You know," she said cheerfully, "I heard about one oceanographer who used to go out and get seasick every time. When he'd have his students out on deck, he would start lecturing, stop to throw up over the side, and then go right back to teaching his classes."

"Such dedication," muttered Nixon, taking yet another
puff on his cigarette. Joe, on the other hand, was hurrying
about, taking pictures of all the living and dead creatures in
our boxes before their colors faded.

As I watched the hard round sea anemones sloshing back
and forth in the boxes, I wondered how they were reacting to
this unfamiliar wave action. Down where the *Chimaera* glide
like fairy creatures over the bottom, and giant sea roaches
tunnel through the soft ooze, there is no wave action. Yet royal
red fishermen spoke of tremendous undersea currents that
were so strong they could catch the nets and nearly flip a shrimp
boat over.

The ecology of the depths is little understood. The bottom is
often formed from the nearly microscopic shells of billions
upon billions of single-celled organisms that have died and sunk
down into the depths over millions of years. So stable are the
bottom temperature, salinity, and other environmental conditions that the shells never deteriorate. Consequently they
cover the seafloor with many feet of fine sediments.

We were now at latitude twenty-nine degrees, twenty-
three minutes North, longitude eighty-seven degrees, twenty-
five minutes West, dragging the nets in 280 fathoms of water.
Three more hours had passed, and it was time for the net to
surface again. It was getting cold: the vanguard of the front
had come through, and the winds were starting to swing around
to the north just as the captain had feared. Those puffy gray
skies were now hard on the move, pushed along by the wind,
leaving a cold empty sky behind.

By the time Thomas Lee said, "All right, let's get her," we
were wearing heavy jackets beneath our foul-weather gear.
Then once again, we waited and watched the winch spools
growing fatter as the incoming cable coiled endlessly around
them. Finally the otter doors rose from the surface and dangled

from the davits with their usual loud noisy crash. The net followed lightly behind, and for just an instant we saw a shark follow it. The crew hoisted up the bag, and it fairly whipped out of the sea it was so light. "God damn!" Thomas Lee snorted and began a stream of angry cussing. "We made a water haul!"

Thomas Lee opened the flaccid bag, and roughly fifty spiny sea robins poured out onto the deck, every one of them dead. The second net also had dead sea robins.

"What happened?" asked Nixon, stooping down to examine the fish.

"Shoot, we weren't even on the bottom; that's what happened. I'll bet we caught those fish in midwater when we were hauling it in. We didn't even tiptoe out here. Damn, I hate that!" He stared up over his shoulder. "This front is coming right along. We've got to go in."

Everyone looked depressed. All that anticipation, all that waiting for nothing. The skipper was embarrassed—he had underestimated the amount of cable to put out—and we were waiting in silence as he cursed and muttered. "All right, I'll tell you what. We'll make one more tow. This time she'll damn sure be on the bottom, I promise you. But the way these seas are building, it's gonna be miserable. If you can take it, that's fine with me. Frankie," he said, turning to the young deckhand, "tie the bags. Let's go fishing. You all watch you don't get washed overboard, you hear me?" he warned unnecessarily, and as he started back to the wheelhouse, he declared for all to hear, "I ain't got no damn sense. We ought to be heading in before we get drowned."

Hours later, the sun was beginning to set, a cold, orange ball sinking into the horizon, when the nets were once again on their way up. This time, as they came out of the depths, there was a heavy solid look about them. The very angle of the trawls pulling solidly down into the water foretold that it was gorged with creatures. Then, suddenly, there was trouble.

"Sharks! Goddamn it, there's sharks all over," yelled Frankie. "They're eating the nets up!" All the dead fish we had culled overboard hours before, all the slime, had attracted them. Suddenly the sea was boiling with sharks, big ones, long sleek grayish-blue bodies whipping in just under the waves and attacking the nets.

Thomas Lee hurried to the pilothouse, shoved down the throttle, and tried to outrun them. But there was no way to buck those twelve- and fifteen-foot waves. Even as the two gorged bags were pulled to the surface in the foaming white wake of the churning boat, we could see the sharks lunging in, biting out mouthfuls of the brightly colored chafing gear that protected the webbing, and violently shaking their bodies to and fro. Fish were spilling out.

"Hold it!" Frankie shouted to the skipper. "We've got to get these nets on deck before they eat the webbing down to the hanging lines."

Desperately the crew began to hoist up the heavy bags, but even as they did so, the sharks streaked in for more. As the first bag lifted clear of the water, fish began spilling out into the sea, and the sharks greedily rammed in and gulped them down. A moment later, the second net was lifted clear and dumped heavily on the deck, riddled with gaping holes. Another minute or two and only shredded webbing would have remained.

The sharks hung back, trailing behind the trawler now, waiting. Frankie put three wraps of rope around the revolving brass cathead, and the big teardrop-shaped net was hoisted up to the lifting boom, where it dangled like an enormous webbed sock. Fish and shrimp began spilling out of the holes, but then they compressed, plugging the leaks. The heavy polyethylene rope groaned under the tremendous weight and stretched taut.

But all that weight didn't prevent the net from swinging

back and forth with the ever-increasing rocking motion of the fifty-ton steel trawler. With each wave bigger than the last, the bag went wild, gyrating like a huge punching bag. Thomas Lee stepped forward like a wrestler about to tackle his opponent with a grim expression. The big man grabbed the two release ropes that hung beneath the swollen bag and tried jerking them open. The net crashed into him, knocking him down.

Joe and I started forward to help him. "Get back," he shouted. "This damn thing will knock you overboard. Get back!"

He rose once again and tried to grab onto the ropes, but they were snatched from his hand. For a moment the giant bag went wild. "Let off on it," he shouted to Frankie, who let the gorged bag crash to the deck with a loud plop. I shuddered to think what that was doing to the specimens; then I glanced at the other shark-torn net bag, also on the deck waiting to be opened with all those fish out of water. If we didn't open the nets and get those creatures into cold seawater soon, what life remained was bound to perish.

"All right, bring it up," said the captain between gritted teeth. This time when it was hoisted back up to the mast, Thomas Lee lunged in, braced himself with all his might, his muscles bulging, and snatched at the trawler knot. The first bit of pressure eased off the constriction, and shrimp and fish began spilling out from the bottom. And suddenly, right in front of us, we were staring at an enormous sea roach on the white deck. It couldn't be confused with anything else in the world, not those slowly flexing, long, needly pointed legs, or that flattened purplish-white segmented body, and those huge triangular eyes and small pointed antennae. I stared at it for an instant, unable to move. It was as if the *Norma Yvonne* had dropped her nets back into time and brought a creature of the Eocene directly into the present.

I rushed over to grab it before it got buried, but the skipper bellowed, "Goddamn it, Jack! Stay the hell clear of them nets, you hear me? I don't want nobody getting killed!"

In a flash our precious *Bathynomus* disappeared beneath the living avalanche. But then, as my eyes took in all those diverse life-forms, those confused flapping or inert creatures with their red, black, or white pale colors, two more prehistoric sea roaches emerged and slid across the deck. Anne grabbed one and we began hollering, "Ya-hoo! We got 'em, we got 'em!"

Nixon had been hanging desperately onto the railing, fighting the rolling seas. He suddenly strode forward and picked up the other one. He was grinning for the first time in hours. An electrical charge went through us at the excitement of discovery. Even the skipper and crew crowded in looking at these creatures. They were strangely beautiful and horrendous at the same time. Their antennae flexed almost mechanically; their segmented bodies, rolled up like armadillos, opened a circular mouth surrounded with cutting plates and spat out a strange brown fluid.

When we dropped them into the waiting boxes of water, they began to swim. I dived back into the pile and dug out yet another *Bathynomus*. By this time the other net was opened and we found three more. The pitching boat and blustering winds made it hard even to squat down now and glean through the catch without falling face-first into the midst of it. But our excitement banished our discomfort. There were more treasures there, sculpins, spiny deep-water crabs, more rattails, large flat skates and gray dogfish sharks with giant green strobe-like eyes. Something great and exciting had just happened aboard the *Norma Yvonne*. The crew was nimbly moving about, shoveling the trash to the sharks, culling out the royal reds, and helping us pick out anything alive.

Suddenly our skipper, Thomas Lee Mills, the man who

had come on this trip only because he wasn't catching any shrimp inshore, was in there with us. His perpetual scowl and doubt were gone. There he was, with his own Brownie camera, cube flashbulbs and all, taking pictures like crazy.

"You know," he declared loudly, "I ain't never paid no attention to this stuff before. I hate to think of all those tons of critters I've shoveled overboard and never bothered to even look at. But damned if there ain't something special about all this. I want pictures so I can show my younguns."

In moments, our boxes were crowded. Then Frankie tossed a mackerellike fish into the box with the largest *Bathynomus*, the one that measured a big muscular twenty-four inches long.

"Good Lord," he said, his eyes wide, "will you look at that!"

The fish lay gasping limply on the bottom of the box as the giant sea roach crawled forward and grasped the hapless creature with its needly hooks. Then it abruptly spun the fish around, its plated round mouth opened, and it clamped down on its tail. Using its ten pairs of claws, before our eyes, it began to shove the entire creature into its mouth. It was almost like seeing a snake swallow a rat, except that the jaws were so powerful that they rapidly sheared away muscle and bone. In no more than a minute the fish was chewed down to bits, leaving only the head and the gnawed-down backbone. Here a creature had come up from fifteen hundred feet of pressure and darkness, up through the warm thermoclines, had been hoisted out of the water, and its first action was to start eating ravenously as if nothing had occurred.

We finished culling under the glare of the deck lights, surrounded by darkness and a star-studded sky. And when the last of the trash was shoveled overboard and the five baskets of royal red shrimp were stashed down below, the crew busied themselves getting the boat ready for the return voyage.

Frankie crawled out on the outrigger, hanging above the rolling black seas, clutching the steel ladder, and looped a rope around the heavy otter doors so they could be hauled up on deck. All it would take was one slip, and we knew these rolling swells of darkness would swallow him up. Even without the sharks, there would be no saving him. But there was no problem. In a moment he was back safely on the boat, manipulating the steel-framed doors back into their brackets.

Then the grueling voyage back to port began. As those twenty-five-foot seas slammed over the bow of the big steel trawler, the sea foamed around the wheelhouse and flooded the rear decks. Periodically we would get up from our exhausted sleep, clutching the railings, and check on our specimens, keeping the heavy, weighted Styrofoam boxes from being swept overboard as we crept back slowly toward shore. We were making scant headway. Soon fifteen hours had passed, then twenty, and still no sight of land. It had only taken ten hours to get out.

It was late in the afternoon before we managed to inch our way back into Mobile Bay. At the dock we worked frantically, repacking the Styrofoam chests, hurrying to meet a six o'clock flight to New York out of the Pensacola airport. Once again darkness fell, only this time Nixon and I were now twenty thousand feet up, watching the lights of the cities down below, the long strings of bioluminescence of the highways.

The full impact of it was beginning to sink in. We had done it: we had captured the monstrous sea roach. Soon many thousands of people would be peering at the deep-water creatures through the glass walls of the aquarium, newly aware, as we had just become, of the incredible diversity and enormity of life in the deeps.

It was eleven o'clock the next morning, and the public relations department of the New York Zoological Society was insistent.

"Hurry up, you've got to get down to the aquarium right away. The press is coming."

Before the blazing lights of television cameras, leathery skinned, prehistoric gooseneck barnacles expanded their scarlet feathery legs. Tan-and-black-striped chain-dogfish sharks glided effortlessly along the bottom. Reporters gazed upon luminous moon snails and deep-sea lobsters. And six *Bathynomus* peered out from behind the glass walls of their new home with their huge triangular eyes, looking more like something from another galaxy rather than creatures of the deep. Nixon Griffis happily held one up for the reporters to see, keeping his fingers well clear of those terrible mandibles. Later viewers across the country awoke to see our giant sea roaches swimming across their television screens on "Good Morning America."

Crowds flocked to the New York Aquarium, even in the dead of winter. The sea roaches had indeed proved to be a monster hit.

Sally Carrighar

THE TWILIGHT SEAS

A WORLD TO DELIGHT

At last the Blue Whale was going to be born. After nearly eleven months in the warm nest of his mother's body he had reached the right time to come forth—but the day itself turned out to be most unfortunate.

He had been ready from long before this; that is, he had been perfectly formed, he had grown through the stages by which his species evolved and had become a complete small Whale, with every organ laid down, when he was only 12 inches long. At that period in his fetal life when most unborn mammals look more like oysters than the adults they eventually become, the Whale had his slender head, tiny flippers, and slim body tapering to the paired flukes of his tail: a perfect whale but in miniature soon after he was conceived—a precocious development it seemed, a mystery even to scientists. All he did from then on was to acquire his enormous size.

In his entire body, finally, the Whale felt a lack. More of something was needed. His mother's blood still came down the umbilical cord. Sped by the pulse of his infant heart, it flowed through his veins; but the blood was no longer satisfying. There was not enough oxygen in it. This emergency was felt by all mammal babies, including the human, when they had grown to the point where they could begin life on their own. And so the Whale did what the others did: in his discomfort he squirmed. His flippers and flukes, which would be floppy and limp until he was born, nevertheless tried to thrash about, and his body turned this way and that. *Help!* his motions were saying, and his mother's muscles were hearing that cry. The walls of his nest, until then so quiet, started to pulse and squeeze. Something was going to happen now.

The nest had produced an opening. It was not large enough, but the infant's own movements and a push by his mother were able to get him through. Next he found himself in a slippery tunnel. The infant himself was such a smooth shape that he should have slid out of it easily. All that was needed was to have his mother's squeeze-push continue. But it didn't, it stopped! This was not right and with the confining tightness here, the Whale could not make his misery known by expressive motions. Motion was the language of whales, all through their lives, but the pressure on every side of him made the infant mute.

At the instant the Whale left her body, his mother would have to be near the surface, for he would need air at once. But the sea just then was a chaos of towering waves plunging over into the seething foam of the troughs. And how would a just-born infant be able to time his breaths so that he took only air into his blowholes, not the tumbling and swirling waters? His mother was staying just under the waves' crashing fall, for now that the birth had commenced she might not be able to hold it back. But if she could she would delay it. And so the infant was finding himself immobilized in the narrow channel.

For the first time he was hearing clearly the sounds from outside his mother. Until then few had penetrated through her thick, insulating blubber. Sounds surrounding him had been only those of his mother's living. Very loud was the *lub-dupp, lub-dupp* of her half-ton heart. Never ceasing, the beat was a lullaby, soothing, an assurance that everything was all right. Already his ears were many times more acute than a human being's; he could also hear the blood rushing along in his mother's veins and arteries. These were sizable streams, some of her arteries being so large that a human child could have crawled through them; and her mighty heart pumped the blood along with enough force to send it quickly down her 96-foot length. Her other organs were doing their work, most of them quietly; no digestion was proceeding because this was the fasting half of her year. The infant could feel her muscles' thrust as she swam about, but the muscles were nearly silent.

Now having progressed almost to his exit, the Whale was hearing the outside sounds of his world-to-come. They warned of violence and peril. From above came crescendos of crashes, heard on all sides but loudest when a wave broke just over him and his mother. A towering storm was raging up there. Lines of waves were gnashing their way ahead, each one gathering itself from below, soaring and then hurling itself down in a long continuous cataract. These were the first sounds that greeted the Whale while he was being born, but to him they were only a great confusion.

The birth would have to take place; there was no turning

back and no further waiting was possible. With the squeeze-push resumed behind him, the Whale reached the end of the tube. It was smallest there but it spread apart suddenly and he was out.

Oh, out! Nothing whatever pressing upon him, nothing surrounding him with protecting walls either! On all sides was a vast noisiness, swing of water, and a pounding down from above. The Whale's mother, having no teeth to bite the umbilical cord in two, made a quick jerk to break it and with the same motion swung under him, lifting him up and out of the sea. Into a nothingness: air! Oxygen! He gasped and took it into his lungs—this that he had been wanting so urgently. More of it, more! But a great wave came down on the two of them. The Whale sensed its fall and closed his blowholes, but water got into them. Instinct would tell him how he could force it out, though not yet. For his mother had sunk away under him, taking him with her below the surface, out of the foam and spray and the battering weight of the waves—too soon, but he couldn't stay out up there, just breathing as he so needed to do, because of the oncoming ridges of water.

Their weight could flatten a newborn whale. His mother must lower him to a safer level, but up they went again into a trough between the waves breaking . . . and down . . . and up, sometimes nearly caught by the waterfall. Oh, this was terrible weather in which to be born!

It was. His mother would never have chosen the height of a storm to introduce her whale baby to life. But here near the Tropics ferocious gales did arise suddenly. Within minutes, it seemed, screaming winds would come over the placid sea, driving the water up into mountainous combers, sending the spindrift flying and worst, slamming down with a crushing force on an infant that needed a quietly rocking cradle of gentle swells. A storm like this was the most dangerous time for a birth, but one never knew when the wind would come in like a ramming enemy. Anyway here the Whale was, born into this turmoil. One could say, though, that it was not an unsuitable welcome, for he belonged to the species of mightiest animals ever to live on earth.

Even below the surface the sea's rolling and tossing frequently pitched the Whale off his mother's back. There was little that he could do then but flounder about. His flippers and flukes, so flabby before he was born, had stiffened immediately; therefore now, when the water threw him around, in his fumbling way he was fighting it. During his very first moments of life he was trying to cope, not successfully but he robustly tried—he was that kind of whale.

The storm lasted too long. Thunder seemed to be splitting the sky wide open. Lightning was streaking out of it, and torrential rain, a new hazard, for the Whale didn't know yet how to turn on his side when he spouted. His blowholes opened directly into the deluge, and each time he breathed he took in some water. But he did sense that he could blow it out when he had a chance. Nevertheless he was a very young baby, with muscles and nerves quickly tiring, and after an hour of being pounded and flung about he was exhausted and made only the feeblest efforts to blow. His flippers and flukes had gone limp again; they were just letting the water swing them. His heart was still beating but with only a small weak flutter. Soon he could be a lost whale.

As the wind and the thunder and lightning went on to savage the farther seas, the Whale's mother, supporting him, kept him up in the air. She sensed that her infant's body was now inert. If he had been dead she would have known; since he wasn't, yet, how could she quicken the life in him, how could she get his flippers to grasp her sides, for he was making no effort to stay on her back? When he started to slide away, she righted him and began swimming, letting him feel the strength of her muscles under him. Nothing helped, and so then, in her fatalistic wild way, she just waited, still holding him out of the gradually diminishing waves.

In the storm's terrible assault many whale infants probably would have relaxed their hold on the cord that tied them to life. The Whale didn't. A moment came when somewhere below thought, below consciousness, he might have been glad to loosen his grasp; when, even, it might have seemed the right thing to do: *Let go now.* And yet he held on. With all his tiny strength he would cling to his life until it was snatched away.

At last, through her back, his mother could feel that he was reviving: the next time he took a breath his chest was deflated suddenly with his blow and then filled out again with new oxygen.

The waves were no longer cascading over. They would advance, roll under the whales, lift them, hold them high, start to pass beyond, let them down in a trough while the next wave advanced, rolled under them. . . . The sea was flattening gradually, the great clouds blew away and the sun came out.

The Whale still was not comfortable. Something new, a different lack, was making him restless. It was a need almost as strong as his hunger for oxygen several hours earlier, and his mouth moved across the back of his mother's head, groping.

She understood and, turning over, slid forward to bring his mouth to the pair of folds on her lower belly. These now spread apart and two immense nipples sprang out. The Whale had no means of sucking, but he closed his large inflatable tongue around one of the nipples, forming the tongue into a funnel with the small end in his throat. And at once, with strong pulses, his mother's breast muscles squeezed a great stream of milk—gallons of it—into him. It was ten times as rich as a cow's milk.

After he'd had enough his mother put one of her flippers under his chin so he wouldn't sink, and both of them lightly slept. At that time, late June, so much oil was contained in her blubber, which was nearly 12 inches thick, that she was a naturally floating object. She had recently finished six months of intensive feeding in the Antarctic where, and only where, the food of southern blue whales could be found. By now, in the antarctic winter, the ice of those seas would have covered the food—but no matter. Her blubber had stored enough nourishment for herself and her young one until the next feeding season would open. During the six months between now and then she would gradually lose weight and buoyancy while her son would be gaining both and would become ever better able to stay on the surface effortlessly. Soon he would be gaining 200 pounds a day and every six and a half days would be a foot longer. Right now on his day of birth he was 24 feet in length and weighed over two tons—4400 pounds. No other infants who ever had lived were as large as blue whales; the adults into which they grew were the all-time giants. But the Whale did not think of his mother as being unique in the earth's history. She was just the right size, a dependable and concerned protector.

The next morning she disappeared—or seemed to; she was not far away. Sudden fright made him splash around helplessly. He was sinking! But no; by lifting his flukes, the large flaps shaped like butterfly wings at the end of his body, he found he could raise his head out of the surface. As his tail lifted and bent up his head, he had gone forward a little—the tail also had thrust. This small emergency was his first lesson in mastery of his element. When he felt that the water was rising along the sides of his head, he tried the maneuver again: he definitely did know how to keep his blowholes out of the surface.

His flukes had no muscles in them, they were not much more than envelopes of skin filled with fibers, now firm, but the hinder part of his body was really a tail, with strong muscles to raise and lower it, and the tail wagged the flukes, partly for extra push and as well to serve as a rudder. That second day he discovered too that depressing his tail sent him *down*. He would not be afraid to do that before very long, especially since he knew how to come up again.

He also was finding out what his flippers, those armlike paddles, were for. They balanced him and prevented his rolling sideward unless he wanted to. What a wonderful body: it could turn in any direction, in this lovely supporting water where he would become almost weightless. How splendid to be a blue whale! Hints of future delights came to the infant while he was discovering what he could do.

More milk! Where was his mother? She was right there. When the folds in her flesh were open, the Whale, failing to grasp a nipple firmly, found himself under a rain of milk, which had spurted up six feet into the air! Next time he took her teat more precisely.

To bear this infant the mother had come to a large bay south of Africa's western shoulder—here because these waters were close to her own warm temperature and small whales needed that external heat. The blue whales that gathered in the Antarctic for summer feeding all came up into subtropical latitudes for the alternate season, for mating one year and birth of a young one the next. And two to four months of each year were spent in migrating between the two areas. This part of the ocean where the small Whale had been born was his mother's home place. She always returned here and knew it well—from the contours of the seabed and flow of the currents.

The storm had been only a brief disturbance and a more congenial world for the temperament of an infant whale followed. The sea lay quiet under a fresh cool breeze, the surface brushed into only the smallest of sun-splashed ripples, which flowed along the Whale's sides in a playful way.

He was enjoying all this when he felt his mother rise under

him, lifting him so that he rode on her back. For a short time they went along at the top of the water and then she dipped below, not far but the infant must hold his breath till she rose into the air again; and then down, a little deeper and going faster and making some shallow turns. The Whale could feel her strong muscles working under him and their motions had as much meaning as spoken words: *You try this*—as soon he was doing, waving his tail up and down as she did but not balancing with his flipper-arms because those were holding onto her tight. When he grew tired from the exercise and too from excitement, she knew and returned with him to the surface where he nursed. And then they lay resting, the Whale's mother no doubt making him feel how pleased she was by his infant self.

During his first independent swim the Whale encircled her as she was idling on top of the water. What he learned on that day was not only what she was like but how he would look when he was grown, though at the time her immense size was just reassurance: she was a great fortress of a mother.

He found that her length was four times his, and she had a head ten feet wide. Being dark blue, she lay on the sea like the shadow of a cloud. Her skin was all glossy smooth, though with a few gray circular scars; the Whale would learn about those when he acquired some himself. She was perfectly streamlined. Her intimate organs were enclosed in those folds on her belly, and no visible ears showed on the sides of her head—tiny holes were all, even though she had phenomenal hearing, as the Whale also had. Her head tapered up to her shoulders and then only very gradually down to her tail. She was the sleekest of whales, wherefore she was one of the best swimmers—unlike the humpbacks, who were shaped some-what like tadpoles, or the sperm whales that breasted the waves with huge cubical heads, seven feet high, wide, and deep, as the infant would see before long. His mother could swim up to 28 miles an hour in an emergency, but most of the others not half as fast.

At the height of the storm no one could tell which was a blown spout from the lungs of a whale and which was spray off the tops of waves; the air was everywhere full of mist. But after the winds calmed down the infant could see what a proud tall fountain was made by his mother's breath. Although coming from two blowholes, close together, the spout rose in a single jet, straight and strong, 20 to 30 feet high. Sadly it would always make her a conspicuous target for anyone wishing to kill a whale. Here in the warm air near the Equator the spout was not steamy, as it would be among the ice of the Antarctic, but was composed of droplets of oil, oil continually gathering in her lungs, as in his, with the life saving effect that the oil absorbed nitrogen, which would be formed in a whale's blood during dives, and the oil, dispersed in spouts, would prevent the bubbles that can be fatal to human divers.

This was only one of several almost miraculous adaptations that allowed whales to live in the sea. They had once been land animals, browsing among the giant ferns and evergreen trees of 40 million years ago; they still had—the young Whale himself had—useless leg bones for walking which were now buried in flippers and flanks. At first the whales could not have survived a dive of half a mile down, but they had developed this oily foam in their lungs, which they did, however, have to get rid of frequently. It was fortunate when the Whale's mother timed her spouts to come after his; otherwise he would gasp in her oily spray and carbon dioxide, the exhausted vapor from a body almost 100 feet long—polluted air until breezes drove it away.

Of all that a bright young whale soon might observe, one fact was overwhelmingly important: none of these others around him had to keep constantly going up to the surface to breathe. All the fishes, even the sharks got their oxygen from the water itself; all had gills, those ruffles along the sides of their throats which drew water in and out and extracted oxygen from it. Why didn't the Whale have them? Almost 300 times every 24 hours, day and night, forever, he would have to come back into the atmosphere of the land animals among whom his ancestors once had lived. Was he still, basically, one of them? Was dry land, even now, his true home? If not why, in 40 million years, had evolution not provided him with a breathing apparatus to make life in the water more practical? While other air-breathing animals could forget their need for a constant supply of oxygen—their lungs sucked it in with unconscious regularity—the Whale would always have to make a voluntary effort to fill his lungs.

There would surely be times when the need to rise to the surface every five minutes would seem a hardship—except that his movement in water, even swimming up to the top for air, was so strong and supple. The Whale, like his mother and all other blues, had a brain with an extra development of the poise and balancing structure, the cerebellum, and so they could manage their huge bodies with silken refinement. It was surely a satisfaction to express such a talent. And being able to bend in three dimensions, which a fish could not do, and very nearly able to ignore gravity, as birds could not, probably made the Whale more physically free than any other creature alive.

He soon showed signs that he would become an excellent swimmer. And he had another skill to perfect: his means of avoiding collisions. He was sharing the sea with countless neighbors, all of them, too, swimming about; indeed if the dry-land world were as fully populated as oceans were, the air would be densely swarming with birds, insects, bats and other mammals like flying squirrels. Marine creatures had to be very adept at weaving their way through the paths of so many others and when they were first born they had to learn how to do that, as did the Whale.

The great problem was that most of the water was turbid,

especially here off the coast of Africa where the Congo River was releasing its load of silt, or in any part of the ocean blurred by clouds of plankton, "the dust of living specks." The Whale's eyes were blue, the ocean's color, and they looked out with a certain knowing expression, seemingly conscious; but they were small in comparison with his size—they would never be larger than grapefruit. And they lacked the cones that would show him color. He never would see the changing hues in the sky and the reflection of these on the water, never the sunrise, the lofty dome of noon, or the splendor of the sun's last light. The Whale would see them but only as grades of pearly radiance.

In the water eyes alone would not allow him to swim about swiftly for danger of hitting other undersea creatures or, later, the seabed and its rocks and ridges; his element would never be one to know visually. Out in the air he could not see a distance more than twice the length of his mother. Under the surface his eyes were more efficient but they were hindered because half the sun's light was lost at a depth of six feet in a sea like this. In his first days of swimming alone, even then in his newborn size, he was so large that those who had better eyes would spray out of his path as he drew near—except the octopuses and their near relatives, squids, which swam backward and couldn't see where they were going. A few times they bumped into *him*, when, alarmed, they would squirt out a cloud of ink in which to hide or escape—a small patch of total night in the ocean.

Lower down it was night everywhere, but the Whale knew that he was among living beings because of the many sounds: fish drumming on their air bladders, tapping their teeth, croaking, squealing, chirping, singing. These and other voices pricked the Whale's interest and he soon discovered how he could know much more about the creatures there in the dark.

Whenever he rode on his mother's back and they went

below to the level where eyes were of little use, she kept sending out probing clicks, small stabs of sound. Was she warning the others that whales were coming? The young one himself was listening to her clicks, which were pitched very high. For that reason those sounds did not spread out in all directions, as sounds in the human range do. Because they went forth at so many cycles a second they could be focused like arrows, precisely aimed. The Whale came to recognize that his mother was feeling ahead with them and then he became aware that the clicks returned echoes from whatever they hit. The echoes were telling the whales that they were approaching something, and by shifting her target about, the mother could learn how large it was; and if it was moving, how fast and in what direction. And by the length of time it took for the return of the echoes she also knew how far away it was. If the object was stationary, maybe a dreaming fish, as she came closer she would investigate with faster clicks. This ear view was actually more clear in deep water than even the sharpest eyes could have shown, because the clicks and their echoes could penetrate roiliness and the dark. And when the young Whale began to grasp what his mother was doing the whole sea came alive for him. He too began to listen for echoes and even could start to interpret them.

It was a special day then as the Whale's mother, with him on her back, was swimming along at a medium level when the Whale felt a swish on his face and, wanting to learn what was causing it, made some clicks of his own. He had not known that he could! It had happened involuntarily and when an echo returned he clicked again and again. He didn't learn what was there because his clicks were not aimed with skill, but having found he could do this, he spent more and more time where there was partial light and he could both see and hear. That was not a deliberate project, rather was due to the fact that the Whale was still timid about going alone into the blacker depths;

but the effect was to match the evidence gained with his eyes and that with his ears and meanwhile have practice in targeting.

He, like most young blues, had a lively interest in everything new, a bright curiosity that would get him into trouble sometimes but also would give him much pleasure. And what a quickening world he had entered, both below and above the surface a world ever changing, diverting, bringing into his days a succession of strangers and mysterious voices, unpredictable weather, light, and movement of water, a world always different and usually harmonious.

Being the largest animal there and eventually the strongest, with the power of 47 horses, it would seem that he must be entirely safe. He was not. A blue whale had several natural enemies, two during his youth and another who might attack when he was of any age. And besides these dangers there was the adversary, man.

As everywhere, some animals in the sea were eaten by other animals—a sad thought but not one to cause despair about nature's program since the deaths were only a way to insure continuing life. It was not until the arrival of men that living creatures were turned into *things*, trivial, insignificant things like machine oil and car wax and shoe polish. In the making of all such products there were inanimate substitutes for the bodies of whales. And the killing of sensate beings, especially intelligent beings, to provide nonessential materials seemed to justify more profound disillusionment than nature's allowing one life to be transformed as food into another life.

Men had the disadvantage of pygmy size but the huge benefit of inventiveness. Any blue whale, then, would need to outwit human technology. But could he? Was there any way that a whale could win such a mismatched conflict?

In any case the young Whale would have an animal's acceptance of the inevitable and would carry on in the ways typical of his kind since long before men had evolved. He would be

great in size and also great in his consideration for other creatures and in helpfulness, as when he would hold a sick or injured
comrade up at the surface to breathe. The Whale himself would
live with no visible sign of bitterness. As his limber body would
make its excursions in the supporting and yielding water, that
beautiful swimming would be his play and perhaps his delight,
since his motions would have more subtlety than even the finest
flyers' among the birds, reaching its climax when he would mate
and his grace would interweave with another's.

When he could avoid the new hideous hazards he would
be one of the most fortunate animals on the earth. If he could
write his autobiography that is possibly what he would say—
at least in the beginning.

EVERYTHING ELSE WAS SMALL

In the early days of his life the Whale spent most of his time
at the surface—for its essential air and also because more things
were happening there. He could see and hear lively splashes
and chases. Dorados were leaping out of the water, always
coming down flat on their sides, noisily smacking up bursts of
spray. They did that for fun, but when flying fish were about,
sailing up into the air and away, the dorados tore after them
at the top of the water, trying to be on the spot where a flyer
would have to come down. Flying fish were their favorite food,
and with many dorados here, more than 40 to share them, they
all went after the flying fish savagely.

The dorados, sometimes called "sailors' dolphins," were
not the small whales now known to most men as dolphins; they
were true fish *(Coryphaena)* and were said to be the most beautiful in the sea. It was not because of their shape that they were
admired, for their heads and necks were so large that they
seemed grotesque. A four-foot dorado could have a head 12
inches high, perhaps a deterrent to getting up speed, for in

chasing a flying fish the dorado swam with his forehead out of the surface. The fame of its beauty was due to something the Whale could not see—its rainbow colors, quick-changing, washing over its scales in a display that was always most dazzling during its death. At men's banquets in Pompeii dorados were sometimes killed at the table to give guests the pleasure of seeing the fires of their dying. But in life too the dorados would blush and go pale in pink, green, yellow and blue, and though the Whale could not see these hues they were visible to him as waves of shine.

The Whale was entertained by the dorados, and as much by the flying fish. They left the water by giving a mighty thrust with the lower blades of their tails, which flung them up into the air where their breast fins opened like wings and carried them 200 yards or more . . . beyond the Whale's sight, but often one ended its glide nearby and the Whale could see how, at the instant it touched the water, another lunge with its tail again sent it up, not so far this time and once more a dip until all its momentum was gone. Especially on sunny days when a lively breeze helped to sustain their flight, the air seemed to be full of the bright little fish, looking as if the waves' sparkles had taken wing. The Whale, stimulated and starting to have some playful urges himself, would make a quick vertical dive down to shivery depths and then dash for the surface, as he came out of it lobbing down hard with his tail—he too could make a fine splash, bigger than any dorado's!

Now what could this possibly be, floating along with the current? It lay near the top of the water, a disc on edge—nine feet wide. Alive? It was not shiny, it had no scales, but on each of two sides it had an appendage that looked like a flipper, narrow and longer than the Whale's infant flippers. The Whale would investigate. He swam close to the object, which made no move. It was just drifting. With his snout the Whale touched the end of the lower appendage.

Splash! The fish—for that's what it was, an ocean sunfish, or headfish *(Mola)*—sprang forward. It had been idling along, soaking up the bright morning's warmth, but now it swam quickly away and was indeed a curious sight. It had no tail, just a slight skirt of ruffled fin from the top to bottom of its rear end, and only those two thin fins, one standing up from its back and the other descending below its belly. To go ahead it was twisting the fins with a sculling motion, surely the strangest swimming arrangement of any fish here, when the usual fish way to progress was by a side-to-side fanning of upright tails —or in the case of a whale, fanning one's tail up and down. And this fish appeared monstrously large, for it weighed a ton. The Whale didn't follow it. Who could tell what damage its mouth with two overlapping teeth, like a beak, might do? (Various jellyfishes, crustaceans, and other marine creatures could have told what damage if they had not already been in the sunfish's stomach.)

It seemed a lazy fish, or perhaps was a female and tired from a recent feat of shedding her eggs in the ocean, of which she laid 300 million at any one time.

The whale's eyes and clicks told him that his fish neighbors were of every size up to that of the giant sunfish. There were schools of tiny ones staying so close together it seemed that the many were one single creature . . . now streaming ahead, now drawing together, a mass of circling bits, a flickering image. Almost all living things in the sea moved continually, as the Whale and his mother did. The water's restlessness was born in them.

Fish in general seemed to be shaped much like himself and his mother. But octopuses and squids were like bags with arms waving out of the tops; and just below the arms, eyes. The bags moved by jet propulsion, taking in seawater and squirting it out to send them along in straight lines very fast —backward, so that even though the creatures had excellent eyes they sometimes hit what was behind them. Very strange was a fish that would blow itself up like a big ball with spines all over it—a porcupine fish. And something puzzling indeed: whether the Whale was up in the water's visual layer or down below, a school of little striped fish swam ahead of him. Why? And how did they know which way he would go? Perhaps they thought they were guiding him. Men called them pilot fish but the Whale certainly knew that he was not following them—or was he? Besides all of these there were other shadowy figures whose blurred outlines approached and then moved away. Many creatures; and there was a wonderful harmony in their flowing about, each of them passing in and out through the companies of his neighbors, going forward or giving way without argument, winding among lives all in motion, most of them with a lovely grace.

With his clicks and echoes the Whale could sense the smooth turnings around him, but there was one fascinating way that his eyes were more useful in even the depths, in fact the darker the better: for eyes could pick up lights, which no ear could do. His ears could not find the millions of tiny animals of the plankton that had luminous spots on their bodies. They were like miniature fireflies, a few of them shining continuously but more that blazed with excitement whenever one of the larger creatures brushed by. They could turn on their lights at will and when some of them brightened, many others would also flash their small beams, kindling together what appeared to be sparks, glowing embers, and sheets of flame in the sea.

It was not an impressive show in the daytime, but at night when the water became black dark, everything moving trailed streamers of light. The Whale's mother did and, being so large, she illuminated the whole sea around her. She was a conflagration compared with the small bright streaks that followed a passing bream or bonito. To those fish and most others the young Whale himself looked immense, defined as he was by pennants of fire.

Why did the plankters light up, for whose benefit were they flaring? For one another no doubt, to warn, attract, be identified or simply to be in touch. Most of the animals in the plankton had eyes and probably recognized their own fellows by the patterns of lights on their bodies. The plankters could not have had any conception of what a whole whale looked like, nor did they see a man as a man, though when one in a boat dropped something over the side, the plankters would flash over a much wider area. Or if the man shone one of his own big lamps on the water, a brightness would spread as if many plankters were answering.

To a human observer the living illumination was always impressive, but there was a greater marvel known only to those with microscopes. Seen under a lens the plankters' lights were of all colors—nearly every observer described them as flashing jewels—and the lights always harmonized with the tints of the owners' bodies: pink, violet, blue, silver-green, scarlet, orange. But after the human eyes became used to the dazzling display, it was the forms of the tiny structures that seemed most amazing.

Some had an exquisite beauty: the radiolarians, for example, which were balls of glass (silicon) pierced in elegant lacy designs and with delicate little antennae, the whole looking like tiny spaceships designed for artistic effect rather than technological traveling. Just visible to a good eye were sapphirinas, minute creatures over which rainbow hues flashed in waves, as over the big dorados; copepods resembling insects, gnat-size but with tails that a peacock might envy; Venus's girdles, long, flat and transparent, edged with bright iridescence; snails with butterfly wings and luminous jellyfish. Those called comb jellies suggested gooseberries, striped with combs which had mobile teeth, a means of moving about; and glass arrowworms had sharp black eyes that could look in every direction, including inside themselves, for what narcissistic purpose one wondered. Some men who thought that surrealistic designs were amusing when made by themselves were a little uncomfortable when confronted by living animals that seemed even less rational. They tried to see the small bodies as like things in the human world: anchors, stars, flowers, seeds borne on thistledown, children's tops, lances. Those descriptions were far-fetched however. In human terms most of the plankton animals looked so nonsensical only one who was daft, it seemed, could have designed them. One observer said the siphonophores, *Physophora hydrostatica*, looked like renditions in glass of jazz solos.

With extraordinary small mobile hairs and other contrivances the plankton animals in the Whale's ocean were migrating many feet up and down every day. They came up in the evenings, fed most of the night on the floating plankton plants that the daytime sun had encouraged to grow; and then as dawn approached, the tiny creatures went back down to the lower levels. But sometimes an individual or a group would linger behind the others; that is, they were not being pushed up and down by some automated force—they had their own way of doing things. And even those that didn't go all the way to the surface usually moved on a vertical course each day. While some were traveling between the top of the water and 200 feet down, others were rising from 600 feet below the surface to 200 feet. Why? Men thought that it was because the various layers of water are often flowing in different directions, and so a change of depth gave the plankters a ride to a new location and maybe to richer foods. Or did they just like mobility? Men seemed unable to imagine any other reason for their moving about than for better nourishment.

A man in a boat off the coast of Guernsey, one of the British Channel islands, or off the coast of California on the other side of the world, puts his hand down in the water, spreads his fingers and moves them back and forth slowly. Between the fingers flow small streams of light. "Phosphorescence," he says and by naming the light diminishes its significance. For phosphorescence isn't composed of one substance, like the phosphorus put in fertilizers. It's a community of living, seeking, responding, flashing little beings, most of them made, literally, of colored glass. The Whale, too, failed to see them as neighbors. They were most important to him as the glow surrounding his mother.

She was not always with him, for she belonged to a species understood to be wanderers and as long as she knew where her infant was by her echo clicks, she would relieve her restlessness by circling about over the area that was her summer home place. It had been chosen ten years before by her and her permanent mate; here in one year she would conceive an offspring and here the next year would bear him—a routine of interludes between their migrations down to the antarctic feeding grounds. As long as her mate was with her it had been an unquestioned habit that they would return to this summer sea; and this year, although she was alone, again it was to here that she had come back.

"Here" was the surface above three small submerged mountain peaks, easy to recognize by a whale's sonar system. The peaks stood at the end of an undersea canyon, a gorge formed by the outflow of Africa's Congo River. The banks of the canyon provided another landmark, and up at the top two familiar currents flowed past. Both currents were swinging westward, side by side but easy to tell apart because one was warm, almost hot, and the other quite cold. Men called the warm one, on the north side of the canyon, the Equatorial Countercurrent because it flowed from the mid-Atlantic doldrums east to Africa where the coast turned it around toward the mid-Atlantic again. It brought the high temperatures of the equatorial region and also a remnant flavor of the Sargasso Sea, of the weeds and the eels that bred there, and some of the human pollution becalmed in that windless area. The Whale's mother, without understanding what caused the flavor, could distinguish it by the taste buds on the back of her tongue.

The other, the cold current, also came from the west and was reversed by the African coast. Called the Benguela Current, it originated in the Antarctic, flowed eastward until it reached the immovable African continent and there, after turning northward along the continent for a way, came as far as the Congo River where it met and was deflected by the Equatorial Countercurrent and had to swing westward, also out toward the mid-Atlantic. Although she had borne her son in the warmer one, the mother, from time to time, made a little foray into the cool, bracing Benguela Current, perhaps because it reminded her of the ice fields around the whales' feeding grounds where, after a few more months, there would come an end to this summer's fasting. The Benguela Current didn't have any particular taste except for the sediment it had scoured off the African shore. It was the less salty current, having been fed by several landward rivers, whereas the current from the Sargasso Sea came from one of the saltiest waters on earth.

The two contrasting flows were along the top of the ocean. Below them there was a southbound layer of water and on the seabed a northward creep of very cold water indeed. The whales wouldn't go all the way down into the lowest one unless they were making very deep dives. That was something the mother and son had to do before long, since it was necessary for him to practice every kind of a whale's swimming.

It was on one of her brief absences in the Benguela Current that she heard a scream from her son—one of fright and perhaps of pain. Instantly she sped toward him.

He had been enjoying himself that day. As always his curiosity inspired him to investigate anything that seemed new and interesting. He was tuned for excitement and his ears scanned the waters around him. What was that? He could hear a prolonged splashing. He listened. The sound came from farther away than the Whale ever had gone alone, but he hurried toward it. As it grew louder he could recognize two kinds of disturbance. The lighter kind was continuous and then there was another, intermittent *bash . . . bash . . . bash*, all the commotion being up at the surface.

When he came near the Whale saw what was happening. A big brown thresher-shark was encircling a school of mackerel, driving them into a closer mass by swimming around and around them and beating the water by slashing into it with a very long pointed tail. The upper lobe of the tail was the same length as the shark's body, ten feet. Like a flail it was, and it whipped into the water in a way that was terrifying the mackerel, now a company of giddily desperate fish.

The Whale, sensing slaughter to come, was mesmerized. The shark would make one more swing around the prey before it attacked them. But—hideous! The shark, catching the scent of the Whale, streaked away from the fish toward the infant! As he flung about to escape, the Whale screamed for his mother. A shriek it was, for the shark's teeth had fastened into his side!

They had ripped out a large bite. Dazzled with pain the Whale sped away. Frantically he was pumping his infant tail, staying up at the surface where the water's resistance was least. *Where was his mother?* Faster, oh, faster! For he could hear the splash of the shark coming on, gaining behind him.

The shark had lost a few seconds in his pursuit because he had not expected a very young whale to dash away so decisively. More often they thrashed about in confused panic. But this time the bite had been deep and its white-hot stab had sent the Whale headlong.

The shark, having swallowed one mouthful of the sweet

mammal flesh, was determined to have another—many more! His snout had come even with the Whale's fibrous flukes. No treat, those, not when six or eight feet ahead was the victim's flank, juicy and trailing provocative blood. The shark's fin was grazing the skin of the Whale and the lofty blade of his thresher tail was flashing into the view of the Whale's right eye. Now breathless and with his energy draining away, the young one put all his remaining strength into a last burst of speed. The hunter's mouth reached his side, opened to clamp again into the living meat when the shark exploded suddenly into scattered lumps of its flesh. For the Whale's mother had rammed it with a force that had killed it instantly, the water here within seconds becoming rank with shattered bones, bloody flesh and half-digested mackerel.

Quickly the Whale's mother slipped under him and carried him back to the home waters where she surfaced to comfort the infant with warm gallons of milk, afterward stroking his wound lightly, tenderly, with her flipper. While he was close to her no other shark would attack, but they must leave this familiar place as soon as the infant's panic was quieted, for the blood in the sea would summon more of these enemies and no infant whale would be safe except in a parent's near and protective presence.

It had been a rare thing for a blue whale to attack. The mother had no proper weapons, no teeth to bite, no claws to rend, but she did have her speed, 41 feet a second when necessary, and her immense size and strength. She would not use these advantages except in an emergency, but then she would hurl herself headfirst at the enemy, and that defense, in the case of a shark, would be enough. There might be occasions when this form of attack could not be used, but this time it had been adequate.

After he finished nursing the young one lay at his mother's side, becoming gradually reassured. The wound would heal.

South of the home place there was a large sink in the seabed, called by men the Angola Basin. Water three miles deep was above it, which made the Basin a good part of the sea for young whales to learn depth diving. In the case of her four other offspring the whale mother had led them there as soon as they started away on their southward migration, but this time, because her infant was wounded, she turned instead toward the coast.

In almost every whale species there was an impulse to go near a shore whenever in trouble—ill or injured. That tendency puzzled men. Why, when one of the greatest hazards to whales was the chance of stranding on beaches, did they deliberately put themselves into this danger at the times when they were least fit? Was there some comfort in being in shallower water? There it was sometimes warmer, and was that the reason they fled toward the land? It didn't have any of the signs of a practical or a learned reaction. In fact their headlong flight shoreward more often suggested panic.

How many, if any, stranded whales that had died in that way had come there with some sort of purpose? Lacking a storm tide to rescue them, stranding was always fatal. Whales did not have a land animal's type of temperature control; out in the air in a sunny atmosphere a whale overheated and died of what essentially was a fever. And yet they did come. It is not necessary, nor possibly reasonable, to think of their motive as a kind of suicide impulse, somewhat as land animals often go off alone when they sense that death is approaching. But in view of the mystery one might wonder whether some very deep instinct draws them toward what, millions of years ago, was their home. Does the land still seem a refuge in spite of all the whales' marine adaptations and long experience? Do the kelp and other seaweeds found only in shallower waters remind them of great terrestrial plants, trees and tall ferns, under which they once foraged? If any such influence sends them toward the land in emergencies, the whales would not, of course, know of it consciously, and these speculations may be too far-fetched to be taken seriously; perhaps their panic flight toward the land must remain unexplained.

Not actually near the beach but in sight of land if they had had better eyes, the mother and son idled slowly along the coast while his bite was healing. There were some unfamiliar small organisms to watch, freshwater fish and crustaceans that had wandered out from the Congo River into this salty element where they would not long survive. That very salt, however, would help speed the Whale's recovery. And so the day came before very long when the Whale's side was well again, and his energy proved that the motions of swimming had ceased to hurt. His interest and curiosity returned and soon led him and his mother into a small and pleasant adventure.

The two were leisurely going along, not far from the African town of Luanda, when the Whale's revived curiosity led him to dash away from her. For he had seen a strange object floating along the surface. Was it a fish? It didn't seem to be swimming and yet it progressed. He must know about this. He swam near it and standing on his tail boldly, he raised his head out of the water. It was a boat, the first he had seen. It had a sail, as large as one of his mother's tail flukes, and was nipping along over the ripples almost silently, the only sound being its whispering bow wave. But if this wasn't an animal there was an animal on it, one long and narrow and small, hardly a quarter as long as the infant Whale. A man. As the Whale came in closer, the man leaned over the side of the boat and made friendly sounds. An amazing thing: he had eyes and his eyes met the interested eye of the Whale and something happened between them, a feeling that passed from one to the other. The Whale, in his excitement at this, turned a somersault in the water. Then he came in still farther. The man splashed a handful of water at his face and the Whale would have laughed if he could.

His mother approached from behind with her great, easy strength. Staying back a few yards, she was watching cautiously to be sure that her infant was safe. She thought that he would be for she knew what these men were like. She had seen the small boats before and the men who sailed them. They did not catch whales, they were not killers. Men were of two kinds, it seemed—like whales. Just as there were safe whales and killer whales, so there were men who were harmless and others that killed. The killer men were usually in the Antarctic, where a whale mother and her infant would soon have to go—unfortunately, because that was the only place she knew that a blue-whale's food could be found. The men there also rode in boats, but very large ones with cruel-sounding machines. Better were tiny boats sailed by only one man, and the mother did not have any animosity toward them. She was too good-natured to feel anger toward all men just because some were killers—as soon dread all whales because there were killer whales.

One would think that the man on this day might have been afraid of her, for one brush of her flipper could have upset his boat. But he, for his part, knew these whales and understood that they managed their mammoth bodies in the most careful way. He was dark, even blacker than she was, and when he stood up and walked to the back of the boat, he moved with a natural sort of grace, like a whale's.

To the east there were palm trees along a low shoreline. The whales could not see that far, but the mother kept sending down sonar signals which told her the seabed was rising beneath them. Soon she and her son must turn back, but they stayed with the boat as long as there was no danger of being stranded upon the beach. For this little incident was diverting—three creatures, all gentle, meeting, exchanging their interest in one another, all three, it might be, having their emotions warmed a little by contact with another so different. Two years later, on the other side of the ocean, the Whale would remember this occasion dimly and recognize that some men were not enemies, as he would have grown in the meantime to think of them.

ONLY THE HARMLESS CAN PLAY

The Whale had been born at the end of June. Since then two weeks had passed. In five and a half more months, at the end of November, he should be weaned and start feeding himself. Most young animals had to learn how to find and, in the case of carnivores, attack and kill what they eat. The food preferred by blue whales, the plankton euphausids or krill, would be so abundant where the Whale and his mother were going that he would need to do no more than open his mouth and they would crowd in. His mother knew where that fortunate concentration was: around South Georgia Island, east of Cape Horn and in the same icy-cold latitude. The island was several thousand miles from this African coast but the route was familiar to the Whale's mother and there was a purposeful steadiness in her swimming as she and her infant started away.

Ordinarily she was not so near land at this early stage of her annual journey. She had come here only because her son's wound had made her wish to be close to a shore. Now that he appeared to be well, they would go back into much deeper water. In all other years when she and her mate had set out from their summer sea, they had gone straight south into the Basin. There they had both enjoyed the greater buoyancy all that volume of water under them gave; and in alternate years, when an infant was with them, they had introduced him to the colder and darker depths. Beyond the Basin there would be no other place as good for a young one to learn depth diving.

The Whale would need that skill when they would encounter killer men later. As they left the coast, then, he was starting into an important new phase of his growing up.

They hadn't gone far when the pulse of the passing waves began to carry a new and astonishing sound. Present everywhere, always, were the mumbling movements of waters stirred by the earth's turning and, when winds were whipping the surface, the crash of rollers and hiss of collapsing foam. These, the seas' impersonal protests, were too familiar to attract attention, but now there was something different: a voice clearly of flesh and blood and, remarkably, not calling but seeming a voice raised for the pure pleasure of making these sounds, which were expressive and musical, what men might describe, did describe, as a song.

The young Whale, hearing something so strange, slowed the beat of his tail. But then he sensed that his mother was not alarmed and he dashed ahead. He must investigate this, and his mother seemed willing to come along.

The singing was under water. The Whale couldn't hear it when he surfaced to blow or nurse but when he submerged the voice was clear and definite even though it was, in the beginning, more than ten miles away. As the Whale and his mother drew near he fell back to swim at her side and she slowed their pace, for—as she expected—a group of humpback whales were ahead and it was not the wild way to rush impulsively into a new situation.

Only one humpback was singing but he was using two voices, one a falsetto, the other a bass voice much lower. The upper voice sounded like that of a very young animal, even the Whale himself if he had been as articulate. It was light and thin, seeming youthfully flexible as it glided around in a high register, plaintive and softly shrill.

And then, as if answering, came low gruff tones that better expressed what the singer was: an animal large in size and weight—and a little impatient perhaps. For a time the two voices suggested a dialogue between an infant and parent.

On a new theme then the humpback began to sing in the baritone range of a human voice. There the singing sounded most musical. One often repeated phrase would slide upward and finish with three puffed staccatoes. There was also a fast staccato, all on the same note, that could almost have been the beat of an outboard motor. Somewhat more throaty were liquid gulping and gurgling tones, like water tumbling along over stones, and with those were quite resonant sounds, as if the stones faintly chimed. Single clear tones were as pure as those from a flute or horn, those, especially, bouncing back from the shallow seabed with echoes so real it was hard to tell which were the original voice. Usually the singer sent out those notes with pauses between them, and an echo would blend with the tone following in a very harmonious way. Finally, sometimes it seemed as if the humpback sang in a large, long tunnel where echoes themselves were echoing, so that the voices came from many directions and made the ocean sound very spacious.

There had been much variety in the humpback's song, splashes of sound that seemed as unplanned as images chasing one another through a whale's mind; and so the singing appeared to be random. Actually it wasn't. During the time the Whale and his mother were on the way coming, the musical themes had been repeated several times, which made them a real song and like nothing else the Whale ever had heard, or ever would hear, in the ocean.

Incidentally, the music that sailors used to think they heard in the hulls of their wooden boats probably was the singing of humpback whales. Most sounds cannot pass from water to air, or from air into water, but wood partly submerged will transmit through that sound barrier. Some primitive fishermen put their ears against the ends of their paddles and say they can hear the voices of fish below. The reason that divers can't hear the underwater sounds is that bubbles of air in their ears won't let the sounds pass.

Echo clicks told the two who approached that the singer had stationed himself a little apart from the rest of the humpbacks. There were ten or twelve of them. He was the first to sense that two blue whales were near, and he fell silent. The blue whales might have gone on then but there was plenty of other interest here. For all the humpbacks were moving about, not in any concerted way but with something like growing excitement: something, it seemed, was about to break into action. It did! As the Whale and his mother listened and watched, the others began a great splashing romp. They had immensely long flippers, white, which were bashing down on the water everywhere, almost concealing the frolicking whales in spray. But this didn't release all their exuberant spirits. They commenced leaping out of the water—entirely out, sometimes to cut an arc in the air and then smoothly, gracefully knife back into the waves, giving a flippant toss of their flukes as they disappeared. Then more and more of them chose to make high dives up into the sunny air and with a twist of their bodies, let themselves, the great expanse of their 40-foot length, fall flat on the surface. Their tremendous weight—30 tons for each whale—sent up geysers of water that rained back over a very wide area—on the Whale and his mother—with a crashing so loud that it was heard 25 miles away by the man in his little boat.

The humpbacks were breaching in every way possible to a whale, rolling out of the water, somersaulting, and again and again throwing themselves out completely and coming down on their backs or bellies. Then gradually all but two played less boisterously. They gave way to a pair whose game had been the most active and who were now slapping each other wildly but soon more tenderly, with short pauses then, to draw together until finally, lying face to face, they had a moment of obliterating all else in delight.

Not long after that the festivities came to an end and the humpbacks idled around or lay resting up at the surface. And the Whale and his mother left, to continue east on their way to the Angola Basin.

Except for his mother, the humpbacks had been the only whales the young one had seen. They were not even half her size but were nearly twice as large as himself—the first creatures to make him seem small. A little tired from the excitement of the humpbacks' display and also perhaps slightly subdued by this revelation of other whales' bigness, he swam forward beside his mother in a way docile for him. What he would not have recognized was that he had seen all that activity, that spectacular leaping, and had heard bewildering volumes of sound without even a tremor of fear. His world on this day had been truly a blue-whale's world, one without animosity. In a few months the humpbacks would have gone south to the Antarctic, to Australia where they too would grow fat on a diet of plankton. The song and the high leaping were not in a blue-whale's temperament, but in one important way blue whales and humpbacks were alike: both could enjoy themselves in easy, relaxed play because their kind of food did not require them to stalk and kill other mammals, creatures akin to themselves. A hunter must keep himself on the alert, ready to pounce. He can't let himself go in a good-natured mood that is not looking for victims but for a chance to have fun.

The Whale might have become a depth diver more quickly if he himself had not been so playful. He often swam at his mother's side, but not for long before he would suddenly somersault and if she turned back to find him, would make a chase of it. He loved chases, the kind that were only games. If his mother started away from him with some private intention, he sometimes would race ahead of her, or try to keep up behind her, nudging her flukes, meaning, *Catch me!* He never jumped all the way out of the water, as the clowning humpbacks did, but he splashed his mother by lobtailing when they were up at the surface; or he splashed with his flipper and once, very daring, he slapped her side as he was finishing nursing and then she whirled around and slapped him, but he ducked down in the water where slaps don't work, and that time she did chase him, which was the best ever.

So far she had not had much inclination to play with this little one as she had with her other infants, because she was depressed by the loss of her mate. But the determined Whale was arousing the playfulness in herself and she began to respond to his splashes and nudges. Besides their fun they were doing some intricate swimming together, first one then the other leading in weaving tumbles and rolls. It wouldn't be long now before this came closer to practice in diving deeply, but they were distracted one morning by seeing a different kind of frolic.

A spanking wind blew from the southeast and here, as if they were the living spirits of tossing waves, came a band of 80 or 90 dolphins *(Steno bredanensis)*, at play with the breakers and with each other. They were vaulting up out of the foaming crests, racing among themselves, proceeding in groups of six or eight that crossed and recrossed their lines of advance. The striking thing, aside from the dolphins' quick and vivacious movements, was that they seemed to be playing as teams. There was an obvious amiability in the whole company, but they divided into small ranks for maneuvers that 90 dolphins together could hardly manage. They turned their gambols into games and, whether or not the teams were competing (who could tell?), the whole object of progress through waves appeared to be lively sport. The humpbacks, except for the two who were mating, had seemed to leap and splash as they individually wished; blue whales frolicked as families, but the dolphins were playing as squads and meanwhile were chirping and calling to one another. The Whale and his mother had not submerged. The dolphins separated to pass around them, for the mother, nearly ten times their size, seemed as large as an island. They knew that she was no threat and for their part the dolphins would not inconvenience blue whales, just go by in their cheerful conviviality. If they had been hungry they might have stopped to eat squids, their favorite food, but squids were everywhere and so dolphins could spend most of their time in their jolly play. (If men had been there they might have called these dolphins the handsomest whales anywhere in the ocean, for they were gray above, white beneath, with the white and gray dappled across their chests against a background of delicate rose.)

Dolphins were not the only ones coming and going in the Angola Basin. Here the blue whales were in the shipping lane between Europe and South Africa, and the grind of the vessels' engines, transmitted through their reverberating steel hulls, filled the sea with metallic clatter. While a ship was passing the din made it harder to hear the delicate echoes from sonar clicks; and the ships' sounds were distressing in a more general way. They destroyed a little the right and comfortable naturalness of a whale's world. When men and men's inventions came into that world, they didn't fit into the flow of normal wild happenings. The ships' unchanging thrust was alien to blues' impulsiveness, and a whale had no intuitive way to adjust to it.

The ships were not always passing and when they weren't, the Whale could get on with learning to make a new kind of dive. His mother had no organized plan for teaching him—no compulsion to take him down at any particular time or in any particular way. Without consciously thinking about it, she would rely on his developing wish to try this new thing. He was as sure to grow into that stage as he was to keep putting on weight from her milk. Some day soon he would want to go far below with her, and when he did she would be with him to make sure that he didn't dive down too far or come up too fast; as events would unfold she would do everything necessary. Her sense of what would be needed and her willingness to carry it out could be depended upon. That was the way of an animal mother.

She herself liked to sound here because in other years echo clicks had told her that no obstructions were there below—no submerged mountain peaks, no canyons with high banks, nothing but a flat seabed so deep that a whale would be in no danger of reaching it. She would make a few dives for her own pleasure and those would be her infant's first lesson.

He would see how she emptied her lungs for this: with four or five short quick spouts so close together that almost no carbon dioxide and no oily froth had time to accumulate between them. In this way her lungs were cleared more completely than they ever could be with one spout. And then the young one would see that she dived vertically, so straight down that the flukes of her tail flipped out of the water as she submerged. He would sense the direction in which she dived, and with a greater thrust too; when he was near he would feel the push of the water. His mother would have done something different and therefore interesting.

He could observe a new way of diving and learn that she would be down, then, longer than usual but that did not alarm him. He played around on the surface until she would appear again, not beside him but breaking out of the water so far away that he could not see her distinctly, though he heard that she lay there, up in the air for a while, panting, before she swam back to him.

When she had sounded enough times that the long absences were becoming familiar, he felt inclined one morning to try expelling everything from his own lungs with a series of fast blows close together. On that occasion she was beside him, spouting in the same way, and then when she flung up her tail for a vertical dive, he did too and found himself head-down with all the light going quickly out of the water; but his mother was right there with her fiery streamers and her great reassuring presence.

He had not gone very far down, actually, before his mother, turning so that her back was against his belly, swung him into an oblique upward course. A strangely gradual rise— the Whale might have gone back to the top as fast as he could . . . but perhaps not after all, for it was in his own instinct to return from a vertical dive on a slant. In that way the oily froth now again entering his lungs had a chance to take up the nitrogen in his blood, formed in the water's depths, before it became bubbles and endangered his life. Having oil in his lungs the Whale could cope with the gas better than human divers could, but to be safe he should allow time for the absorption of gas to take place. The slow rise had been tedious and had postponed the taking in of new oxygen, which would be welcome. But at last here they were out of the water, and the young Whale, like his mother beside him, felt inclined to lie at the surface for a few minutes, breathing deeply.

At his age he should not be allowed to go down to the maximum that a blue-whale adult might dive, which would be about half a mile. That far below the surface a whale would need the full means of adapting to the water's tremendous pressure. The food the young Whale would seek when he stopped nursing was usually not more than 600 feet down, but even at 600 feet the pressure would be 270 pounds on every square inch of his body, contrasting with 15 pounds at the surface. At half a mile there would be a pressure of 1200 pounds per square inch. There it would be necessary for the Whale's ribs to push farther and farther inward until, it might be, his lungs would actually collapse, but without damage. The lungs themselves would yield because they would have been thoroughly emptied before the dive; and the ribs would give because they were not attached to a breastbone, as most mammals' ribs are; in the front they were "floating" ribs. He would not have gasped in oxygen for the dive because he would already have an abundance of it for a dive of about 20 minutes—an excessive number of oxygen-laden red corpuscles in his blood and besides, oxygen stored in his muscles which were even reddish in color with it. Automatically the Whale would draw on those supplies when submerged; automatically too the flow of his blood would change so that most of the blood would circulate in his heart and brain while the rest of his body would suspend its demands and go into something like hibernation during the dive. Up in his head, heavy muscles would protect his eyes from the water's pressure and very dense foam would hold it away from his delicate inner ears.

Even as he went down a short way in his play or restlessness—perhaps no more than 100 feet—all these adaptations were functioning, these and others were being protection against changing pressures as he dived or came back to the surface.

During the present days of play and diving experiments the Whale and his mother had continued to migrate toward the Antarctic. They would soon go beyond the Angola Basin, passing over its southern rim into a sea where there were enough submerged highlands that the mother would stay in the water's safe upper levels. And the Whale, still guiding his actions by hers, would try no more depth dives for a while; but before the pair came away from the Basin they would go far below once again, this time the deepest yet, perhaps to a thousand feet.

On an afternoon of bright sun, when the youngster had emptied his lungs very thoroughly and his mother hers, they turned straight down together.

The water, which was warm, 60 degrees F. at the surface,

dropped to 53 degrees at 600 feet. No matter; they wouldn't be there very long. Their tails were giving their strongest strokes and soon the whales had reached the 700-foot level, 25 times the son's length, then 800 feet, then 900. It was nearly time to reverse their dive and start the gradual upward rise.

But what terrifying pulse in the water is this! The mother knew. On a wide line approaching above was a beat with a deadly regularity. It was the pace of creatures who hunt. The Whale himself had not heard that particular sound before, but his heart gave a great thump of fear. For he had an innate dread of that forward sweeping, the typical driving advance of a band of killer whales at the surface.

His mother could visualize how they were coming: a dozen or more of them stretched out horizontally, the same distance apart, the great black fins on their backs all rising together, submerging, rising with a precision, determination, in which there would never be any impulse to play. They had a motion frightfully like the propellers of whale-catching boats at South Georgia.

The mother had seen and heard killer whales there many times. They went to the feeding grounds for the same reason the killer men did: because a predictable gathering of the immense baleen whales would have been drawn to the island by the concentration of plankton. From December through April the killer men and the killer whales were certain to find a throng of victims there, although now, at the height of the antarctic winter, the island would be deserted. The men would be loafing and the killer whales would be seeking and slaughtering their prey—mammals the same as themselves—throughout other oceans.

The two kinds of killers were rivals down at South Georgia. After one of the boats had sent its fatal explosive harpoon into a mammoth whale, there would be a delay before its dead body was towed to the factory ship and drawn up inside to be "processed." During that time a band of killer whales might have an easy feast. They would snatch greedy bites out of the lips of the carcass, eat its tongue away, then the flesh of its head, its blowholes, the parts with the finest oil, most valued by men, and so the men hated the killer whales, whom they called robbers, and as well as slaughtering baleen whales they tried also to kill the killers.

The killer whales were not only after the carcasses. If none were available they pursued living whales and, being very fast swimmers, could overtake many of them. The killers' best speed was about 14 miles an hour. A fin whale could swim faster and a full-grown blue whale could also escape, but not one who was only a little more than a month old; nor could a humpback or one of the smaller species of baleen whales. And so relentless were killers that, swimming alongside one of the slower giants they would tear out its mouth even while it was fleeing. Therefore the great ones often gave up at the start of a hopeless chase. The mother of the young Whale had seen a humpback female stop trying to get away and, rolling over onto her back, had simply waited for death as a band of killer whales ate her face and tongue. From her stillness after the first terrible moments, it seemed that she died of the pain and agony, as her wounds would not have been fatal at once.

These memories undoubtedly were revived in the whale mother's mind as she heard the recognizable menace draw near. The direction from which the sound of the killers was coming was south. Therefore she and her son would start their rise also southward, hoping the killers would pass overhead and be farther north before the two had to surface.

She slid under her son so he would ride on her back. For she would determine the speed and slant of their going up, which she would make as slow as the infant could possibly stand the lack of breath and the water's pressure. The Whale himself sensed an emergency but only vaguely grasped what his mother would be attempting. As yet he was not too un-comfortable. They had not been submerged for more than five or six minutes. Ordinarily an oblique rise would get them to the top in ten minutes more, which was about the limit that he could wait at his age. But his mother was not starting up! Why was she taking so long to tilt into a slant? Why were they going along at this great depth on a level course? Instinct told him that this was not right.

The steady and fearful beat of the killer whales' swimming had grown louder and louder. Would they have heard the blue whales below? Would they dive down and attack? The possi-bilities were not clear in the infant's mind. He only knew that there was an ominous strangeness in the way his mother was acting.

The bright fish, bathed in glowing light, were swimming around the two whales with an indolent grace. A squid shot by. His mother should rise through the water fast, like the squid! Instead, though he could feel from her muscles that she was pushing with all her strength, there didn't seem to be any lessening of the water's pressure. They couldn't go on like this!

The sound of the killers' swimming continued its mur-derous hunting momentum. As they rhythmically dipped up and down through the surface, their splash filled the sea with a clamor. How different they were from the frolicking hump-back whales! Or the sport of the dolphins! Or indeed of the play of the Whale himself. The killers' deliberate movement seemed fateful. Nothing could stop their coming. Unless they broke ranks, scattered their terrible purpose, this onward drive surely would end in attack.

The sound was the loudest now. But the Whale's fear was becoming dulled in the agony of his need to breathe. His lungs were full of oil and carbon dioxide and nitrogen and these were pushing against the sea's heavy pressure. He could wait no longer. He must leave his mother and dash to the top. As he started to lift away however, she rose beneath him so firmly that her motion was a command: *Stay!* But he was desperate. He must get up out of these depths. Squirming, he lifted away and she tried to relieve his distress by swimming faster.

It was a frantic ride. But the killers had gone beyond them. Slowly, oh, very slowly, the Whale and his mother were rising now. They were coming nearer the surface. But hurry! As the water's pressure grew lighter, the Whale's lungs felt as if they would burst. Then he and his mother were out, and with a great blow the Whale stretched along the top, prepared to lie there and pant as they had always done after their other deep dives. Yet this time his mother insisted that he must come down again, for even now the sounds of the killer whales were too near. They were receding but she knew how suddenly killers could swing around and come back.

Had they seen the two spouts? Apparently not, since there was no change in the frightful rhythm. She took her son underwater again, not far, keeping him out of the killers' sight until the retreating sounds diminished to almost nothing. And then they were gone, quite out of hearing, in the direction the playful dolphins had taken another day.

At last the two could relax on the surface. Gently the Whale's mother stroked her son with her flipper. He was a very dear Whale and the ending so easily could have been different.

Ann Haymond Zwinger

THE MYSTERIOUS LANDS

OF SAND FOOD AND PALM TREES

The Cabeza Prieta stretches westward almost to Yuma, astride one of the few crossings of the main desert river, the Colorado. Although in 1538 the Spanish navigator, Francisco de Ulloa, laid claim for Spain to all the area drained by the Colorado River, not until 1774 was this desert crossed on foot when Juan Bautista de Anza, commandant at Tubac, scouted a route west, forded the Colorado River, and trekked across the Salton Basin as far as San Gorgonio Pass. The next year he took this route all the way to San Francisco Bay and established a path used for years. Seven years later Don Pedro Fages made the exceedingly difficult southern crossing to San Diego Bay, a route little used until Kearny and Cooke brought Army contingents through in 1846–47. Present-day California highways across the desert largely follow their routes.

Lieutenant Emory crossed here in November 1846, and "encountered an immense sand drift, and from that point until we halted, the great highway between Sonora and California lies along the foot of this drift, which is continually but slowly encroaching down the valley." That great drift is called the Algodones Dunes, a massive dunefield of two hundred square miles that runs roughly north and south athwart the travel path, paralleling the western edge of the river.

I walk the Algodones Dunes late one companionable spring afternoon with Anita Williams, a gracious woman who knows and writes about the Cocopa Indians of northern Mexico. Late light softens the flowing flanks of the dunes and low light pushes our Giacometti shadows ahead of us. The fluted white trumpets of desert lilies luminesce in the low-shafting sunlight. The plants are large, the flowers opulent, much larger than the usual calf-high lilies I just saw in the Cabeza Prieta, as are some of the creosote bushes and evening stars, a gigantism that frequently occurs on dunes because of the lenses of deeply held moisture these plants can tap.

In the hollows of the dunes, yellow evening primroses trace volutes, sand verbena lays deep pink serpentines, and hundreds of birdcage evening primroses unfurl. In 1846, when he brought Mormon troops across the desert, Philip St. George Cooke noted the last as a plant "which grows into a pear-shaped basket frame, the stems or branches all uniting as if tied above. . . . The bushes furnish a small fire for making tea and frying meat."

Cooke was referring to the skeletons of the evening primrose, which endure from season to season, some of which lie fragmented in the sand beside us. As the plant grows, it does so by adding stalks around the periphery of the original rosette. As they dry they curve upward to meet toward the top, forming an open basket of seed pods, giving rise to its common name of birdcage primrose. When the tubular capsules pop open at the top, some seeds spill out while others remain attached. The drifting sand eventually buries the skeletons with their cache of half-filled capsules. With rains, the capsule-enclosed seeds germinate in place, often in a circle following the outline of the buried skeleton.

The white evening primroses, like the pearly desert lily, luminesce, their luster intensified by the tawny color of the dunes themselves. Red iron-bearing minerals are common in sand, and some dune sands acquire a coating of ferric oxide concentrated in pits on the grain surface, becoming darker with age. The Algodones Dunes have acquired, over the centuries, a lovely resonance of color.

One dune crest looms behind another, a heavenly disorder of line and flow. To see order I have to remember how they looked when Herman and I flew over them: nearly parallel ridges defining the western edge, long, almost straight ridges bending slightly eastward, extending for eight or nine miles, five hundred to six hundred feet apart. Or to recall the even greater simplicity of a Landsat photograph, where the western edge is merely wrinkled, and the main body of the dunes themselves mere ripples, facing neatly southeast. To the south the ridges curve into immense barchans, some of which may have horns up to a mile apart. The blended shapes flow in elegant sinuous patterns with a purity of line absent in the surrounding piled-up and turreted mountains.

The prevailing winds blow from the west-southwest, but the orientation of the dunes—both in their slipfaces and general trend—indicates that the preponderance of dune-forming winds comes from the north-northwest. Such strong northwesterlies probably blew during the Pleistocene, and drove large amounts of sand inshore from the beaches of Lake Cahuilla, the much larger Pleistocene predecessor of the Salton Sea, and built the Algodones Dunes to their present configuration. Now there is little movement. Plotted on aerial maps, the present-day dunes travel some three inches a year.

I had no reading of magnitude when I first saw them from the air, no sense of scale. They might as well have been ripples on a beach. But on the ground, this floating evening, the dunes are impressive and massive, unforgettable. Wherever I trace a horizon, I follow an elegant line that protects and defines, brings order and serenity into a chaotic world.

Anita and I are meeting Gary Nabhan to look for sand food. The name of this well-camouflaged parasitic plant comes from its intensive use for food by local Sonoran Desert Papago, the Hiach-eD O'odham: the Sand People. The easiest way to find sand food is to find the plants upon which it is parasitic, especially pleated tiquilia. This tiquilia, noted by William Emory in 1846 as "a low bush with small oval plaited leaves, unknown," is itself not all that easy to find. The sand-drifted rosettes, splayed flat on the ground, with its spare quarter-inch leaves, tend to disappear in the vastness of the dunes.

We scuffle a lot of dune, looking for sand food. Gary's eye is better attuned to subtle changes in sand texture and infinitesimal shadows, and soon he cups sand away from a flattened golf ball-size head, exposing the stem. Beset with small bracts, the stem is the diameter of a man's thumb and exactly the color of the sand. The flavor of the stem is described as melonlike, and early travelers, as well as Indians, relished its freshness of flavor and texture. Potential hosts are close by: six feet separate it from a pleated tiquilia, ten feet from another, five feet from a dune peabush. A sand-food root need barely touch the host plant with only a few root hairs to tap into its moisture and sugar.

How sand-food roots locate a host, without which they cannot survive, no one knows for sure. Possibly desert kangaroo rats, the only species present here in any numbers, bury the seed heads near a host plant, where contact can be made on germination. Or sand food's dry seed heads break off and tumble, peppering the sand with thousands of seeds, and with odds that some of this number will catch around the base of a potential host and work into the sand. Or vice versa, seeds from other plants snag in sand food's dried head and germinate there, bringing the mountain to Mohammed. Or perhaps they are simply part of the vast reservoir of seeds in the top few inches of desert soil, galvanized to germinate by exudations given out by germinating seeds of potential host plants.

Given the good health of its hosts, sand food's classification as a parasite may need to be revised. Sand food imbibes and stores moisture from infrequent rainfall in its succulent stem, moisture then possibly available to its host plant, providing it with a better supply than it could have achieved on its own. Excepting parts of the Mojave Desert and Death Valley, the 250 square miles of the Algodones Dunes are the driest part of the United States. Rain falls occasionally in the winter and in spasmodic torrential summer rains, but the mean annual precipitation here is less than three inches.

The plump disk of the two-inch flowering head is lost in the vastness of sand around it. Just beginning to bloom, sand food's few tubular flowers are scarcely a sixteenth of an inch across, nestled in the dense plumose pubescence of the calyx. Since ultraviolet photography picks up the flowers' lilac hue, they must be clearly visible to pollinating bees, and some additional pollination undoubtedly occurs at night when the majority of dune insects are active.

Like sand food, all the sensitive plants of the Algodones Dunes have evolved unusually long stems, forced by the continually blowing and drifting sand. Elongated stems develop shortly after germination in Wiggins' croton, dune sunflower, Pierson's locoweed, and dune buckwheat, as well as in sand food. After buckwheat seeds germinate on moist dunes, only a leaf tuft shows on the surface, while the rapidly growing stem may anchor ten feet or more into sand.

Pierson's locoweed displays another adaptation to dune living. Although it is a perennial that matures into a small shrub, it can flower and fruit just two months after germination, assuring seed survival even if the plant never reaches shrub status. It produces seeds larger and heavier than those of any other southwestern locoweed. Housed in inflated seed pods, they tumble and bounce across the sandy hillocks, gathering in hollows or piling up against other plants, confined by their weight to the place they grow best.

All these plants are threatened. Sand food has almost totally disappeared from Arizona because of a combination of agriculture (desert soils can be very fertile when irrigated) and off-the-road-vehicles (ORVs); dune buggies also destroy the habitat and seedlings of croton, sunflowers, and locoweed.

Gary fills in the hole and sprinkles some sand across the top and, if I didn't know better, I'd swear he murmured some Hiach-eD O'odham incantation for its longevity.

The Coachella Dunes pile up on the east side of a narrow valley extending some sixty miles from the Salton Sea on the southeast to the summit of San Gorgonio Pass to the northwest. The booming winds that pour out of San Gorgonio Pass and funnel down the Coachella Valley are legendary (attested to by the numerous wind-generator "farms" at the throat of the pass). Desert areas of lower barometric pressure draw relatively cool air in off the Pacific, creating an inversion at around three thousand feet over the pass. The narrow configuration of the pass creates a Venturi effect: air pressure drops and wind accelerates, unleashing high-velocity and extremely desiccating winds into the basin below, and whisking sand into low, rolling dunes against the Coachella Mountains.

The day has built to full blaze when I reach the dunes at noontime. I loathe hats but clamp one on my sweating forehead anyway, and check to see that my water bottle is full, knowing the water will soon be as unappetizing as warm dishwater. I roll down my sleeves and turn up my shirt collar to keep off the sun—loose clothing that does not hinder perspiring can halve the radiant heat load and cut water loss by two-thirds. I pick up my notebook, take a deep, reluctant breath, and push my way into a desert blanched with heat.

The dunes are so light in color that they reflect and focus the heat on any object above ground. Small dunes, the largest but twenty-five feet or so above the interdunal flats, face mostly south and east. I tromp through a rich growth of dune plants stabilizing small hummocks on the dunes' apron, laced with rabbit and fox tracks. Ladybugs clamber on desert yellow primrose and over desert four-o'clocks.

I bend over to admire a clump of yellow evening primroses and find myself contemplating instead a desert iguana tucked away neatly in the shadows. A stocky medium-size lizard with a five-inch body plus a five-inch tail, its buff-colored body is patterned with brown spots that form varied patterns across its back and coalesce to form lines around the tail, pied patterns that render it part and parcel of the broken shadows. A diagnostic row of beaded scales runs all the way down its back and onto its tail. It has a narrower range than that of creosote bush, which makes up almost all of its diet. It often climbs into the bushes to feed on the blossoms although it avoids the unpalatable leaves.

Desert iguanas are beautifully adapted to life on the desert.

They may drink water if available, but can exist with only the water contained in their food. As they are primarily vegetarians, desert iguanas must have some extrarenal mechanism to get rid of the sodium and potassium salts they pick up in their diet. This they do in much the same way sea birds do, through a pair of glands opening into the nasal cavity that act as accessory kidneys and remove salt ions with a minimum loss of water.

In this afternoon heat the iguana operates very close to its tolerable limit. It regulates its body temperature within rather narrow limits by behavior, as it is doing now. It began its day in its burrow, waiting as long as two hours after sunup before emerging to forage, remaining nearby its burrow until its body temperature reached its optimum, a high 111 degrees F. Unlike most lizards, which tend to take over old rodent burrows, desert iguanas usually dig their own.

I lean closer to see it better and it bolts off across sand so hot I feel the heat through my soles. It runs with such a peculiar shuffling, swimming gait that I think it has a bad leg, before I realize it is brushing the hotter surface sand aside to bring its body into contact with cooler sand beneath. As the day cools —and I fervently wish it would—it will spend more time in the sun, orienting its body to receive the rays at a more perpendicular angle, until it finally tucks into its burrow for the night.

When I reach the dunes proper, I find them well used by ORVs, tire tracks superimposed on the chained imprints of lizards. Although it is windy now, it is nothing like what it can be, for which I am thankful. Fast-moving hot air quickly adds to heat stress. Because the winds from San Gorgonio Pass are so powerful, accumulations of larger grains and pebbles in the troughs emphasize the spectacular ripple patterns. Smaller-scale patterns crosshatch larger ripples going in another direction, as if the wind constantly wearied and shifted, quixotically rearranging the patterns to its liking every hour or so.

Small cylinders of damp sand dot the top of one dune, an inch or so high, as if tubes of sand had been piped up. Smaller ones are sealed. Several dozen stud the area, always several feet apart, probably belonging to large female wolf spiders, who find happy hunting on the dunes.

The distinctive and extraordinarily even track of a sidewinder loops across a sand flat. The impressions in the sand sit at an oblique angle to the rattlesnake's direction of travel, arcs with little walled ridges of heaped-up sand on the outer arc of the curve. Each impression has a hook made by the head and neck and a T by the tail, which is to say that this snake was heeding Horace Greeley's advice.

Sidewinders are specialized for living in dunes by this peculiar type of locomotion, not by any anatomical or physiological alterations. The principle of sidewinding is that of a sidewise rolling screw, touching ground with only two sections of its length at a time, efficient for fast travel in a loose substrate like sand because the force exerted is vertical, better than a horizontal force, which is inefficient in loose sand. I follow the tracks until they peter out, only to emerge a few feet farther, going in the other direction, looping around a bush, setting out straight again, and finally disappearing. They have no rhyme or reason and there are interruptions where I think this crazy snake must have been airborne. Where tamarisks have formed copses in the dunes, the hummocks are riddled with holes, oases for the sidewinder's favorite prey of small mammals and lizards.

What I have come here for, am enduring this fiery furnace for, is a glimpse of a fringe-toed lizard, one of the most highly specialized vertebrates of North America sand-dune habitats, a lizard endowed with distinctive anatomical adaptations to living in this flowing, unstable medium. Its sloping snout is wedge-shaped, with a countersunk lower jaw to slip easily into the sand. Its eyelids are thick and overlap and interlock to keep out sand. Scales form protective flaps over the ears, pressing back as the animal burrows. Elongated scales fringe its toes, which increase the surface area of the feet and therefore the traction, allowing greater purchase in loose sand by doubling back when its leg bends forward and flaring outward as its leg extends. Fringe-toed lizards are the only reptiles limited to and totally dependent upon dunes.

Fringe-toed lizards' ability to bury themselves on the instant is legendary. They wiggle their heads sideways and push strongly with muscular hind legs, holding their front legs close to their sides to reduce resistance as a scuba diver, arms held back, powers through the ocean with his legs. A U-shaped nasal passage prevents sand grains from penetrating its head. A cavity of air under its body keeps sand from falling in about it and constricting its next inhalation.

I follow endless tracks chaining the dune rims and skidding down over the edge into loose sand, where there are dark burrow holes on the slope face. But not a lizard is in sight. Possessing exceptionally keen eyesight, they may have spotted me long before I could see them. Or, being so closely matched to the sand that sometimes sophisticated optical equipment cannot differentiate between lizard and background, perhaps I couldn't see them even if they were out. I scan the dune crest for movement. Nothing.

Even though I know they bask only in early morning and late evening, the eternal optimist in me bids me wait. I sit down on the dune. Perspiration streams down my nose and drips on my notebook. I drain my water bottle. The sun notches another degree to the west, cleaning out the shadows, encouraging the wind, blow-torching my shoulders. If I don't get up I may be glued to the sand. After an hour, it takes great effort to heave myself to my feet and admit defeat. No lizard in its right mind would be out.

Disappointed, I shuffle my way out of the dunes, pondering

the impossibility of fringe-toed lizards. I can see where one went to market, where they held the Olympic games, where one had another to tea, where the chorus line formed, but they themselves have sensibly retired beneath the surface, waiting for a respite from the heat. I envision dozens of them studding the lee side of the dunes, all neatly tucked beneath the sand, front legs folded close to their sides, hind limbs flexed for action, pineal gland shuttered, eyelids locked closed, all perversely, unreasonably, and inconsiderately keeping out of sight.

The 13,000-acre Coachella Valley Preserve protects not only fringe-toed lizards, but also a palm oasis, the anomaly of the Sonoran Desert. The flickering shade of the palm oasis is the complete antithesis of the unmitigated heat of the dunes. The shade is as much aural as visual. The fronds converse quietly, rustling like taffeta, from an air flow that moves through treetops thirty feet above but scarcely reaches me below. I can almost hear the whir of cameras as Rudolph Valentino gesticulated through *Son of the Sheik* or Cecil B. De Mille filmed *King of Kings* in this very palm oasis.

The only way in which early travelers survived crossing the desert was to know where there were tinajas, springs, or oases. On the first major crossing, when de Anza trekked the desert in the 1770s, he took with him an Indian as guide to direct him to water, and survived the trip. These oases are still marked on the map but now they are towns, among them Twentynine Palms, Thousand Palms, Palm Springs.

Native desert palms have been present in western North America for 100 million years, and at this time may be actively expanding their range. The genus developed in the Pliocene, probably in Baja California, and spread northward, retreating southward during the cooler, rainier times of the Pleistocene. When climate warmed after the Pleistocene, they again migrated northward, increasing their range, able to establish where there were springs and seeps. Every canyon with enough water to support a desert palm grove is on a fracture or fault that allows, or forces, water to rise to the surface.

The kind of soil in which the palms root is not as important as permanent water. If runoff and seepage increase, so do the number of desert palms; likewise, successive drought years decrease grove size. These groves look as if they are all the same height and the same age, but after reaching thirty feet the palms slow in growth, limited by the ability of the vascular system to pump water any higher; between seventy and eighty feet seems to be the limit. While close in height, they may actually be varied in age.

Since the seeds are so heavy, long-range dispersal depends on coyotes, which eat the whole fruit, passing the seeds through their digestive tracts unharmed—those seeds germinate better than uneaten ones. In the fall, coyote scat is filled with desert palm seeds, most of which are ready to sprout. Robins have been seen with seeds in their beaks, and possibly other birds may help in distributing palm seeds. Over eighty species of migratory birds frequent palm oases.

Fire is the most important factor, other than water, in maintaining desert palm oases. Palm oases are very flammable from the large skirts of dry fronds that thatch the trunks. Without burning, palms eventually decline in number, crowded out by plants with more extensive root systems. Fire eliminates the above-ground portion of competing plants so that palms can outgrow even the grasping tamarisk with its incredibly dense and shallow water-usurping root system. Although tamarisks will return by sprouting from rhizomes and roots, at least three years pass before they reach their former growth. Meanwhile, the industrious desert palms, growing one to two feet a year, will have produced multitudes of new seeds, which will have enjoyed ideal germinating conditions.

The cambium layer, which produces new growth, in deciduous trees is near the surface and thus vulnerable to fire; in palms, such transport tissue is scattered throughout the trunk. Even though the green leaves are killed when the dead hanging skirts catch fire, the palm generally survives and puts out new fronds in two to three weeks.

With no plants except desert palms transpiring, the soil actually becomes more moist, creating a perfect situation for palm seeds to germinate. In response, desert palms produce twice as much fruit after a fire. The survival rate of new palms is enhanced by arming the petioles of tender new fronds with spines, which discourage browsing. The large grove at Palm Springs burned in 1980; now, half a dozen years later, it has regained its lushness. Only blackened bark recalls the blaze. Nevertheless, each fire removes a little more wood, destroys a few more vascular bundles, eventually weakening the tree and making it vulnerable to boring beetles.

Nearly every tree has thumb-size exit holes, made by the two-inch larvae of the giant palm borer beetle on their way out. Once a desert palm is well-started, few insects attack it, but this beetle can literally reduce the core of a tree to sawdust. The fat, pale-yellow larvae spend up to six years eating their way through the trunk. Seventy to 90 percent of the palms are done in by these beetles, either directly or indirectly. The only control is fire, which decimates beetles close to the surface. The palm borer occurs in every palm oasis except the two remote ones in Arizona to which the beetles have not yet spread.

My cultural belief that fires are one of the evils of nature has required some massive rethinking on my part to discard, in order to accept that fire has always been a natural part of certain of the North American biomes, even a necessary part. The burning of the prairies fertilized and kept the tall grasses vigorous. The Indians who lived near the palm oases—and each oasis had its own group—often fired them to promote new growth. Plant and animal populations are adjusted to fire. Man is not, and has contrived to remove fire as much as possible from the ecosystem, frequently to its detriment.

But before there were careless campers there were, and still are, careless storms with careless lightning. Nature has its own cadence, in which there is little tolerance for houses built in floodplains or flammable chaparral, little consideration for cities built on fault zones, or irrigation works that turn the desert green. Nature acts without calling in a consultant or submitting an environmental impact statement.

OF OLD SEAS AND NEW SEAS

From one of the tufa-coated boulders that makes up Travertine Point on the western edge of the Sonoran Desert, I gaze out over a panoramic view of the Salton Sea and the Chocolate Mountains beyond. This place invites looking back and looking forward, for there is so much visible evidence of past seas, past plants, past animals.

There was a time when the land around me was not desert. At my feet are the familiar Sonoran Desert plants—tiquilia and pincushion plant, four-o'clocks and brown-eyed primrose, California chicory and burrobush, whose bracts cast a satiny sheen. These plants did not grow here fifteen million years ago for it was much too wet. What did grow here was a rich woodland with numerous live oaks and walnut, mountain mahogany and sumac, all of which require frost-free winters and twenty to twenty-five inches of annual rainfall (today's flora gets along with barely three inches). In fact, woodlands covered the whole vast area of the Southwest until recently when the area now called Mojave Desert was uplifted as a block, breaking the continuity.

Nearly three hundred feet above the northwest corner of the Salton Sea, Travertine Point was once connected to the Santa Rosa Mountains by a tombolo, a sandy spit on the shoreline of the Salton Sea's ancient predecessor, Lake Cahuilla. Actually, Travertine Point is misnamed, since travertine forms from calcium carbonates precipitated around hot springs and is different in appearance and formation from tufa, which coats the rocks here. When the waves of Lake Cahuilla lapped these rocks, they were covered with algae. In summertime algae take in great amounts of carbon dioxide in photosynthesis. Tufa forms when turbulence and wave agitation rob the algae mat of carbon dioxide, leaving calcium carbonate that spontaneously precipitates out and coats, in this instance, the beige quartzites that mark the high-water level of Lake Cahuilla.

This Pleistocene sea exceeded the Salton Sea in area and depth, being some three hundred feet deeper. An ancient beachline beneath me is drawn but a few feet above present sea level and likely represents the former northern limit of the Gulf of California, before the delta of the Colorado River diked the basin. This lake did not fluctuate with climate, as other pluvial lakes did during the Pleistocene; rather its water level depended on the Colorado River—how its channels ran and whether or not there was an outlet through to the gulf. The lake existed long enough to produce beach deposits of considerable extent, particularly those along the northeast shore that supplied the sand for the Algodones Dunes.

Today tufa wraps the rocks with softly pillowed shapes, as abrasive as coarse sandpaper, hot to the touch, full of centuries of sunshine. Encrustations coat the rocks to a depth of 20 inches in a 120-foot vertical band. Where the tufa erodes away, frond patterns ice the granite surface or form imbricated patterns on the sides of the rock, leaving lacy dendritic tapestries. On others the tufa looks like spongy dough, risen full of huge air holes. Sometimes the coating resembles dried shaving cream, the foam dried into little holes and thousands of little empty wiggly tunnels. Lichens, black as iron filings, darken the interstices of the tuberculated surface. Big rinds of tufa lie on the ground, some nearly a foot thick. I work out a tiny snail shell trapped within the tufa, an algae grazer that plied its trade thousands of years ago, and think of its modern cousins who have adapted to survive in a place without water.

The rocks lie like old lizards piled on top of one another. Bulldog faces guard crouched frogs. All kinds of eyes look out from the rocks, baleful, slanted, downcast, deepset, smiling, and in one profile a Janus face, looking back and looking forward.

In reading a landscape no longer here, it is as if I have access to an understanding of change beyond the span human eyes can reach to scan the future. When Lake Cahuilla dried up around 500 years ago, it left a desert valley some 110 miles long and 34 wide, a structural trough bounded by two mountain ranges and deeply filled with erosional debris from their granites, gneisses and schists—sediments so deep that borings 1,700 feet through interbedded sands and clays do not reach bedrock. So much salt accumulated here that it was an item of trade for the Cahuilla Indians, and the conspicuous salt beds were worthy of note in the Pacific Railroad Survey reports. The desert stretched from the Colorado River west to the coastal mountains, north from northern Baja California to the end of the Coachella Valley.

To cross it was a nightmare. In 1849, the *New Orleans Picayune* sent a reporter, John Durivage, to write about the rigors of this desert crossing. When Durivage's group reached the desert, they waited until the morning to set off:

through the dreariest road imaginable. The heat was intense, and reminded one in many essential particulars of a portion of Goldsmith's "Deserted Village." . . . By ten o'clock in the morning the rays of the sun poured down upon our devoted heads with the utmost intensity. The animals faltered and staggered in their tracks; one half of our little party were on foot; and the signs of the times around us were such to alarm the most intrepid. The scorching, seething sun provoked the most intolerable thirst, and none had that with which to allay it; those who had supplied themselves most liberally with water having exhausted their precious store. The dejected countenances, the unnatural brilliancy of the eye, and the inflamed veins in the face gave token of the sufferings of the men.

Afterward he acknowledged that "until one has crossed a barren desert without food or water, under a burning tropical sun at the rate of three miles an hour, he can form no conception of what misery is."

William Blake, geologist on this segment of the Pacific Railroad Survey in 1853, recognized that there had once been a very large amount of water here and named the ancient Pleistocene lake Lake Cahuilla because it lay in the Cahuilla Valley, in turn named after the local Indian tribe of itinerant farmers and fishermen who lived there when the lake still filled the basin. Blake named the desert the Colorado Desert for its proximity to the river, a confusing name that must be qualified with "of the Sonoran Desert" since there is now a state of Colorado.

Blake returned in 1905 to find a desert he had named now inundated, found his "old-traveled trail across the desert lay 15 fathoms deep under water, where before not a drop could be found." He saw what I now see below me: a placid body of water sparkling in the sunshine, a shimmer of liquid blue over a desert pavement.

And how the Salton Sea came to be is a product of "making the desert bloom," one of those patriotic slogans like "God, mother, and apple pie," tied up with hopes and dreams, avarice and ignorance.

In the 1880s and 1890s the empty Colorado Desert provided impoverished grazing. It was a dry, empty basin where a meager living could be made if you liked sun and wind and heat and living by yourself. The first Southern Pacific Railroad train crossed the Colorado Desert to Indio in 1876. Within a few years, more than twelve thousand people had moved into the desert, determined to have agriculture by building an irrigation canal from the Colorado River to the valley. Capital was so difficult to come by that the canal was not finished until 1902, but it was, from the start, a powerful impetus to settlement, and was built despite a 1902 United States Department of Agriculture circular in which a soil expert reported that the land was too full of alkali to grow anything.

By 1909 the canal had silted up. After two years of heavy rains and inadequate canal maintenance, the Colorado River breached the canal's headworks, rampaged into the lowest part of the desert, the Salton Sink, and defied desperate efforts to stanch the flow. Tons of rock and heroic effort finally contained the river, but not before millions of gallons of water had poured into the Salton Sink, creating the Salton Sea. Today the Salton Sea has a net evaporation loss of five feet per year, and a salinity quadruple that of the Pacific Ocean.

A section of the Colorado Desert south of the Salton Sea was reclaimed for agriculture and named Imperial Valley, a name that would become synonymous with an almost contin-uous growing season that produces fine fruits and vegetables. In 1928, Congress passed the Boulder Canyon Project Act creating Hoover Dam, Imperial Dam, and the All-American Canal System, and put the Colorado River behind cement. By 1940 the All-American Canal conveyed water 125 miles from the Colorado River, irrigating 450,000 acres of desert. When the Coachella Branch was finished in 1948, more than 80,000 desert acres were added, much of the valley lying below sea level.

"Making the desert bloom" is not a happy-ever-after story. It blooms, but at a price. Successful reclamation is possible in inland salt deserts but tremendous "soil amendment" is necessary. Once a crop is planted, it must be irrigated. An efficient drainage system is essential to avoid the buildup of subterranean salts, which gradually work up to the surface, making the land unfit for crops. The natural flora of the Salton Sink is differentiated from the general flora of the Colorado Desert by the preponderance of saltbushes and other halophytes, diagnostic of highly saline soils.

Yesterday I involuntarily watched crop spraying, trying not to inhale the acrid fumes. Little planes like smart yellow dragonflies flew to the end of a field, pulled into a tight left turn, tilted their wings, then buzzed back down the next row, leaving a mist of chemicals hanging pungent in the air. Along with its fruits and vegetables, California also grows rapidly adapting, voracious, quick-breeding insects, encouraged by the seldom-rotated single-crop plantings of large agribusiness. Although some farmers are abandoning synthetic fertilizers and pesticides, information on crop yields and the costs of conversion are not yet widely available.

Even with good irrigation practices, many of the fields in the Imperial Valley that I see both from the ground and from the air have empty moth-eaten spots shaved out of the green. Salt is not only present in the soil but is carried in solution by the Colorado River, and desalinization is becoming a major concern in the valley. The gloomiest estimates are for only twenty years more life before the valley is no longer fit for crops and the desert takes back its own. Once there was desert here, and here there may be desert again.

At the turn of the century, John Van Dyke, an eastern art

historian who came to the desert for his health, was captivated by the dry heat and clear air, and became concerned for its preservation. In *The Desert* he wrote prophetic words: "The deserts should never be reclaimed. They are the breathing-spaces of the west and should be preserved forever."

Rachel L. Carson

SUMMER'S END

From *Under the Sea Wind*

It was September before the sanderlings, now in whitening plumage, ran again on the island beach or hunted Hippa crabs in the ebbing tide at the point of land called Ship's Shoal. Their flight from the northern tundras had been broken by many feeding stops on the wide mud flats of Hudson Bay and James Bay and on the ocean beaches from New England southward. In their fall migration the birds were unhurried, the racial urge that drove them northward in the spring having been satisfied. As the winds and the sun dictated, they drifted southward, their flocks now growing as more birds from the north joined them, now dwindling as more and more of the migrants found their customary winter home and dropped behind. Only the fringe of the great southward wave of shorebirds would push on and on to the southernmost part of South America.

As the cries of the returning shorebirds rose once more from the frothy edge of the surf and the whistle of the curlews sounded again in the salt marshes, there were other signs of the summer's end. By September the eels of the sound country had begun to drop downstream to the sea. The eels came down from the hills and the upland grasslands. They came from cypress swamps where black-watered rivers had their beginnings; they moved across the tidal plain that dropped in six giant steps to the sea. In the river estuaries and in the sounds they joined their mates-to-be. Soon, in silvery wedding dress, they would follow the ebbing tides to the sea, to find—and lose—themselves in the black abysses of mid-ocean.

By September, the young shad, come from the eggs shed in river and stream by the spawning runs of spring, were moving with the river water to the sea. At first they moved slowly in the vaster currents as the sluggish rivers broadened toward their estuaries. Soon, however, the speed of the little fish, no longer than a man's finger, would quicken, when the fall rains came and the wind changed, chilling the water and driving the fish to the warmer sea.

By September the last of the season's hatch of young shrimp were coming into the sounds through the inlets from the open sea. The coming of the young was symbolic of another journey which no man had seen and no man could describe— a journey taken weeks before by the elder generation of shrimp. All through the spring and summer more and more of the grown shrimp, come to maturity at the age of a year, had been slipping away from the coastal waters, journeying out across the continental shelf, descending the blue slopes of undersea valleys. From this journey they never returned, but their young, after several weeks of ocean life, were brought by the sea into the protected inside waters. All through the summer and fall the baby shrimp were brought into the sounds and river mouths—seeking warm shallows where brackish water lay over muddy bottoms. Here they fed eagerly on the abundant food and found shelter from hungry fish in the carpeting eel grass. And as they grew rapidly, the young turned once more to the sea, seeking its bitter waters and its deeper rhythms. Even as the youngest shrimp from the last spawning of the season came through the inlets on each flood tide of September, the larger young were moving out through the sounds to the sea.

By September the panicles of the sea oats in the dunes had turned a golden brown. As the marshes lay under the sun, they glowed with the soft greens and browns of the salt meadow grass, the warm purples of the rushes, and the scarlet of the marsh samphire. Already the gum trees were like red flares set in the swamps of the river banks. The tang of autumn was in the night air, and as it rolled over the warmer marshes it turned to mist, hiding the herons who stood among the grasses at dawn; hiding from the eyes of the hawks the meadow mice who ran along the paths they had made through the marshes by the patient felling of thousands of marsh-grass stems; hiding the schools of silversides in the sound from the terns who fluttered above the rolling white sea, and caught no fish until the sun had cleared away the mists.

The chill night air brought a restlessness to many fish scattered widely throughout the sound. They were steely gray fish with large scales and a low, four-spined fin set on the back like a spread sail. The fish were mullet who had lived throughout the summer in the sound and estuary, roving solitary among the eel grass and widgeon grass, feeding on the litter of animal and vegetable fragments of the bottom mud. But every fall the mullet left the sounds and made a far sea journey, in the course of which they brought forth the next generation. And so the first chill of fall stirred in the fish the feeling of the sea's rhythm and awakened the instinct of migration.

The chilling waters and the tidal cycles of the summer's end brought to many of the young fish of the sound country, also, a summons to return to the sea. Among these were the young pompano and mullet, silversides and killifish, who lived in the pond called Mullet Pond, where the dunes of the barrier island fell away to the flat sands of the Ship's Shoal. These young fish had been spawned in the sea, but had found their way to the pond through a temporary cut earlier that year.

On a day when the full harvest moon sailed like a white balloon in the sky, the tides, which had grown in strength as the moon swelled to roundness, began to wash out a gully across the inlet beach. Only on the highest tides did the torpid pond receive water from the ocean. Now the beat of the waves and the strong backwash that sucked away the loose sand had found the weak place in the beach, where a cut had been made before, and in less time than it took a fishing launch to cross from the mainland docks to the banks a narrow gully or slough had been cut through to the pond. Not more than a dozen feet across, it made a bottleneck into which the surf rolled as the waves broke on the beach. The water surged and seethed as in a mill race, hissing and foaming. Wave after wave poured through the slough and into the pond. They dug out an uneven, corrugated bottom over which the water leaped and tobogganed. They spread out into the marshes that backed the pond, seeping silently and stealthily among the grass stems and the reddening stalks of the marsh samphire. Into the marshes they carried the frothy brown scud thrown off by the waves. The sandy foam filled the spaces between the grass stalks so closely that the marsh looked like a beach thickly grown with short grass; in reality the grass stood a foot in water and only the upper third of the stalks showed above the froth.

Leaping and racing, foaming and swirling, the incoming flood brought release to the myriads of small fishes that had been imprisoned in the pond. Now in thousands they poured out of the pond and out of the marshes. They raced in mad confusion to meet the clean, cold water. In their excitement they let the flood take them, toss them, turn them over and over. Reaching mid-channel of the slough they leaped high in the air again and again, sparkling bits of animate silver, like a swarm of glittering insects that rose and fell, rose again and fell. There the water seized them and held them back in their wild dash to the sea, so that many of them were caught on the slopes of the waves and held, tails uppermost, struggling help-lessly against the might of the water. When finally the waves released them they raced down the slough to the ocean, where they knew once more the rolling breakers, the clean sandy bottoms, the cool green waters.

How did the pond and the marshes hold them all? On they came, in school after school, flashing bright among the marsh grasses, leaping and bounding out of the pond. For more than an hour the exodus continued, with scarcely a break in the hurrying schools. Perhaps they had come in, many of them, on the last spring tide when the moon was a pencil stroke of silver in the sky. And now the moon had grown fat and round and another spring tide, a rollicking, roistering, rough-and-ready tide, called them back to the sea again.

On they went, passing through the surf line where the white-capped waves were tumbling. On they went, most of them, past the smoother green swells to the second line of surf, where shoals tripped the waves coming in from the open sea and sent them sprawling in white confusion. But there were terns fishing above the surf, and thousands of the small migrants went no farther than the portals of the sea.

Now there came days when the sky was gray as a mullet's back, with clouds like the flung spray of waves. The wind, that throughout most of the summer had blown from the southwest, began to veer toward the north. On such mornings large mullet could be seen jumping in the estuary and over the shoals of the sound. On the ocean beaches fishermen's boats were drawn up on the sand. Gray piles of netting lay in the boats. Men stood on the beach, with eyes on the water, patiently waiting. The fishermen knew that mullet were gathering in schools throughout the sound because of the change in the weather. They knew that soon the schools would run out through the inlet before the wind and then would pass down along the coast, keeping, as the fishermen had told it from one generation to the next, "their right eyes to the beach." Other mullet would come down from the sounds that lay to the north and still others would come by the outside passage, following down along the chain of barrier islands. So the fishermen waited, confident in their generations of tested lore; and the boats waited with the nets that were empty of fish.

Other fishers besides the men awaited the runs of mullet. Among them was Pandion, the fish hawk, whom the mullet fishermen watched every day as he floated, a small dark cloud, in wide circles in the sky. To pass the hours as they stood watch on the sound beach or among the dunes, the fishermen wagered among themselves when the osprey would dive.

Pandion had a nest in a clump of loblolly pines on the shore of the river three miles away. There he and his mate had hatched and reared a brood of three young that season. At first the young had been clothed in down that was the color of old, decaying tree stumps; now they had grown their pinions and had gone away to fish for themselves, but Pandion and his mate, who had been faithful to each other throughout life, continued to live in the nest which they had used year after year.

The nest was six feet across at its base and more than half as wide at the top. Its bulk would have overflowed any of the farm carts that were drawn by mules along the dirt roads of the sound country. The two ospreys had repaired the nest and added to it during the years anything they could find washed up on the beaches by the tides. Now practically the whole top of a forty-foot pine served as support for the nest, and the great weight of sticks, branches, and pieces of sod had killed all but a few of the lower branches. In the course of years the ospreys had woven or worked into the nest a twenty-foot piece of haul seine with ropes attached that they had picked up on the shore of the sound, perhaps a dozen cork floats from fishing gear, many cockle and oyster shells, part of the skeleton of an eagle, parchmentlike strings of the egg cases of conchs, a broken oar, part of a fisherman's boot, tangled mats of seaweed.

In the lower layers of the huge, decaying mass many small birds had found nesting places. That summer there had been three families of sparrows, four of starlings, and one of the Carolina wren. In the spring an owl had taken up quarters in the osprey nest, and once there had been a green heron. All these lodgers Pandion had suffered good-naturedly.

After the third day of grayness and chill, the sun broke through the clouds. Watched by the mullet fishermen, Pandion sailed on set wings, riding the mounting columns of warm air that shimmered upward from the water. Far below him the water was like green silk rippling in a breeze. The terns and skimmers resting on the shoals of the sound were the size of robins. The black, glistening backs of a school of dolphins, diving and rolling, moved, a dark serpent, over the face of the sound. The amber eyes of Pandion flickered as a whipper ray leaped three times from the water, coming down with a sharp spat that was carried away on the wind and lost.

A shadow took form on the green screen beneath the osprey and the surface dimpled as a fish nosed at the film. In the sound two hundred feet below the fish hawk, Mugil, the mullet—the leaper—gathered his strength and flung himself in exhilaration into the air. As he was flexing his muscles for the third leap a dark form fell out of the sky and viselike talons seized him. The mullet weighed more than a pound, but Pandion carried him easily in his taloned feet, bearing the fish across the sound and to the nest three miles away.

Flying up the river from the estuary the osprey carried the mullet head first in his talons. As he neared the nest he relaxed his grip with the left foot and, checking flight, alighted on the outer branches of the nest with the fish still gripped in his right foot. Pandion lingered over his meal of fish for more than an hour, and when his mate came near he crouched low over the mullet and hissed at her. Now that the nesting was done, every bird must fish for itself.

Later in the day, as he returned down the river to fish, Pandion swooped low to the water and for the space of a dozen wing beats dragged his feet in the river, cleansing them of the adhering fish slime.

On his return Pandion was watched by the sharp eyes of a large brown bird perched in one of the pines on the west bank of the river, overlooking the marshes of the estuary. White Tip, the bald eagle, lived as a pirate, never fishing for himself when he could steal from the ospreys of the surrounding country. When Pandion moved out over the sound the eagle followed, mounting into the air and taking up a position far above the fish hawk.

For an hour two dark forms circled in the sky. Then from his high station White Tip saw the body of the osprey suddenly dwindle to sparrow size as he fell in a straight drop, saw the white spray mount from the water as the fish hawk disappeared. After the passing of thirty seconds Pandion emerged from the water, mounting straight for fifty feet with short, heavy wing beats and then leveling out into straight flight toward the river's mouth.

Watching him, White Tip knew that the osprey had caught a fish and was taking it home to the nest in the pines. With a shrill scream that fell down through the sky to the ears of the osprey, the eagle whirled in pursuit, keeping his elevation of a thousand feet above the fish hawk.

Pandion cried out in annoyance and alarm, redoubling the force of his wing beats in an effort to reach the cover of the pines before his tormentor should attack. The speed of the hawk was retarded by the weight of the catfish that he carried and by the convulsive struggles of the fish, held firmly in the strong talons.

Between the island and the mainland and several minutes' flight from the mouth of the river the eagle gained a position directly over the hawk. On half-closed wings he dropped with terrific speed. The wind whined through his feathers. As he passed the osprey he whirled in air, back to the water, presenting his talons to the attack. Pandion dodged and twisted, eluding the eight curved scimitars. Before White Tip could recover himself Pandion had shot aloft two hundred, five hundred feet. The eagle hurtled after him, mounted above him. But even as he began the stoop, the fish hawk, in another upward soaring, surmounted the position of his enemy.

Meanwhile the fish, drained of life by separation from the water, grew limp as all its struggles ceased. Like a mist gathering on a clear glass surface, a film clouded its eyes. Soon the iridescent greens and golds that made its body, in life, a thing of beauty had faded to dullness.

By turns rising and swooping, hawk and eagle rose to a great height, into the empty places of the air, of which the sound and its shoals and white sands had no part.

Cheep! Cheep! Chezeek! Chezeek! screamed Pandion in a frenzy of excitement.

A dozen white feathers, ripped from his breast as he barely evaded White Tip's talons on the last stoop, fluttered earthward. Of a sudden the osprey bent his wings sharply and dropped like a stone toward the water. The wind roared in his ears, half blinded him, plucked at his feathers as the sound rushed up to meet him. It was his final effort to outwit a stronger and more enduring enemy. But from above, the relentless dark form fell even faster than Pandion, gained on him, passed him as the fishing boats on the sound grew big as gulls afloat, whirled and tore the fish from his grasp.

The eagle carried the fish to his pine-tree perch to rend it, muscle from bone. By the time he reached the perch Pandion was beating out heavily over the inlet to new fishing grounds at sea.

Henry Beston

NIGHT ON THE GREAT BEACH

I

Our fantastic civilization has fallen out of touch with many
aspects of nature, and with none more completely than with
night. Primitive folk, gathered at a cave mouth round a fire,
do not fear night; they fear, rather, the energies and creatures
to whom night gives power; we of the age of the machines,
having delivered ourselves of nocturnal enemies, now have a
dislike of night itself. With lights and ever more lights, we drive
the holiness and beauty of night back to the forests and the
sea; the little villages, the crossroads even, will have none of

it. Are modern folk, perhaps, afraid of night? Do they fear that vast serenity, the mystery of infinite space, the austerity of stars? Having made themselves at home in a civilization obsessed with power, which explains its whole world in terms of energy, do they fear at night for their dull acquiescence and the pattern of their beliefs? Be the answer what it will, to-day's civilization is full of people who have not the slightest notion of the character or the poetry of night, who have never even seen night. Yet to live thus, to know only artificial night, is as absurd and evil as to know only artificial day.

Night is very beautiful on this great beach. It is the true other half of the day's tremendous wheel; no lights without meaning stab or trouble it; it is beauty, it is fulfilment, it is rest. Thin clouds float in these heavens, islands of obscurity in a splendour of space and stars: the Milky Way bridges earth and ocean; the beach resolves itself into a unity of form, its summer lagoons, its slopes and uplands merging; against the western sky and the falling bow of sun rise the silent and superb undulations of the dunes.

My nights are at their darkest when a dense fog streams in from the sea under a black, unbroken floor of cloud. Such nights are rare, but are most to be expected when fog gathers off the coast in early summer; this last Wednesday night was the darkest I have known. Between ten o'clock and two in the morning three vessels stranded on the outer beach—a fisherman, a four-masted schooner, and a beam trawler. The fisherman and the schooner have been towed off, but the trawler, they say, is still ashore.

I went down to the beach that night just after ten o'clock. So utterly black, pitch dark it was, and so thick with moisture and trailing showers, that there was no sign whatever of the beam of Nauset; the sea was only a sound, and when I reached the edge of the surf the dunes themselves had disappeared behind. I stood as isolated in that immensity of rain and night as I might have stood in interplanetary space. The sea was troubled and noisy, and when I opened the darkness with an outlined cone of light from my electric torch I saw that the waves were washing up green coils of sea grass, all coldly wet and bright in the motionless and unnatural radiance. Far off a single ship was groaning its way along the shoals. The fog was compact of the finest moisture; passing by, it spun itself into my lens of light like a kind of strange, aërial, and liquid silk. Effin Chalke, the new coast guard, passed me going north, and told me that he had had news at the halfway house of the schooner at Cahoon's.

It was dark, pitch dark to my eye, yet complete darkness, I imagine, is exceedingly rare, perhaps unknown in outer nature. The nearest natural approximation to it is probably the gloom of forest country buried in night and cloud. Dark as the night was here, there was still light on the surface of the planet. Standing on the shelving beach, with the surf breaking at my feet, I could see the endless wild uprush, slide, and withdrawal of the sea's white rim of foam. The men at Nauset tell me that on such nights they follow along this vague crawl of whiteness, trusting to habit and a sixth sense to warn them of their approach to the halfway house.

Animals descend by starlight to the beach. North, beyond the dunes, muskrats forsake the cliff and nose about in the driftwood and weed, leaving intricate trails and figure eights to be obliterated by the day; the lesser folk—the mice, the occasional small sand-coloured toads, the burrowing moles— keep to the upper beach and leave their tiny footprints under the overhanging wall. In autumn skunks, beset by a shrinking larder, go beach combing early in the night. The animal is by preference a clean feeder and turns up his nose at rankness. I almost stepped on a big fellow one night as I was walking north to meet the first man south from Nauset. There was a scamper, and the creature ran up the beach from under my feet; alarmed he certainly was, yet was he contained and continent. Deer are frequently seen, especially north of the light. I find their tracks upon the summer dunes.

Years ago, while camping on this beach north of Nauset, I went for a stroll along the top of the cliff at break of dawn. Though the path followed close enough along the edge, the beach below was often hidden, and I looked directly from the height to the flush of sunrise at sea. Presently the path, turning, approached the brink of the earth precipice, and on the beach below, in the cool, wet rosiness of dawn, I saw three deer playing. They frolicked, rose on their hind legs, scampered off, and returned again, and were merry. Just before sunrise they trotted off north together down the beach toward a hollow in the cliff and the path that climbs it.

Occasionally a sea creature visits the shore at night. Lone coast guardsmen, trudging the sand at some deserted hour, have been startled by seals. One man fell flat on a creature's back, and it drew away from under him, flippering toward the sea, with a sound "halfway between a squeal and a bark." I myself once had rather a start. It was long after sundown, the light dying and uncertain, and I was walking home on the top level of the beach and close along the slope descending to the ebbing tide. A little more than halfway to the Fo'castle a huge unexpected something suddenly writhed horribly in the darkness under my bare foot. I had stepped on a skate left stranded by some recent crest of surf, and my weight had momentarily annoyed it back to life.

Facing north, the beam of Nauset becomes part of the dune night. As I walk toward it, I see the lantern, now as a star of light which waxes and wanes three mathematic times, now as a lovely pale flare of light behind the rounded summits of the dunes. The changes in the atmosphere change the colour of the beam; it is now whitish, now flame golden, now golden red; it changes its form as well, from a star to a blare of light, from a blare of light to a cone of radiance sweeping a circumference of fog. To the west of Nauset I often see the apocalyptic flash of the great light at the Highland reflected on the clouds or even on the moisture in the starlit air, and, seeing it, I often think of the pleasant hours I have spent there when George and Mary Smith were at the light and I had the good fortune to visit as their guest. Instead of going to sleep in the room under the eaves, I would lie awake, looking out of a window to the great spokes of light revolving as solemnly as a part of the universe.

All night long the lights of coastwise vessels pass at sea, green lights going south, red lights moving north. Fishing schooners and flounder draggers anchor two or three miles out, and keep a bright riding light burning on the mast. I see them come to anchor at sundown, but I rarely see them go, for they are off at dawn. When busy at night, these fishermen illumine their decks with a scatter of oil flares. From shore, the ships might be thought afire. I have watched the scene through a night glass. I could see no smoke, only the waving flares, the reddish radiance on sail and rigging, an edge of reflection overside, and the enormous night and sea beyond.

One July night, as I returned at three o'clock from an expedition north, the whole night, in one strange, burning instant, turned into a phantom day. I stopped and, questioning, stared about. An enormous meteor, the largest I have ever seen, was consuming itself in an effulgence of light west of the zenith. Beach and dune and ocean appeared out of nothing, shadowless and motionless, a landscape whose every tremor and vibration were stilled, a landscape in a dream.

The beach at night has a voice all its own, a sound in fullest harmony with its spirit and mood—with its little, dry noise of sand forever moving, with its solemn, overspilling, rhythmic seas, with its eternity of stars that sometimes seem to hang down like lamps from the high heavens—and that sound the piping of a bird. As I walk the beach in early summer my solitary coming disturbs it on its nest, and it flies away, troubled, invisible, piping its sweet, plaintive cry. The bird I write of is the piping plover, *Charadrius melodus*, sometimes called the beach plover or the mourning bird. Its note is a whistled syllable, the loveliest musical note, I think, sounded by any North Atlantic bird.

Now that summer is here I often cook myself a camp supper on the beach. Beyond the crackling, salt-yellow driftwood flame, over the pyramid of barrel staves, broken boards, and old sticks all atwist with climbing fire, the unseen ocean thunders and booms, the breaker sounding hollow as it falls. The wall of the sand cliff behind, with its rim of grass and withering roots, its sandy crumblings and erosions, stands gilded with flame; wind cries over it; a covey of sandpipers pass between the ocean and the fire. There are stars, and to the south Scorpio hangs curving down the sky with ringed Saturn shining in his claw.

Learn to reverence night and to put away the vulgar fear of it, for, with the banishment of night from the experience of man, there vanishes as well a religious emotion, a poetic mood, which gives depth to the adventure of humanity. By day, space is one with the earth and with man—it is his sun that is shining, his clouds that are floating past; at night, space is his no more. When the great earth, abandoning day, rolls up the deeps of the heavens and the universe, a new door opens for the human spirit, and there are few so clownish that some awareness of the mystery of being does not touch them as they gaze. For a moment of night we have a glimpse of ourselves and of our world islanded in its stream of stars—pilgrims of mortality, voyaging between horizons across eternal seas of space and time. Fugitive though the instant be, the spirit of man is, during it, ennobled by a genuine moment of emotional dignity, and poetry makes its own both the human spirit and experience.

II

At intervals during the summer, often enough when the tides are high and the moon is near the full, the surf along the beach turns from a churn of empty moonlit water to a mass of panic life. Driven in by schools of larger fish, swarms of little fish enter the tumble of the surf, the eaters follow them, the surf catches them both up and throws them, mauled and confused, ashore.

Under a sailing moon, the whole churn of sea close off the beach vibrates with a primeval ferocity and intensity of life; yet is this war of rushing mouth and living food without a sound save for the breaking of the seas. But let me tell of such a night.

I had spent an afternoon ashore with friends, and they had driven me to Nauset Station just after nine o'clock. The moon, two days from the full, was very lovely on the moors and on the channels and flat, moon-green isles of the lagoon; the wind was southerly and light. Moved by its own enormous rhythms, the surf that night was a stately incoming of high, serried waves, the last wave alone breaking. This inmost wave broke heavily in a smother and rebound of sandy foam, and thin sheets of seethe, racing before it up the beach, vanished endlessly into the endless thirst of the sands. As I neared the surf rim to begin my walk to the southward, I saw that the beach close along the breakers, as far as the eye would reach, was curiously atwinkle in the moonlight with the convulsive dance of myriads of tiny fish. The breakers were spilling them on the sands; the surf was aswarm with the creatures; it was indeed, for the time being, a surf of life. And this surf of life was breaking for miles along the Cape.

Little herring or mackerel? Sand eels? I picked a dancer out of the slide and held him up to the moon. It was the familiar sand eel or sand launce, *Ammodytes americanus*, of the waters between Hatteras and Labrador. This is no kin of the true eels, though he rather resembles one in general appearance, for his body is slender, eel-like, and round. Instead of ending bluntly, however, this "eel" has a large, well-forked tail. The fish in the surf were two and three inches long.

Homeward that night I walked barefooted in the surf, watching the convulsive, twinkling dance, now and then feeling the squirm of a fish across my toes. Presently something occurred which made me keep to the thinnest edge of the foam. Some ten feet ahead, an enormous dogfish was suddenly borne up the beach on the rim of a slide of foam; he moved with it unresisting while it carried him; the slide withdrawing and drying up, it rolled him twice over seaward; he then twisted heavily, and another minor slide carried him back again to shore. The fish was about three feet long, a real junior shark, purplish black in the increasing light—for the moon was moving west across the long axis of the breakers—and his dark, important bulk seemed strange in the bright dance of the smaller fish about him.

It was then that I began to look carefully at the width of gathering seas. Here were the greater fish, the mouths, the eaters who had driven the "eels" ashore to the edge of their world and into ours. The surf was alive with dogfish, aswarm with them, with the rush, the cold bellies, the twist and tear of their wolfish violence of life. Yet there was but little sign of it in the waters—a rare fin slicing past, and once the odd and instant glimpse of a fish embedded like a fly in amber in the bright, overturning volute of a wave.

Too far in, the dogfish were now in the grip of the surf, and presently began to come ashore. As I walked the next half mile every other breaker seemed to leave behind its ebb a mauled and stranded sharklet feebly sculling with his tail. I kicked many back into the seas, risking a toe, perhaps; some I caught by the tails and flung, for I did not want them corrupting on the beach. The next morning, in the mile and three quarters between the Fo'castle and the station, I counted seventy-one dogfish lying dead on the upper beach. There were also a dozen or two skates—the skate is really a kind of shark—which had stranded the same night. Skates follow in many things, and are forever being flung upon these sands.

I sat up late that night at the Fo'castle, often putting down the book I read to return to the beach.

A little after eleven came Bill Eldredge to the door, with a grin on his face and one hand held behind his back. "Have you ordered to-morrow's dinner yet?" said he. "No." "Well, here it is," and Bill produced a fine cod from behind his back. "Just found him right in front of your door, alive and flopping. Yes, yes, haddock and cod often chase those sand eels in with the bigger fish; often find them on the beach about this time of the year. Got any place to keep him? Let me have a piece of string and I'll hang him on your clothesline. He'll keep all right." With a deft unforking of two fingers, Bill drew the line through the gills, and as he did so the heavy fish flopped noisily. No fear about him being dead. Make a nice chowder. Bill stepped outside; I heard him at the clothesline. Afterward we talked till it was time for him to shoulder his clock and Coston case again, pick up his watch cap, whistle in his little black dog, and go down over the dune to the beach and Nauset Station.

There were nights in June when there was phosphorescence in the surf and on the beach, and one such night I think I shall remember as the most strange and beautiful of all the year.

Early this summer the middle beach moulded itself into a bar, and between it and the dunes are long, shallow runnels into which the ocean spills over at high tide. On the night I write of, the first quarter of the moon hung in the west, and its light on the sheets of incoming tide coursing thin across the bar was very beautiful to see. Just after sundown I walked to

Nauset with friends who had been with me during the afternoon; the tide was still rising, and a current running in the
pools. I lingered at the station with my friends till the last of
sunset had died, and the light upon the planet, which had been
moonlight mingled with sunset pink, had cleared to pure cold
moon.

Southward, then, I turned, and because the flooded runnels were deep close by the station, I could not cross them and
had to walk their inner shores. The tide had fallen half a foot,
perhaps, but the breakers were still leaping up against the bar
as against a wall, the greater ones still spilling over sheets of
vanishing foam.

It grew darker with the westing of the moon. There was
light on the western tops of the dunes, a fainter light on the
lower beach and the breakers; the face of the dunes was a unity
of dusk.

The tide had ebbed in the pools, and their edges were wet
and dark. There was a strange contrast between the still levels
of the pool and the seethe of the sea. I kept close to the land
edge of the lagoons, and as I advanced my boots kicked wet
spatters of sand ahead as they might have kicked particles of
snow. Every spatter was a crumb of phosphorescence; I walked
in a dust of stars. Behind me, in my footprints, luminous patches
burned. With the double-ebb moonlight and tide, the deep-
ening brims of the pools took shape in smouldering, wet fire.
So strangely did the luminous speckles smoulder and die and
glow that it seemed as if some wind were passing, by whose
breath they were kindled and extinguished. Occasional whole
breakers of phosphorescence rolled in out of the vague sea—
the whole wave one ghostly motion, one creamy light—and,
breaking against the bar, flung up pale sprays of fire.

A strange thing happens here during these luminous tides.
The phosphorescence is itself a mass of life, sometimes protozoan its origin, sometimes bacterial, the phosphorescence I

write of being probably the latter. Once this living light has seeped into the beach, colonies of it speedily invade the tissues of the ten thousand thousand sand fleas which are forever hopping on this edge of ocean. Within an hour the grey bodies of these swarming amphipods, these useful, ever hungry sea scavengers (*Orchestia agilis; Talorchestia megalophthalma*), show phosphorescent pin points, and these points grow and unite till the whole creature is luminous. The attack is really a disease, an infection of light. The process had already begun when I arrived on the beach on the night of which I am writing, and the luminous fleas hopping off before my boots were an extraordinary sight. It was curious to see them hop from the pool rims to the upper beach, paling as they reached the width of peaceful moonlight lying landward of the strange, crawling beauty of the pools. This infection kills them, I think; at least, I have often found the larger creature lying dead on the fringe of the beach, his huge porcelain eyes and water-grey body one core of living fire. Round and about him, disregarding, ten thousand kinsmen, carrying on life and the plan of life, ate of the bounty of the tide.

III

All winter long I slept on a couch in my larger room, but with the coming of warm weather I have put my bedroom in order—I used it as a kind of storage space during the cold season—and returned to my old and rather rusty iron cot. Every once in a while, however, moved by some obscure mood, I lift off the bedclothing and make up the couch again for a few nights. I like the seven windows of the larger room, and the sense one may have there of being almost out-of-doors. My couch stands alongside the two front windows, and from my pillow I can look out to sea and watch the passing lights, the stars rising over ocean, the swaying lanterns of the anchored fishermen, and the white spill of the surf whose long sound fills the quiet of the dunes.

Ever since my coming I have wanted to see a thunderstorm bear down upon this elemental coast. A thunderstorm is a "tempest" on the Cape. The quoted word, as Shakespeare used it, means lightning and thunder, and it is in this old and beautiful Elizabethan sense that the word is used in Eastham. When a schoolboy in the Orleans or the Wellfleet High reads the Shakespearean play, its title means to him exactly what it meant to the man from Stratford; elsewhere in America, the terms seems to mean anything from a tornado to a blizzard. I imagine that this old significance of the word is now to be found only in certain parts of England and Cape Cod.

On the night of the June tempest, I was sleeping in my larger room, the windows were open, and the first low roll of thunder opened my eyes. It had been very still when I went to bed, but now a wind from the west-nor'west was blowing through the windows in a strong and steady current, and as I closed them there was lightning to the west and far away. I looked at my watch; it was just after one o'clock. Then came a time of waiting in the darkness, long minutes broken by more thunder, and intervals of quiet in which I heard a faintest sound of light surf upon the beach. Suddenly the heavens cracked open in an immense instant of pinkish-violet lightning. My seven windows filled with the violent, inhuman light, and I had a glimpse of the great, solitary dunes staringly empty of familiar shadows; a tremendous crash then mingled with the withdrawal of the light, and echoes of thunder rumbled away and grew faint in a returning rush of darkness. A moment after, rain began to fall gently as if someone had just released its flow, a blessed sound on a roof of wooden shingles, and one I have loved ever since I was a child. From a gentle patter the sound of the rain grew swiftly to a drumming roar, and with the rain came the chuckling of water from the eaves. The tempest was crossing the Cape, striking at the ancient land on its way to the heavens above the sea.

Now came flash after stabbing flash amid a roaring of rain, and heavy thunder that rolled on till its last echoes were swallowed up in vast detonations which jarred the walls. Houses were struck that night in Eastham village. My lonely world, full of lightning and rain, was strange to look upon. I do not share the usual fear of lightning, but that night there came over me, for the first and last time of all my solitary year, a sense of isolation and remoteness from my kind. I remember that I stood up, watching, in the middle of the room. On the great marshes the lightning surfaced the winding channels with a metallic splendour and arrest of motion, all very strange through windows blurred by rain. Under the violences of light the great dunes took on a kind of elemental passivity, the quiet of earth enchanted into stone, and as I watched them appear and plunge back into a darkness that had an intensity of its own I felt, as never before, a sense of the vast time, of the thousands of cyclic and uncounted years which had passed since these giants had risen from the dark ocean at their feet and given themselves to the wind and the bright day.

Fantastic things were visible at sea. Beaten down by the rain, and sheltered by the Cape itself from the river of west wind, the offshore brim of ocean remained unusually calm. The tide was about halfway up the beach, and rising, and long parallels of low waves, forming close inshore, were curling over and breaking placidly along the lonely, rain-drenched miles. The intense crackling flares and quiverings of the storm, moving out to sea, illumined every inch of the beach and the plain of the Atlantic, all save the hollow bellies of the little breakers, which were shielded from the light by their overcurling crests. The effect was dramatic and strangely beautiful, for what one saw was a bright ocean rimmed with parallel bands of blackest advancing darkness, each one melting back to light as the wave toppled down upon the beach in foam.

Stars came out after the storm, and when I woke again before sunrise I found the heavens and the earth rainwashed, cool, and clear. Saturn and the Scorpion were setting, but Jupiter was riding the zenith and paling on his throne. The tide was low in the marsh channels; the gulls had scarcely stirred upon their gravel banks and bars. Suddenly, thus wandering about, I disturbed a song sparrow on her nest. She flew to the roof of my house, grasped the ridgepole, and turned about, apprehensive, inquiring . . . *'tsiped* her monosyllable of alarm. Then back toward her nest she flew, alighted in a plum bush, and, reassured at last, trilled out a morning song.

Ernest Thompson Seton

JOHNNY BEAR

Johnny was a queer little Bear cub that lived with Grumpy, his mother, in the Yellowstone Park. They were among the many Bears that found a desirable home in the country about the Fountain Hotel.

The steward of the Hotel had ordered the kitchen garbage to be dumped in an open glade of the surrounding forest, thus providing throughout the season a daily feast for the Bears, and their numbers have increased each year since the law of the land has made the Park a haven of refuge where no wild thing may be harmed. They have accepted man's peace-offering, and many of them have become so well known to the Hotel men that they have received names suggested by their looks or ways. Slim Jim was a very long-legged thin Blackbear; Snuffy was a Blackbear that looked as though he had been singed; Fatty was a very fat, lazy Bear that always lay down to eat; the Twins were two half-grown, ragged specimens that always came and went together. But Grumpy and Little Johnny were the best known of them all.

Grumpy was the biggest and fiercest of the Blackbears, and Johnny, apparently her only son, was a peculiarly tiresome little cub, for he seemed never to cease either grumbling or whining. This probably meant that he was sick, for a healthy little Bear does not grumble all the time, any more than a healthy child. And indeed Johnny looked sick; he was the most miserable specimen in the Park. His whole appearance suggested dyspepsia; and this I quite understood when I saw the awful mixtures he would eat at that garbage-heap. Anything at all that he fancied he would try. And his mother allowed him to do as he pleased; so, after all, it was chiefly her fault, for she should not have permitted such things.

Johnny had only three good legs, his coat was faded and mangy, his limbs were thin, and his ears and paunch were disproportionately large. Yet his mother thought the world of him. She was evidently convinced that he was a little beauty and the Prince of all Bears, so, of course, she quite spoiled him. She was always ready to get into trouble on his account, and he was always delighted to lead her there. Although such a wretched little failure, Johnny was far from being a fool, for he usually knew just what he wanted and how to get it, if teasing his mother could carry the point.

7

II

It was in the summer of 1897 that I made their acquaintance. I was in the Park to study the home life of the animals, and had been told that in the woods, near the Fountain Hotel, I could see Bears at any time, which, of course, I scarcely believed. But on stepping out of the back door five minutes after arriving, I came face to face with a large Blackbear and her two cubs.

I stopped short, not a little startled. The Bears also stopped and sat up to look at me. Then Mother Bear made a curious short *Koff, Koff*, and looked toward a near pine tree. The cubs seemed to know what she meant, for they ran to this tree and scrambled up like two little monkeys, and when safely aloft they sat like small boys, holding on with their hands, while their little back legs dangled in the air, and waited to see what was to happen down below.

The Mother Bear, still on her hind legs, came slowly toward me, and I began to feel very uncomfortable indeed, for she stood about six feet high in her stockings and had apparently never heard of the magical power of the human eye.

I had not even a stick to defend myself with, and when she gave a low growl, I was about to retreat to the Hotel, although previously assured that the Bears have always kept their truce with man. However, just at this turning-point the old one stopped, now but thirty feet away, and continued to survey me calmly. She seemed in doubt for a minute, but evidently made up her mind that, "although that human thing might be all right, she would take no chances for her little ones."

She looked up to her two hopefuls, and gave a peculiar whining *Er-r-r Er-r*, whereupon they, like obedient children, jumped, as at the word of command. There was nothing about them heavy or bear-like as commonly understood; lightly they swung from bough to bough till they dropped to the ground, and all went off together into the woods. I was much tickled by the prompt obedience of the little Bears. As soon as their mother told them to do something they did it. They did not even offer a suggestion. But I also found out that there was a good reason for it, for had they not done as she had told them they would have got such a spanking as would have made them howl.

This was a delightful peep into Bear home life, and would have been well worth coming for, if the insight had ended there. But my friends in the Hotel said that that was not the best place for Bears. I should go to the garbage-heap, a quarter-mile off in the forest. There, they said, I surely could see as many Bears as I wished (which was absurd of them).

Early the next morning I went to this Bears' Banqueting Hall in the pines, and hid in the nearest bushes.

Before very long a large Blackbear came quietly out of the woods to the pile, and began turning over the garbage and feeding. He was very nervous, sitting up and looking about at each slight sound, or running away a few yards when startled by some trifle. At length he cocked his ears and galloped off into the pines, as another Blackbear appeared. He also behaved in the same timid manner, and at last ran away when I shook the bushes in trying to get a better view.

At the outset I myself had been very nervous, for of course no man is allowed to carry weapons in the Park; but the timidity of these Bears reassured me, and thenceforth I forgot everything in the interest of seeing the great, shaggy creatures in their home life.

Soon I realized I could not get the close insight I wished from that bush, as it was seventy-five yards from the garbage-pile. There was none nearer; so I did the only thing left to do: I went to the garbage-pile itself, and digging a hole big enough to hide in, remained there all day long, with cabbage-stalks, old potato-peelings, tomato-cans, and carrion piled up in odorous heaps around me. Notwithstanding the opinions of countless flies, it was not an attractive place. Indeed, it was so unfragrant that at night, when I returned to the Hotel, I was not allowed to come in until after I had changed my clothes in the woods.

It had been a trying ordeal, but I surely did see Bears that day. If I may reckon it a new Bear each time one came, I must have seen over forty. But of course it was not, for the Bears were coming and going. And yet I am certain of this: there were at least thirteen Bears, for I had thirteen about me at one time.

All that day I used my sketch-book and journal. Every Bear that came was duly noted; and this process soon began to give the desired insight into their ways and personalities.

Many unobservant persons think and say that all Negroes, or all Chinamen, as well as all animals of a kind, look alike. But just as surely as each human being differs from the next, so surely each animal is different from its fellow; otherwise how would the old ones know their mates or the little ones their mother, as they certainly do? These feasting Bears gave a good illustration of this, for each had its individuality; no two were quite alike in appearance or in character.

This curious fact also appeared: I could hear the Woodpeckers pecking over one hundred yards away in the woods, as well as the Chickadees chickadeeing, the Blue-jays blue-jaying, and even the Squirrels scampering across the leafy forest floor; and yet I *did not hear one of these Bears come*. Their huge, padded feet always went down in exactly the right spot to break no stick, to rustle no leaf, showing how perfectly they had learned the art of going in silence through the woods.

III

All morning the Bears came and went or wandered near my hiding-place without discovering me; and, except for one or two brief quarrels, there was nothing very exciting to note. But about three in the afternoon it became more lively.

There were then four large Bears feeding on the heap. In the middle was Fatty, sprawling at full length as he feasted, a picture of placid ursine content, puffing just a little at times as he strove to save himself the trouble of moving by darting out his tongue like a long red serpent, farther and farther, in quest of the tidbits just beyond claw reach.

Behind him Slim Jim was puzzling over the anatomy and attributes of an ancient lobster. It was something outside his experience, but the principle, "In case of doubt take the trick," is well known in Bearland, and settled the difficulty.

The other two were clearing out fruit-tins with marvelous dexterity. One supple paw would hold the tin while the long tongue would dart again and again through the narrow opening, avoiding the sharp edges, yet cleaning out the can to the last taste of its sweetness.

This pastoral scene lasted long enough to be sketched, but was ended abruptly. My eye caught a movement on the hilltop whence all the Bears had come, and out stalked a very large Blackbear with a tiny cub. It was Grumpy and Little Johnny.

The old Bear stalked down the slope toward the feast, and Johnny hitched alongside, grumbling as he came, his mother watching him as solicitously as ever a hen did her single chick. When they were within thirty yards of the garbage-heap, Grumpy turned to her son and said something which, judging from its effect, must have meant: "Johnny, my child, I think you had better stay here while I go and chase those fellows away."

Johnny obediently waited; but he wanted to *see*, so he sat up on his hind legs with eyes agog and ears acock.

Grumpy came striding along with dignity, uttering warning growls as she approached the four Bears. They were too much engrossed to pay any heed to the fact that yet another one of them was coming, till Grumpy, now within fifteen feet, let out a succession of loud coughing sounds, and charged into them. Strange to say, they did not pretend to face her, but as soon as they saw who it was, scattered and all fled for the woods.

Slim Jim could safely trust his heels, and the other two were not far behind; but poor Fatty, puffing hard and waddling like any other very fat creature, got along but slowly, and, unluckily for him, he fled in the direction of Johnny, so that Grumpy overtook him in a few bounds and gave him a couple of sound slaps in the rear which, if they did not accelerate his pace, at least made him bawl, and saved him by changing his direction. Grumpy, now left alone in possession of the feast, turned toward her son and uttered the whining *Er-r-r Er-r-r Er-r-r-r*. Johnny responded eagerly. He came "hoppity-hop" on his three good legs as fast as he could, and joining her on the garbage, they began to have such a good time that Johnny actually ceased grumbling.

He had evidently been there before now, for he seemed to know quite well the staple kinds of canned goods. One might almost have supposed that he had learned the brands, for a lobster-tin had no charm for him as long as he could find those that once were filled with jam. Some of the tins gave him much trouble, as he was too greedy or too clumsy to escape being scratched by the sharp edges. One seductive fruit-tin had a hole so large that he found he could force his head into it, and for a few minutes his joy was full as he licked into all the farthest corners. But when he tried to draw his head out, his sorrows began, for he found himself caught. He could not get out, and he scratched and screamed like any other spoiled child, giving his mother no end of concern, although she seemed not to know how to help him. When at length he got the tin off his head, he revenged himself by hammering it with his paws till it was perfectly flat.

A large sirup-can made him happy for a long time. It had had a lid, so that the hole was round and smooth; but it was not big enough to admit his head, and he could not touch its riches with his tongue stretched out its longest. He soon hit on a plan, however. Putting in his little black arm, he churned it around, then drew out and licked it clean; and while he licked one he got the other one ready; and he did this again and again, until the can was as clean inside as when first it had left the factory.

A broken mousetrap seemed to puzzle him. He clutched it between his forepaws, their strong inturn being sympathetically reflected in his hind feet, and held it firmly for study. The cheesy smell about it was decidedly good, but the thing responded in such an uncanny way when he slapped it that he kept back a cry for help only by the exercise of unusual self-control. After gravely inspecting it, with his head first on this side and then on that, and his lips puckered into a little tube, he submitted it to the same punishment as that meted out to the refractory fruit-tin, and was rewarded by discovering a nice little bit of cheese in the very heart of the culprit.

Johnny had evidently never heard of ptomaine poisoning, for nothing came amiss. After the jams and fruits gave out he turned his attention to the lobster and sardine-cans, and was not appalled by even the army beef. His paunch grew quite balloon-like, and from much licking his arms looked thin and shiny, as though he was wearing black silk gloves.

IV

It occurred to me that I might now be in a really dangerous place. For it is one thing surprising a Bear that has no family responsibilities, and another stirring up a bad-tempered old mother by frightening her cub.

"Supposing," I thought, "that cranky little Johnny should wander over to this end of the garbage and find me in the hole; he will at once set up a squall, and his mother, of course, will think I am hurting him, and without giving me a chance to explain, may forget the rules of the Park and make things very unpleasant."

Luckily, all the jam-pots were at Johnny's end; he stayed by them, and Grumpy stayed by him. At length he noticed that his mother had a better tin than any he could find, and as he ran whining to take it from her he chanced to glance away up the slope. There he saw something that made him sit up and utter a curious little *Koff Koff Koff Koff*.

His mother turned quickly, and sat up to see "what the child was looking at." I followed their gaze, and there, oh, horrors! was an enormous Grizzly Bear. He was a monster; he looked like a fur-clad omnibus coming through the trees.

Johnny set up a whine at once and got behind his mother. She uttered a deep growl, and all her back hair stood on end. Mine did too, but I kept as still as possible.

With stately tread the Grizzly came on. His vast shoulders sliding along his sides, and his silvery robe swaying at each tread, like the trappings on an elephant, gave an impression of power that was appalling.

Johnny began to whine more loudly, and I fully sympathized with him now, though I did not join in. After a moment's hesitation Grumpy turned to her noisy cub and said something that sounded to me like two or three short coughs—*Koff Koff Koff*. But I imagine that she really said: "My child, I think you had better get up that tree, while I go and drive the brute a-way."

At any rate, that was what Johnny did, and this was what she set out to do. But Johnny had no notion of missing any fun. He wanted to *see* what was going to happen. So he did not rest contented where he was hidden in the thick branches of the pine, but combined safety with view by climbing to the topmost branch that would bear him, and there, sharp against the sky, he squirmed about and squealed aloud in his excitement. The branch was so small that it bent under his weight, swaying this way and that as he shifted about, and every moment I expected to see it snap off. If it had been broken when swaying my way, Johnny would certainly have fallen on me, and this would probably have resulted in bad feelings between myself and his mother; but the limb was tougher than it looked, or perhaps Johnny had had plenty of experience, for he neither lost his hold nor broke the branch.

Meanwhile, Grumpy stalked out to meet the Grizzly. She stood as high as she could and set all her bristles on end; then, growling and chopping her teeth, she faced him.

The Grizzly, so far as I could see, took no notice of her. He came striding toward the feast as though alone. But when Grumpy got within twelve feet of him she uttered a succession of short, coughy roars, and charging, gave him a tremendous blow on the ear. The Grizzly was surprised; but he replied with a left-hander that knocked her over like a sack of hay.

Nothing daunted, but doubly furious, she jumped up and rushed at him.

Then they clinched and rolled over and over, whacking and pounding, snorting and growling, and making no end of dust and rumpus. But above all their noise I could clearly hear Little Johnny, yelling at the top of his voice, and evidently encouraging his mother to go right in and finish the Grizzly at once.

Why the Grizzly did not break her in two I could not understand. After a few minutes' struggle, during which I could see nothing but dust and dim flying legs, the two separated as by mutual consent—perhaps the regulation time was up—and for a while they stood glaring at each other, Grumpy at least much winded.

The Grizzly would have dropped the matter right there. He did not wish to fight. He had no idea of troubling himself about Johnny. All he wanted was a quiet meal. But no! The moment he took one step toward the garbage-pile, that is, as Grumpy thought, toward Johnny, she went at him again. But this time the Grizzly was ready for her. With one blow he knocked her off her feet and sent her crashing on to a huge upturned pine-root. She was fairly staggered this time. The force of the blow, and the rude reception of the rooty antlers, seemed to take all the fight out of her. She scrambled over and tried to escape. But the Grizzly was mad now. He meant to punish her, and dashed around the root. For a minute they kept up a dodging chase about it; but Grumpy was quicker of foot, and somehow always managed to keep the root between herself and her foe, while Johnny, safe in the tree, continued to take an intense and uproarious interest.

At length, seeing he could not catch her that way, the Grizzly sat up on his haunches; and while he doubtless was planning a new move, old Grumpy saw her chance, and making a dash, got away from the root and up to the top of the tree where Johnny was perched.

Johnny came down a little way to meet her, or perhaps so that the tree might not break off with the additional weight. Having photographed this interesting group from my hiding-place, I thought I must get a closer picture at any price, and for the first time in the day's proceedings I jumped out of the hole and ran under the tree. This move proved a great mistake, for here the thick lower boughs came between, and I could see nothing at all of the Bears at the top.

I was close to the trunk, and was peering about and seeking for a chance to use the camera, when old Grumpy began to come down, chopping her teeth and uttering her threatening cough at me. While I stood in doubt, I heard a voice far behind me calling:

"Say, Mister! You better look out; that ole B'ar is liable to hurt you."

I turned to see the cowboy of the Hotel on his Horse. He had been riding after the cattle, and chanced to pass near just as events were moving quickly.

"Do you know these Bears?" said I, as he rode up.

"Wal, I reckon I do," said he. "That there little one up top is Johnny; he's a little crank. An' the big un is Grumpy; she's a big crank. She's mighty onreliable gen'relly, but she's always strictly ugly when Johnny hollers like that."

"I should much like to get her picture when she comes down," said I.

"Tell ye what I'll do: I'll stay by on the pony, an' if she goes to bother you I reckon I can keep her off," said the man.

He accordingly stood by as Grumpy slowly came down from branch to branch, growling and threatening. But when she neared the ground she kept on the far side of the trunk, and finally slipped down and ran into the woods, without the slightest pretence of carrying out any of her dreadful threats. Thus Johnny was again left alone. He climbed up to his old perch and resumed his monotonous whining:

Wah! Wah! Wah! ("Oh, dear! Oh, dear! Oh, dear!")

I got the camera ready, and was arranging deliberately to take his picture in his favorite and peculiar attitude for threnodic song, when all at once he began craning his neck and yelling, as he had done during the fight.

I looked where his nose pointed, and here was the Grizzly

coming on straight toward me—not charging, but striding along, as though he meant to come the whole distance.

I said to my cowboy friend: "Do you know this Bear?"

He replied: "Wal I reckon I do. That's the old Grizzly. He's the biggest B'ar in the Park. He gen'relly minds his own business, but he ain't scared o' nothin'; an' today, ye see, he's been scrappin', so he's liable to be ugly."

"I would like to take his picture," said I; "and if you will help me, I am willing to take some chances on it."

"All right," said he, with a grin. "I'll stand by on the Horse, an' if he charges you I'll charge him; an' I kin knock him down once, but I can't do it twice. You better have your tree picked out."

As there was only one tree to pick out, and that was the one that Johnny was in, the prospect was not alluring. I imagined myself scrambling up there next to Johnny, and then Johnny's mother coming up after me, with the Grizzly below to catch me when Grumpy should throw me down.

The Grizzly came on, and I snapped him at forty yards, then again at twenty yards; and still he came quietly toward me. I sat down on the garbage and made ready. Eighteen yards—sixteen yards—twelve yards—eight yards, and still he came, while the pitch of Johnny's protests kept rising proportionately. Finally at five yards he stopped, and swung his huge bearded head to one side, to see what was making that aggravating row in the treetop, giving me a profile view, and I snapped the camera. At the click he turned on me with a thunderous

G—R—O—W—L!

and I sat still and trembling, wondering if my last moment had come. For a second he glared at me, and I could note the little green electric lamp in each of his eyes. Then he slowly turned and picked up—a large tomato-can.

"Goodness!" I thought, "is he going to throw that at me?"

But he deliberately licked it out, dropped it, and took another, paying thenceforth no heed whatever either to me or to Johnny, evidently considering us equally beneath his notice.

I backed slowly and respectfully out of his royal presence, leaving him in possession of the garbage, while Johnny kept on caterwauling from his safety-perch.

What became of Grumpy the rest of that day I do not know. Johnny, after bewailing for a time, realized that there was no sympathetic hearer of his cries, and therefore very sagaciously stopped them. Having no mother now to plan for him, he began to plan for himself, and at once proved that he was better stuff than he seemed. After watching, with a look of profound cunning on his little black face, and waiting till the Grizzly was some distance away, he silently slipped down behind the trunk, and despite his three-leggedness, ran like a hare to the next tree, never stopping to breathe till he was on its topmost bough. For he was thoroughly convinced that the only object that the Grizzly had in life was to kill him, and he seemed quite aware that his enemy could not climb a tree.

Another long and safe survey of the Grizzly, who really paid no heed to him whatever, was followed by another dash for the next tree, varied occasionally by a cunning feint to mislead the foe. So he went dashing from tree to tree and climbing each to its very top, although it might be but ten feet from the last, till he disappeared in the woods. After perhaps ten minutes, his voice again came floating on the breeze, the habitual querulous whining which told me he had found his mother and had resumed his customary appeal to her sympathy.

V

It is quite a common thing for Bears to spank their cubs when they need it, and if Grumpy had disciplined Johnny this way, it would have saved them both a deal of worry.

Perhaps not a day passed that summer without Grumpy getting into trouble on Johnny's account. But of all these numerous occasions that most ignominious was shortly after the affair with the Grizzly.

I first heard the story from three bronzed mountaineers. As they were very sensitive about having their word doubted, and very good shots with the revolver, I believed every word they told me, especially when afterward fully endorsed by the Park authorities.

It seemed that of all the tinned goods on the pile the nearest to Johnny's taste were marked with a large purple plum. This conclusion he had arrived at only after most exhaustive study. The very odor of those plums in Johnny's nostrils was the equivalent of ecstasy. So when it came about one day that the cook of the Hotel baked a huge batch of plum tarts, the telltale wind took the story afar into the woods, where it was wafted by way of Johnny's nostrils to his very soul.

Of course Johnny was whimpering at the time. His mother was busy "washing his face and combing his hair," so he had double cause for whimpering. But the smell of the tarts thrilled him; he jumped up, and when his mother tried to hold him he squalled, and I am afraid—he bit her. She should have cuffed him, but she did not. She only gave a disapproving growl, and followed to see that he came to no harm.

With his little black nose in the wind, Johnny led straight for the kitchen. He took the precaution, however, of climbing from time to time to the very top of a pine-tree lookout to take an observation, while Grumpy stayed below.

Thus they came close to the kitchen, and there, in the last

tree, Johnny's courage as a leader gave out, so he remained aloft and expressed his hankering for tarts in a woebegone wail.

It is not likely that Grumpy knew exactly what her son was crying for. But it is sure that as soon as she showed an inclination to go back into the pines, Johnny protested in such an outrageous and heartrending screeching that his mother simply could not leave him, and he showed no sign of coming down to be led away.

Grumpy herself was fond of plum jam. The odor was now, of course, very strong and proportionately alluring; so Grumpy followed it somewhat cautiously up to the kitchen door.

There was nothing surprising about this. The rule of "live and let live" is so strictly enforced in the Park that the Bears often come to the kitchen door for pickings, and on getting something, they go quietly back to the woods. Doubtless Johnny and Grumpy would each have gotten their tart but that a new factor appeared in the case.

That week the Hotel people had brought a new Cat from the East. She was not much more than a kitten, but still had a litter of her own, and at the moment that Grumpy reached the door, the Cat and her family were sunning themselves on the top step. Pussy opened her eyes to see this huge, shaggy monster towering above her.

The Cat had never before seen a Bear—she had not been there long enough; she did not know even what a Bear was. She knew what a Dog was, and here was a bigger, more awful bobtailed black dog than ever she had dreamed of, coming right at her. Her first thought was to fly for her life. But her next was for the kittens. She must take care of them. She must at least cover their retreat. So, like a brave little mother, she braced herself on that doorstep, and spreading her back, her claws, her tail, and everything she had to spread, she screamed out at that Bear an unmistakable order to

STOP!

The language must have been "Cat," but the meaning was clear to the Bear; for those who saw it maintain stoutly that Grumpy not only stopped, but she also conformed to the custom of the country and in token of surrender held up her hands.

However, the position she thus took made her so high that the Cat seemed tiny in the distance below. Old Grumpy had faced a Grizzly once, and was she now to be held up by a miserable little spike-tailed skunk no bigger than a mouthful? She was ashamed of herself, especially when a wail from Johnny smote on her ear and reminded her of her plain duty, as well as supplied his usual moral support.

So she dropped down on her front feet to proceed.

Again the Cat shrieked, "STOP!"

But Grumpy ignored the command. A scared mew from a kitten nerved the Cat, and she launched her ultimatum, which ultimatum was herself. Eighteen sharp claws, a mouthful of keen teeth, had Pussy, and she worked them all with a desperate will when she landed on Grumpy's bare, bald, sensitive nose, just the spot of all where the Bear could not stand it, and then worked backward to a point outside the sweep of Grumpy's claws. After one or two vain attempts to shake the spotted fury off, old Grumpy did just as most creatures would have done under the circumstances: she turned tail and bolted out of the enemy's country into her own woods.

But Puss's fighting blood was up. She was not content with repelling the enemy; she wanted to inflict a crushing defeat, to achieve an absolute and final rout. And however fast old Grumpy might go, it did not count, for the Cat was still on top, working her teeth and claws like a little demon. Grumpy, always erratic, now became panic-stricken. The trail of the pair was flecked with tufts of long black hair, and there was even bloodshed (in the fiftieth degree). Honor surely was satisfied, but Pussy was not. Round and round they had gone in the mad race. Grumpy was frantic, absolutely humiliated, and ready to make any terms; but Pussy seemed deaf to her cough-like yelps, and no one knows how far the Cat might have ridden that day had not Johnny unwittingly put a new idea into his mother's head by bawling in his best style from the top of his last tree, which tree Grumpy made for and scrambled up.

This was so clearly the enemy's country and in view of his reinforcements that the Cat wisely decided to follow no farther. She jumped from the climbing Bear to the ground, and then mounted sentry-guard below, marching around with tail in the air, daring that Bear to come down. Then the kittens came out and sat around, and enjoyed it all hugely. And the mountaineers assured me that the Bears would have been kept up the tree till they were starved, had not the cook of the Hotel come out and called off his Cat—although this statement was not among those vouched for by the officers of the Park.

VI

The last time I saw Johnny he was in the top of a tree, bewailing his unhappy lot as usual, while his mother was dashing about among the pines, "with a chip on her shoulder," seeking for someone—anyone—that she could punish for Johnny's sake, provided, of course, that it was not a big Grizzly or a Mother Cat.

This was early in August, but there were not lacking symptoms of change in old Grumpy. She was always reckoned "on-sartain," and her devotion to Johnny seemed subject to her characteristic. This perhaps accounted for the fact that when the end of the month was near, Johnny would sometimes spend half a day in the top of some tree, alone, miserable, and utterly unheeded.

The last chapter of his history came to pass after I had left the region. One day at gray dawn he was tagging along behind his mother as she prowled in the rear of the Hotel. A newly hired Irish girl was already astir in the kitchen. On looking out, she saw, as she thought, a Calf where it should not be, and ran to shoo it away. That open kitchen door still held unmeasured terrors for Grumpy, and she ran in such alarm that Johnny caught the infection, and not being able to keep up with her, he made for the nearest tree, which unfortunately turned out to be a post; and soon—too soon—he arrived at its top, some seven feet from the ground, and there poured forth his woes on the chilly morning air, while Grumpy apparently felt justified in continuing her flight alone. When the girl came near and saw that she had treed some wild animal, she was as much frightened as her victim. But others of the kitchen staff appeared, and recognizing the vociferous Johnny, they decided to make him a prisoner.

A collar and chain were brought, and after a struggle, during which several of the men got well scratched, the collar was buckled on Johnny's neck and the chain made fast to the post.

When he found that he was held, Johnny was simply too mad to scream. He bit and scratched and tore till he was tired out. Then he lifted up his voice again to call his mother. She did appear once or twice in the distance, but could not make up her mind to face that Cat, so disappeared, and Johnny was left to his fate.

He put in the most of that day in alternate struggling and crying. Toward evening he was worn out, and glad to accept the meal that was brought by Norah, who felt herself called on to play mother, since she had chased his own mother away.

When night came it was very cold; but Johnny nearly froze at the top of the post before he would come down and accept the warm bed provided at the bottom.

During the days that followed, Grumpy came often to the garbage-heap, but soon apparently succeeded in forgetting all about her son. He was daily tended by Norah, and received all his meals from her. He also received something else; for one day he scratched her when she brought his food, and she very properly spanked him till he squealed. For a few hours he sulked; he was not used to such treatment. But hunger subdued him, and thenceforth he held his new guardian in wholesome respect. She, too, began to take an interest in the poor motherless little wretch, and within a fortnight Johnny showed signs of developing a new character. He was much less noisy. He still expressed his hunger in a whining *Er-r-r Er-r-r Er-r-r*, but he rarely squealed now, and his unruly outbursts entirely ceased.

By the third week of September the change was still more marked. Utterly abandoned by his own mother, all his interest had centered in Norah, and she had fed and spanked him into an exceedingly well-behaved little Bear. Sometimes she would allow him a taste of freedom, and he then showed his bias by making, not for the woods, but for the kitchen where she was, and following her around on his hind legs. Here also he made the acquaintance of that dreadful Cat; but Johnny had a powerful friend now, and Pussy finally became reconciled to the black, woolly interloper.

As the Hotel was to be closed in October, there was talk of turning Johnny loose or of sending him to the Washington Zoo; but Norah had claims that she would not forgo.

When the frosty nights of late September came, Johnny had greatly improved in his manners, but he had also developed a bad cough. An examination of his lame leg had shown that the weakness was not in the foot, but much more deeply seated, perhaps in the hip, and that meant a feeble and tottering constitution.

He did not get fat, as do most Bears in fall; indeed, he continued to fail. His little round belly shrank in, his cough became worse, and one morning he was found very sick and shivering in his bed by the post. Norah brought him indoors, where the warmth helped him so much that thenceforth he lived in the kitchen.

For a few days he seemed better, and his old-time pleasure in *seeing things* revived. The great blazing fire in the range particularly appealed to him, and made him sit up in his old attitude when the opening of the door brought the wonder to view. After a week he lost interest even in that, and drooped more and more each day. Finally not the most exciting noises or scenes around him could stir up his old fondness for seeing what was going on.

He coughed a good deal, too, and seemed wretched, except when in Norah's lap. Here he would cuddle up contentedly, and whine most miserably when she had to set him down again in his basket.

A few days before the closing of the Hotel, he refused his usual breakfast, and whined softly till Norah took him in her lap; then he feebly snuggled up to her, and his soft *Er-r-r Er-r-r* grew fainter, till it ceased. Half an hour later, when she laid him down to go about her work, Little Johnny had lost the last trace of his anxiety to see and know what was going on.

Thomas Bledsoe

BROWN BEAR SUMMER

LATE SPRING

McNeil Falls thundered in the distance for a moment and then the roar faded softly on the sea breeze until all was quiet and still. This simple sound, perhaps the most vivid in my mind, brought memories of the two previous summers rushing back to me. I stood alone on the beach of McNeil Cove and scanned the vast wilderness scene around me. The country was immense, quiet, and empty.

It was June 1, 1975, and I had one day to be completely alone at McNeil. Tomorrow, Jim Taggart, my co-worker, would arrive on the high tide with our second and final load of summer supplies. I had spent enough time here and I loved it so much that I felt a closeness with the country that was intensified by being alone. The familiar hum of Bill deCreeft's "Beaver" float plane had already vanished out over Kamishak Bay, having deposited me and a planeload of gear on the beach near our camp. In the deep silence of early evening, I slowly began to perceive the subtle sights and sounds of McNeil.

I decided to put off setting up camp so that I could have a first look at the country in the evening light. I grabbed a chair from the tiny cabin and climbed a ladder to the flat roof and sat down, with the panorama of McNeil before me in a misty gold sunset. The sun shot between the mountains and clouds northwest of camp and poured down over McNeil Cove with the kind of deep intensity that can appear only at sunset from beneath dark clouds. The jagged peaks of the Aleutian Range lay buried in late spring snow and formed a steep frozen wall to the south and west. Patterns of deep purple and gold undulated across the slopes as the clouds formed and rolled, then disappeared beyond the pinnacles.

The source of the McNeil River plunged from one of the many small glaciers that lay hidden in protected bowls in these mountains. From its icy beginning, the McNeil flows north and then east until it emerges from the steep bluffs, winds across the tidal flats, and slips into the sea through a deep channel near our camp. The long days of early June were rapidly melting the snow and filling the river with a racing, foaming torrent of icy blue water. Again, the roar of McNeil Falls 1½ miles from camp drifted in and out on the sea breeze and alternated with the plaintive cry of the glaucous-winged gulls.

The lower mountain slopes were streaked with snowdrifts and greening alder brush all the way down to the tidal flats, where 20- and 30-foot drifts remained in the lee of the bluffs. East of camp the massive cliffs of McNeil Head jutted out into the sea and ended abruptly with sheer 700-foot cliffs plunging directly into the surf. The cool ocean breeze rose over the cliffs and formed soft wisps of fog that curled and billowed like giant breakers as it flowed softly down the slopes toward camp.

These were the sights and sounds that I remembered most from McNeil. It was a good welcome back for the summer. I felt a belonging, as though McNeil were a part of me and I had become a part of it.

The last streaks of sunlight were slanting across the tidal flats when I noticed two bears out on the flats about a half mile south of camp. They appeared to be together and one was quite large, so I assumed that it was an estrous sow with a large boar in mating consort. June is the most common month for mating so it was not unusual to see boars and sows in consort at this time. I was curious to know their identities, so I grabbed my binoculars and headed down to inspect them.

I followed the well-used trail that led down the beach, across the tidal flats, and eventually on to the falls. Six-foot snowdrifts remained on the beach, but they were rapidly melting, and already pale yellow shoots of wild celery and beach grass were reaching around the edges of the crusted snow. The plants had no patience with this late spring. Out on the tidal flats sharp points of sprouting sedge peppered the slimy mud on the scattered areas that were not still covered by a thick mantle of muddy ice. Soon I reached Mikfik Creek where it joins the McNeil and then I hiked along its banks until I had a clear view of the bears.

While en route I had noticed a few familiar characteristics of these bears and so I was almost certain I knew the pair. They undoubtedly were both McNeil residents because neither was concerned with my presence. This is because the resident McNeil bears have been habituated to the presence of humans, whereas nonresidents are usually frightened of people and will run when they encounter them. Now that I had a good close look, I could verify that they were Patches, a twenty-one-year-old boar, and White, a seven-year-old sow. Patches and White were both familiar to me and vice versa, so I moved closer to observe the pair.

White still wore her thick, creamy blond winter coat that made her look larger than her 300 pounds (all weights are estimates). She had a full "teddy bear" face and a fat, rounded belly that gave her the typical look of a sow. Patches' blond coat was marred by a large black bald spot that saddled across his hips and gave him his name. He weighed about 750 pounds, which is not particularly large for a twenty-one-year-old boar.

I was never quite sure if Patches had noticed me observing him that day. His apparent senility had grown worse each year and I was certain he also had very poor vision. I doubt if he would have acknowledged my presence, even if he had noticed me, because, of all the McNeil bears, he was the most nonchalant toward people.

I had never noticed any evidence that White had been in estrous in previous years and her behavior now suggested that this might be one of her first estrous cycles. Patches obviously frightened her, for she moved away from him constantly while cautiously glancing back at him. Her ears were laid back against her neck and she cowered slightly, which were signs that she was either under stress or, in this case, just plain afraid. White's reaction to Patches was typical, because any young bear is afraid of an older or larger bear. When she was older, White would be less afraid of a boar consort.

White was moving fast and soon the pair disappeared in the alders on the edge of the flats. Patches would have to pursue her for at least a day or two and probably longer before White would begin to become habituated to his presence.

While walking back to camp, I reflected a bit on White's estrous condition. She was one of the six McNeil sows I had counted the summer before that potentially could have cubs this year, although realistically I hoped that two or three would show up with young. Several litters of cubs would make the summer very exciting, so I was especially interested when I noticed the tracks of a sow and two spring cubs on the beach near my trail. The smooth wet beach sand, freshly washed by tidal waters, was broken in a wandering line by the unequal tracks of the family. The large determined prints of the sow were braided unevenly by the tiny marks of two inquisitive cubs. Their tangled line of travel continued across the slick tidal mud flats toward the river and disappeared in the glare of the lingering sunset. Little did I know that those tracks were just the beginning of yet another summer filled with unbelievable observations of maternal behavior, some of which were previously undocumented.

The next morning I made a quick tour of the five structures in camp and found that everything, including the gear we had left in storage, had made it through the winter in good condition. I untied the nylon lines that had held the roofs on the two small corrugated steel cabins during the 80-knot winter storms, and then I swept out the dust and debris that had accumulated in the cabins during the past nine months. The two canvas wall tents that Jim and I would live in for the next three months were clean and dry inside the thick-walled cedar-shake sauna, which was located beside a pond 200 feet south of the two cabins. An elevated food cache and a cedar-shake outhouse were situated closer to the cabins.

Our little camp was literally on the beach of McNeil Cove. During high tides, the waves lapped within 10 feet of my wall tent. The location of the camp was a discreet 1½ miles from McNeil Falls, but its position was primarily a practical necessity, as it was the only protected beach area where float planes could land and unload people and supplies. The camp dated back over twenty years to the early 1950s, when the U.S. Fish and Wildlife Service first began research on the bears. The area was

closed to hunting in 1955, and in 1967 the Alaska state legislature provided permanent protection by making approximately 72 square miles around the river a state sanctuary. Near
the mouth of the river we found the ruins of a small native
village from an unknown era and also a few broken-down cabins
of a small mining town that had a brief history in the 1920s.
No one knows how these early inhabitants affected the bears,
but today the McNeil brownies use the immediate area during
the summer with very little disturbance.

The bears congregate each summer on McNeil Falls to
fish for chum salmon. The falls is not a true falls in the usual
sense; rather it is a 300-foot stretch of violent white water
located a mile from the mouth of the river. This falls is created
by a descending series of conglomerate rock slabs in the riverbed, which also create many excellent fishing spots for the
bears along the riverbank. The chum salmon are not powerful
swimmers, so they are easy prey for the bears. I have seen as
many as thirty-five bears fishing on the falls at one time, and
sixty to seventy different bears will fish during July and August.

This concentration of fishing bears makes McNeil Falls an
ideal outdoor laboratory for the study of brown bear behavior.
Each summer these solitary, asocial animals are brought together into an intensely competitive situation and forced to
deal with one another in a social context. Aggression runs high
and bears interact constantly with one another. Every day it is
possible to observe and document behaviors at the falls that
would never be seen among solitary bears in open country.
Furthermore, the same bears return every year, which makes
it possible to follow the behavioral development of many
individuals.

We had the perfect location for a behavioral study that
was also a very timely undertaking. There was a total lack of
field studies on brown bear behavior. There had been ecological
field studies that dealt with home range, seasonal movements,

denning, reproduction, and food habits, but their behavior in the wild remained essentially unexplored. The importance of behavioral study was apparent in newspaper headlines of the bear-human conflicts in national parks and elsewhere. Clearly there was an urgent need for more understanding of the behavior of wild brown bears.

The coastal brownies at McNeil are the same species (*Ursus arctos*) as the interior grizzly bear. In fact, this single species includes all North American, European, and Asian brown bears, as well as all North American grizzlies. There are subtle morphological differences between the grizzly and brown bear, but size is the most obvious variable. Brown bears attain a size easily twice as large as the interior grizzlies because of their habitat and diet. The longer growing season, lush vegetation, and, most important, the abundant salmon of the coastal areas are all factors that contributed to the evolution of the large size of brown bears. In contrast, the short growing season and leaner diet of the interior grizzlies mandate a smaller body size.

The range of brown bears in Alaska follows roughly the coastal areas where salmon runs are plentiful in the summer. Grizzlies range over most other interior areas. The famous Kodiak Island brown bears are often thought to be the largest of all brown bears, but this reputation is partly due to the well-established trophy hunting on the island. Bears on the Alaska Peninsula, which includes McNeil River, reach sizes equal to those on Kodiak Island.

The beautiful weather on my first day added to my excitement about the summer. It was difficult to keep my mind on setting up and organizing camp when I could have been beachcombing or hiking on the alpine tundra of McNeil Head. I was thankful to hear Bill deCreeft's plane in the distance the next afternoon with Jim and the rest of our gear onboard. The plane was unloaded quickly, and when Bill took off he was our last contact with the outside world for the next five weeks. We had no radio equipment, and no planes were due until after the first of July, when an Alaska Department of Fish and Game technician would arrive in advance of the first tourists. These visitors would begin arriving the first or second week in July, when the bears began fishing on the falls. The number of both tourists and fishing bears peaks in late July and then dwindles by mid-August. Jim and I always relished our June isolation because all the subtle activities of McNeil carried on more naturally when few people were around.

The final leg of our annual trip from Logan, Utah, to McNeil was finally completed. We had flown from Salt Lake City to Anchorage with most of our essential gear. I'm sure the airline agents hated to see us approach with thirty-five odd-shaped boxes and packs. In Anchorage we rented a van and spent three days assembling food and more supplies before driving 220 miles down the Kenai Peninsula to Homer, a small, picturesque fishing village that would be our final stop. The only access to McNeil from Homer was by chartered float plane. We always used Bill deCreeft's Kachemak Air Service for the 140-mile flight across Cook Inlet. Bill's "Beaver" float plane could carry about one-half of our summer gear, so two trips timed on the high tide were necessary. Much of our life at McNeil was timed by the tides, which could rise and fall as much as 28 feet.

I couldn't have found a better man than Jim Taggart to work with me. He was a mountain climber, rock climber, runner, hiker, wilderness skier, naturalist, and a real adventurer. Jim explored the McNeil country more than anyone had ever done and was responsible for finding trails and routes into previously unexplored areas. Hikes and climbs with Jim gave me much more insight into the country than I would have had without him. Sometimes, while trying to keep up with him, I felt like five feet of his blond six-foot frame was all leg.

It was sunny and calm, so we wasted no time in unpacking and putting up our wall tents and cook tent. Only the occasional hammering of a nail or stake broke the silence and resounded off the bluffs. Now and then we were startled by a cracking explosion as the ice and snow covering the flats began to break up. Twice each day the warm tide advanced on the icy mantle and slowly tore small black icebergs from its edges. These rafts of ice, which were often 20 feet across and two feet thick, floated about on the currents of the tide and rammed into other ice formations and massive cornices of snow and accelerated the breakup of the ice-clogged cove. As the tide receded, the crashing and splintering of ice ended and the floating pieces silently rode the tide out to sea.

Early evening found us finished and sitting on top of our cabin with spotting scopes to observe the flats and bluffs. We saw no bears, but this was expected as very few bears arrive on the flats until mid- to late June. An occasional young bear or estrous sow with a boar consort might stop briefly on the flats, but generally we saw foxes, bald eagles, and very rarely moose, wolves, and wolverines.

The summertime concentration of salmon-fishing bears around McNeil gives the impression that there is an enormous permanent population of bears in the immediate area, but this is not the case. For most of the bears, McNeil is only one stop in their annual range, which may extend over 60 miles of country. From late August until late June of each year most of the bears are in other areas and very few are found around McNeil.

The annual activities of the McNeil bears center primarily on feeding on salmon, berries, and, to a lesser degree, the sedge that grows on the tidal flats. Their year usually begins in April when most bears emerge from their dens with the first signs of spring. This is the leanest time for the bears because no reliable food is available from the snow-covered land. They wander in search of the carcasses of animals that died during the long winter, and sometimes they prey on animals weakened by the winter and the old, sick, or newborn animals. Some beachcomb for beached whales, walrus, seals, and sea lions, and for clams and even seaweed. The sprouting grasses, forbs, and particularly the protein-rich sedges in June provide the first dependable food for the hungry bears. Then the vital salmon runs begin in late June, peak in July and August, and continue sporadically through the fall. Finally, a variety of berries are very important in late summer and fall until the bears den in November.

NEW BEGINNINGS

During the first two weeks of June the McNeil bears were still in the springtime wandering and searching stage of their annual schedule, so we rarely spotted any of our ursine friends. This was our time to relax and enjoy all aspects of McNeil from beachcombing and hiking to digging for clams and fishing for king salmon. Late evening always found us sitting on top of our cabin to take in the spectacular scenery and to watch for arriving bears.

I remember one evening in mid-June 1973 that was unusually enjoyable. A strong west wind had blown afternoon showers over the area, and straight curtains of rain had fallen in the dead calm that followed the wind. Heavy wetness hung in the air long after the rain had stopped and the drenched landscape was clean and crisp. The tide shimmered like a glowing mirror with sharp reflections of snow-covered mountains. The higher peaks held wispy remnants of the showers in protected bowls as streams of setting sun shot randomly through the clouds and set the lush green flats ablaze with color. McNeil was good to us and we were glad to be there.

The day was just fading into twilight when I noticed three bears ambling along the beach toward the sedge flats. A quick glance with my spotting scope revealed them to be a sow and a pair of two-year-old cubs. They were too far away for identification, so we scrambled off the cabin roof and headed down the beach for a closer look. We stopped at the end of the beach behind a sandbar to avoid being seen. From this closer vantage point, I could identify Goldie, a familiar McNeil sow of unknown age, and her two cubs. The two cubs lagged behind and played while Goldie strolled along toward the flats. Goldie was a shy sow who usually avoided people, so I decided we should remain hidden to avoid spooking her.

Goldie's cubs were average size for their age. They were great golden balls of fur with large puffs of shimmering hair obscuring their facial features and giving them teddy bear faces. The nose of one was a bit longer than the other, so they were temporarily dubbed Long Nose and Short Nose, but they were later named Clara and Rama. Early in the summer all cubs the age of Clara and Rama have dense fluffy coats, which make them appear larger than they really are. From a distance, they may appear almost as large as their mother, though she may weigh three or four times as much. (Many of the names of McNeil bears dated back to the early 1960s when personnel of the Alaska Department of Fish and Game chose names for various reasons. The majority of names originated in the first years of the Utah State University study [1970–1972] and were derived from names of friends, relatives, celebrities, and by what just seemed to fit a bear's personality. I continued this naming tradition that had no rhyme or reason.)

Goldie and her cubs grazed alone on the green sedge that rippled and swirled as the sea breeze danced patterns around them. Clara and Rama ate the protein-rich sedge voraciously, since they were no longer nursing from Goldie. As spring cubs (first year, newborn cubs) they depended mostly on milk but also ate sedge, fish, and other solid foods. The next summer, as yearlings, they utilized more solid foods and nursed much less. If Goldie now came into estrous, this would mean that she had not nursed the cubs recently, because lactation inhibits the onset of estrous in bears. When cubs have been weaned, the sow ceases to lactate and comes into estrous, thereby allowing a sow to leave her cubs only after they have become independent of her as a food source.

A sow retains her cubs until coming into estrous, but when a boar first approaches her, the cubs must leave immediately, because the boar may possibly kill them if given the chance. The sow will initially attempt to avoid the boar, but his persistent following soon forces her frightened cubs to flee, and even though they may rejoin their mother briefly, this forced separation is soon complete.

Occasionally a sow does not wean her cubs on schedule and retains them for a third summer. Because two sows who did this during my study had large litters of three or more cubs, I concluded that the cubs, with less milk and fish to go around, had remained dependent on their mother's milk and thus inhibited her estrous cycle and their weaning. However, this occurs sometimes even with single cubs, so perhaps other factors are involved.

Goldie's cubs would soon leave her, as a large dark boar appeared on the flats and approached the trio, which was conclusive evidence that Goldie was in estrous. Jim and I froze and crouched below the sandbar in excitement. I had never observed a sow leaving her cubs before, so we wanted to be careful and not let the boar detect our presence. Goldie became uneasy and began to move away after realizing that the boar was approaching, but she was no longer the excessively aggressive mother who had protected Clara and Rama for the previous two years. Before coming into estrous she would have run frantically or fought fiercely when confronted with a mature boar, but now she only halfheartedly avoided the boar and let her cubs run and fend for themselves. Goldie undoubtedly would have chased her cubs away from the danger of the dark boar had they not fled on their own.

Clara and Rama headed straight for the cliffs that overlook the sedge flats, running as only young bears can run: hind paws placed in unison in front of and on either side of where the front paws had landed together, rocking back and forth, a blur of flying paws and legs, a jolt and shimmer of fur with each planting of the paws. Once to the cliffs, they ran up one of the many bear trails that crisscross every slope in the area and settled on a spot with a good view and watched as Goldie meandered about the sedge flats with the boar in tow.

Goldie's behavior was typical of a mature sow's reaction to a boar consort. She was tolerant of his presence, but moved constantly as if attempting to escape. Her erratic movements eventually led them into the alders and out of sight. The cubs, alone for perhaps the first time, held their position on the bluff and eagerly surveyed the scene for their mother, as though not sure of a course of action without her direction.

The late-night sunset had given way to midnight twilight when Goldie returned without the boar and searched around for a few minutes before locating her cubs' outpost. Unlike younger cubs, who can seldom recognize their mother from a distance, Clara and Rama identified Goldie and descended to her. After some smells of recognition, all three grazed undisturbed in the gathering darkness.

The cubs were on their own when I next sighted them two days later. Even though I did not witness their final separation, the temporary split two days earlier proved to me that the process had been gradual. I had no idea how many times the cubs left and rejoined Goldie before making a permanent break, but based on other observations, I suspect that eventually a boar became persistent and separated Goldie from her cubs for a long enough period of time to effect a permanent separation. Whenever a sow and cubs of any age are separated, there is a possibility that they may simply fail to find one another again. This hypothesis is certainly a plausible explanation for the final separation and it is supported by the fact that Goldie had no active role in the temporary separation I witnessed. Nonetheless, there is a point at which a sow will have nothing more to do with her cubs. For instance, approximately three weeks after their departure, Goldie ran into her cubs at the falls and aggressively chased them away from the fishing area just as she would have done with any young bears. At no time during my study did I observe any familial attachments between a sow and her cubs after their final separation.

After two and a half years of consummate maternal protection, the sudden reality of being alone is a drastic change in a cub's life. During the years it spent with its mother, it became more and more confident as every new element in its environment was discovered, investigated, and, with the influence of the sow, most likely dominated. The loss of the sow's support destroys most of a cub's confidence in dealing with other bears. Years of learning by trial and error will be needed to rebuild the confidence it had had only one day before that final separation.

Goldie's cubs retained a little confidence by staying together after being weaned. Siblings, like Clara and Rama, invariably remain together for their first summer and sometimes a second summer, and in extreme cases for a third and fourth summer. This sort of sibling relationship is advantageous to the young cubs, who can present a common front and thus dominate single bears their own age or older. This is particularly important at the falls, where dominance can control fishing success.

Clara and Rama remained on and around the flats for several days. One afternoon, while surveying the sedge flats with my spotting scope, I noticed the golden pair roaming around the bluffs overlooking Mikfik Creek, where sockeye salmon were still running. I suspected that they were fishing the many shallow areas along the creek, so we gathered up our gear and set out in hopes of observing their first attempts at catching fish.

Mikfik Creek has a number of small waterfalls where the sockeye are vulnerable to eagles and bears. The shallow waters of the creek are ideal for eagle and bear predation, but the run is too small and sporadic to provide a reliable food source, so the brownies fish the creek only in passing. However, the eagles fish intensively for two or three weeks and often as many as twenty of the majestic birds can be seen on the bluffs overlooking the creek.

We spooked several eagles on the trail leading to our overlook at the first waterfall, where we settled down in an opening and waited for some action to develop. The deep pool below the falls was plugged with salmon that surged in schools every few seconds and jumped from the pool, flip-flopping about in the thin veil of rushing water. Hearing all this commotion, Clara and Rama came splashing upstream toward the falls with salmon splashing in every direction as they approached. Both cubs jumped around sporadically, then stood on hind legs to better view the concentrations of schooling salmon that swirled all around them, then they continued splashing and jumping here and there in total confusion. Clara stood with her head jerking wildly around not knowing which direction to go, for every frantic chase ended with the salmon escaping to deep water. Soon the shallow rapids were so full of fish that experience and technique were less important, and

Clara pinned a fish with her paws and grabbed at its back several times before securing it. Rama became aware of the success and ran to share the catch, but Clara had something else in mind and headed for the creek bank with her sister in pursuit.

It was a close race to dry ground with Rama alongside Clara, snapping incessantly at her catch. Once ashore, Clara crouched closely over her fish, pushing her back to her new adversary, who began the low guttural growl or deep-pitched infantile bawling that characterizes fish stealing in all young bears. It is similar to the insistent cry of hunger or begging that spring cubs emit when ready to nurse. Rama's method of stealing from her sister was considerably more brazen than it would have been with any other bear, as she did not fear her sister and made few of the appeasing gestures that accompany stealing from a higher ranking or less familiar bear. Clara held her fish to the ground and pivoted around it, while Rama clawed and snapped at the salmon. The grunting and roaring continued for several seconds before Clara tried to escape up a steep bluff with her salmon flopping so unguarded in her mouth that Rama easily grabbed the tail and began a short tug of war, which split the salmon and provided both cubs with a portion.

The cubs acted out this same stealing episode after catching two more salmon, but as they returned for a fourth try, both cubs suddenly lifted their noses to the wind and scanned the bluffs. They had detected our scent or that of an approaching bear. In order to keep from being seen, we froze in our places and moved only our eyes, knowing that the cubs could spot moving objects from great distances but their vision made it difficult for them to discern stationary forms. The cubs were unable to spot the source of an apparently strange scent and so they set out to find it. I was fairly certain that we were the objects of their curiosity, as the cubs would have run away in fear had they smelled another bear approaching. Their interest in us was not surprising because young cubs are often curious about humans, particularly cubs like Clara and Rama, who had had little previous contact with people. Their mother, Goldie, was a shy, retiring sow who kept her distance from people.

We were downwind from the cubs, but the breeze had apparently swirled around in the gorge and reached their super-sensitive noses. They began to circle around us and disappeared into the alders and were gone for several minutes as they moved downwind to locate our position. I had almost decided that they had become frightened and left, when suddenly the pair appeared on our right scarcely 15 feet away with their mouths frothing heavily as they held their noses high, sniffing and chewing the air closely, while meandering around and observing us for several seconds. This brief inspection apparently satisfied their curiosity, for they left quickly. While sauntering away, Clara stopped to roll around on a low-growing alder bush as though scent marking it or herself. The purpose of this behavior was not clear, but perhaps it was somehow connected with our meeting.

Jim and I were somewhat surprised at the boldness of the newly weaned cubs, because such behavior is more typical of yearlings and two-year-olds that still have their sows to back them up. Although a sow has enough experience to steer clear of people, yearlings and two-year-olds, who are accustomed to their mother's protective shield, are confident and very inquisitive about people. Occasionally, they may approach in a playful romp or even a menacing stalk, while smelling the air and sizing up the strange beings. This daring behavior by cubs can force a sow into protective action and cause her to be more aggressive than she would be in other situations with humans.

After returning to the creek, neither cub fished with the same wild abandon that had characterized their hungry attempts before detecting the two of us. Our presence made them so uneasy that frequent glances in our direction interrupted their efforts and reduced their success. Before another fish was caught, both cubs once again lifted their black noses to the wind and deciphered the breeze. Almost immediately they climbed the bluff on the far side of the creek and stopped on the summit long enough to survey the drainage below and to take another whiff of breeze before continuing their retreat. Their fearful behavior almost certainly meant that another bear was approaching. A moment later, Big Ears, a five-year-old male, came lumbering downstream with a spray of salmon parting in his wake. Big Ears smelled the cubs and, pointing his nose to the wind, he set out to find them. Clara and Rama had already traversed two small hills in their retreat, while stopping frequently and standing on their hind legs for a better view. Meanwhile, Big Ears loped up the drainage below them and made an unseen approach. The cubs obviously were afraid that the intruder would attempt to track them down but, unfortunately, by the time they had detected his presence, the pair had unwittingly cornered themselves on a dangerous precipice overlooking the creek sixty feet below. Suddenly looming out of the alders directly in front of them, Big Ears stood and casually scrutinized the pair, who were beside themselves with terror as they backed to the edge and scrambled to maintain their footing on the rocky cliff. Loose rocks broke free and showered into the water below as their hind feet slipped from the edge and scratched frantically for a new hold on the crumbling ledge. Then, just as one or both cubs seemed certain to fall, their antagonist turned and slowly retreated to the salmon stream and began to fish eagerly as the frightened cubs cautiously withdrew into the brush.

Big Ears had interrupted what was certain to be a successful fishing bout just to investigate two young cubs. This is a typical reaction, for despite their asocial nature, most bears are immensely curious about other bears. I inferred from his swift, undaunted approach to the cubs that Big Ears was capable of discerning from smell alone that they were young bears, for he certainly would not have approached a mature boar or sow with such confidence. In fact, he most likely would have avoided such bears in the same way Clara and Rama had avoided him. Critical perception of this sort may be hard to appreciate until one realizes that bears depend on their sense of smell for their very livelihood and can detect strong odors from miles away. Discriminating young bears from sows or boars by smell alone may well be routine.

Following their traumatic encounter with Big Ears, Clara and Rama disappeared and were not seen around the flats for several days. During their absence other newly weaned cubs appeared around the flats and provided us with continuing insights into the activities of weaned cubs. Two new arrivals were M.J. and Miss Kitty. Their mother was a sow named Leeland P., who never returned to McNeil during my study and was killed in the fall hunt of 1976. Her two-year-old cubs were incredibly beautiful. The light female cub, Miss Kitty, was a dazzling platinum blond with a small black bald spot that resembled a beauty mark below her right eye. The dark male cub, M.J., wore a deep auburn coat that parted in deep creases as the breeze whipped around him. They were not playful cubs and seemed quite businesslike as they grazed methodically on the lush sedge.

When these cubs spent long afternoons foraging on the flats, I often sat on the beach very near the pair and watched them for hours. Two summers at the falls with their mother had so habituated them to people that they totally ignored me. I tossed stones into a nearby stream to see if they might venture after the splashes, hoping for fish, but quick glances toward the splashes were their only reactions.

M.J. and Miss Kitty remained on and around the flats until late July. They preferred to stay on the coastal areas where mature bears seldom venture and where visibility is excellent, making it possible effectively to isolate themselves from their greatest danger—other bears. The proximity of our camp to the sedge flats created an atmosphere of potential conflict between people and young bears like M.J. and Miss Kitty, because the latter have no fear of humans and will venture boldly into camp if they smell food.

One day in late July several persons were roasting several fresh chum salmon on an open fire in the middle of camp, when M.J. and Miss Kitty came sauntering up the beach. As it happened, I had just climbed on top of the cabin to scan the flats, so I noticed the pair as they approached on the beach with their noses high in the air. Suddenly the smell of salmon hit Miss Kitty and she broke into a dead run and headed straight for the roasting salmon. Fortunately Jim was talking to campers near the fire, so I yelled at him and pointed out the approaching cubs. He ran at them waving his arms wildly in the air in an aggressive display that surprised the cubs, ended their foray, and sent them running in the opposite direction. Had Jim not been there the cubs might have run head-on into the midst of the campers and plunged into the roasting salmon to create a scene that would have been unbelievable, though I doubt if anyone would have been hurt.

That was the last time I saw M.J. and Miss Kitty in 1973, and as they did not return the next year, I was both surprised and glad to see them arrive on the flats two summers later in 1975. They grazed near each other but stayed apart and had nothing to do with each other. Close observation soon revealed that Miss Kitty was more ill-tempered and asocial than her brother and was probably responsible for the dissolution of their sibling bond. After playfully approaching Miss Kitty several times and being aggressively rebuffed each time, M.J. soon found more playful bears on the flats and abandoned all attempts to join his sister.

When a play partner was unavailable, M.J. often found

inventive ways of entertaining himself. He was particularly fond of body sledding on the thirty-foot snowbanks that remained under the bluffs surrounding the flats. After climbing to the top of a large drift, he usually flipped over onto his back and slid down feet first to the bottom, but later he contrived several different methods of sledding: feet first, head first, on his back, on his stomach, tumbling, rolling, and combinations of these. After several runs he took a short break to straddle the peak of the snowdrift and scratch his belly for several seconds.

M.J. found several suitable play partners on the flats, but most often he played with Flashman, a four-year-old male. They were the same age, about the same size, and very playful. While on the flats, the pair normally stayed apart except for prolonged play bouts, but later in the summer they became closer when they entered the competitive situation at the falls. They usually arrived at the falls together and would remain together and watch the fishing action from a distance if the falls was crowded with bears. M.J. and Flashman closely resembled brothers in a sibling bond, but unlike siblings, they did not cooperate with each other at the falls in fishing, stealing, or in agonistic encounters with other bears. They were primarily amiable play partners.

The history of M.J. and Miss Kitty leads me to believe that the temperament of siblings largely controls the length of time they remain together after being weaned. This is supported by the history of Red, White, and Blue, who were weaned in 1970. White, who was more asocial, left her siblings by 1971. In contrast, Red and Blue were amiable and playful and remained together in 1971, and rejoined in 1972 and again in 1973. White's early independence from her sisters was not made possible by greater dominance, because Red was more dominant than either White or Blue. White was simply a very solitary animal who preferred to be alone.

The difference in dominance and the temperaments of these sisters illustrates that there is no clear-cut similarity between a sow and her cubs. It is logical to assume that a sow might transmit, either genetically or through learning, some aspects of her disposition to her offspring, but such transference is not apparent. I was never able to detect any familial similarities. Instead, the observable aspects of dispositions are extremely variable in different bears and become more so after weaning.

The Mikfik sedge flats south of camp were created by the constant intertidal action at the mouth of the McNeil. At each high tide the ocean water advances into the cove and dams up the river, filling the entire area with fresh river water. The flats are alternately dry and flooded twice daily by up to 28-foot tides. This constant intertidal action creates tidal flats covered with deep, slimy sediments deposited during high tide. Dense growths of sedge (*Carex langlii*) cover the tidal flats on all areas protected from the pounding surf by sandspits or gravel bars. The sedge is a grasslike plant that is often called *sedge grass*, even though it is not a member of the grass family. It is a perennial that sprouts from rootstock as soon as the ice and snow melt in the spring. The thin pointed blades grow thick and luxuriant and cover the flats in a rippling green sea of vegetation that reaches a depth of three or four feet before maturing and turning brown in late July or August.

The sedge is important to the bears because in its early stage of growth this plant contains up to 26 percent protein (as percentage of dry weight). It is also the first vegetation that sprouts in the spring, which makes it the first abundant, reliable food source in the bears' meager springtime diet. The Mikfik flats produce an enormous crop of sedge in close proximity to

McNeil Falls, so this area is important in the early summer activities of many McNeil bears.

Young resident McNeil bears, such as Clara, Rama, M.J., and Miss Kitty, are often the first bears to arrive on the flats in June. They are invariably the offspring of McNeil sows that spend each summer fishing on McNeil Falls. This annual exposure to people habituates the young bears to humans and allows them to continue utilizing the area as adults without fear. However, there are wary bears at McNeil. Boars, many mature sows, particularly those with cubs, and many other cautious bears are rarely seen on the wide-open exposed tidal flats. Their avoidance of the coastal areas, which is where most bears are killed by hunters, enables them to survive to maturity. They are big and old because they fear man.

What causes a bear to become so wary of man? There are several arguments that could be proposed to explain this phenomenon. First, the bears learn that man is dangerous and that he can be avoided. Second, intense trophy hunting eliminates all the fearless individuals at a young age and leaves only those that are shy to survive to maturity. Third, bears are initially afraid of man on first encountering him because he is a strange being they have not encountered previously in their natural world. Later, this wary behavior may be reinforced in unprotected areas or, in parks and protected areas, repeated uneventful encounters with man may teach bears to overcome their initial fear of man. I suspect that the wariness of bears may result from a combination of all these factors.

Whatever combination of these factors produces wary behavior, the result is that in general younger bears are less wary than older bears. However, the younger bears occupy the coastal areas for reasons other than their fearless attitude toward man. Since adult bears have historically been the only enemies of young bears, the latter avoid adults as if they were predators. Therefore, younger bears remain near the coast, where they are less likely to encounter adult bears and where good visibility helps them avoid the few adults who do venture near the coast.

Of course the coastal area is prime brown bear habitat that is obviously very desirable for any bear, and so it is not as though young bears are driven into marginal habitat, as is the case with many animals in which the young disperse after weaning. It is possible that the only reason the young bears occupy the coastal area is to utilize the bountiful food sources found there. Nonetheless, I believe they receive these benefits by default, since historically this segregation pattern was probably reversed. That is, before man arrived on the Alaska Peninsula, mature bears probably roamed the coastal areas without fear of any living thing, whereas today they are relegated to inferior, remote habitat in order to survive.

This sort of regional segregation by bears might appear to be simply an efficient division of habitat and food for optimum use of available resources, but even though the habitat is effectively divided by this segregation, the latter does not occur because of necessity. For instance, bears do not divide up their use of sedge flats because there is not enough sedge to go around. On the contrary, only a small fraction of the sedge is eaten on the flats each summer. It is obvious that the Mikfik flats and other sedge flats could easily support many more bears in early summer, but the reluctance of most bears to risk the exposure of grazing on the flats reduces the utilization of sedge.

Because the sedge is the first major stand of vegetation that sprouts from the thawing landscape, one would expect the bears to eagerly welcome this abundant, nutritious addition to their meager springtime diets. Historically, however, they wait until one or two weeks before the start of the fishing season at the falls before grazing heavily on the sedge. By this time the protein content of the sedge has declined considerably because the percentage of crude fiber increases, which makes the plants' texture much ranker and presumably more difficult to digest. Why they wait for two weeks or more before grazing on the sedge, where they are located, and what they are doing during this time was a mystery I intended to explore.

It is common knowledge that brown bears eat very little in the spring, but they have no choice, as very little food is available. McNeil brownies probably wander around in the springtime and pursue a variety of scarce and unreliable foods. Since there are no abundant staple foods, such as the sedge, salmon, and berries, which appear later in the season, I suspect that they literally follow their noses and pursue any smells that might lead to potential food. These rare foods include carrion, beached whales and other marine animals, seaweed, clams, roots, tubers, and perhaps sick or weakened moose or moose calves.

Why do the bears continue to seek these unreliable food resources long after the rich sedge has sprouted on the flats? In order to discover some clues to this intriguing puzzle I needed to observe bear activities in early spring, while the land was still frozen and snow covered. Fortunately the late spring of 1975 provided me with this opportunity.

Each night in mid-June we sat on top of the cabin in the midnight twilight and surveyed the powder-white highlands and wondered where the bears were and what they were doing. We decided to wait until the flats had a good stand of sedge and then ski up into the McNeil drainage in hopes of discovering something about the bears' activities. What were they doing while the sedge grew rapidly and ungrazed on the flats?

By June 20 the sedge was several inches high and covered most of the flats. This was our signal. We planned and prepared for a seven-day ski trip and left camp that morning. The land above

400 feet elevation was still covered with snow, so we carried our cross-country skis strapped to our packs. Our early morning trail led to the falls where the McNeil, swollen by melting snow, thundered over the rocks. As the country upstream was entirely unexplored and unknown to us, we stopped at this familiar place and studied our topographic maps. We knew that picking a route would be tricky because most of the terrain from sea level to an elevation of 600 to 1,000 feet was almost entirely covered with dense alder thickets. Wherever these alders grow in uninterrupted stands they make travel on foot impossible. Even though the region ahead of us appeared to be impenetrable, we found that careful use of the detailed maps allowed us to pick an easy trail along streams and through clearings in the alders.

This first day of sloshing along streams and soggy meadows led us about eight miles up the drainage before we were stopped by an incredibly dense thicket of willows and alders. There appeared to be no direct routes ahead so we scrambled up a nearby ravine that was filled with snow and finally reached the top of a 400-foot ridge. From this vantage point we inspected the lay of the land and planned a route to the snow-covered alpine tundra on the mountains ahead.

The next day we skied up ravines, along ridges, through tiny valleys of snow, and eventually reached the tundra at the 1,000-foot level on the mountains to the south of the river. The sun was beating down directly on the snow and the radiating heat warmed and dried our cold, wet clothes, so we made use of the good conditions and skied another five miles over the alpine tundra and stopped in late evening to establish a base camp at 1,500 feet on the mountain slopes. We planned to make several day trips from this camp in search of bear activities.

The sun was setting through white jagged peaks to the

northwest as we finished anchoring our tent with heavy rocks to guard against the 80-mile-per-hour winds that can rake the area when strong Pacific storms move over the Peninsula.

Clean snowy peaks rose all around us, appearing distinct and somewhat unnatural against a contrasting backdrop of threatening blue-gray clouds made dark by the sunset. McNeil Cove was dwarfed at the base of the drainage by surrounding peaks and massive Kamishak Bay. Forty miles out in the Pacific the icy white cone of Augustine Volcano rose from the sea in solemn isolation.

The next day got off to an exciting start when a wolverine wandered into camp while we were eating breakfast. He made an unseen approach and then suddenly appeared from behind a scrubby alder about 10 feet from us and was calmly inspecting me when I first looked up and met his gaze. He so startled me that I jumped and yelled at Jim and caused the wolverine to spook and run. While we watched, he loped away at a smooth, easy pace, stopping every 50 feet or so to turn and inspect us again. He looked very much like a lanky yearling bear cub with a shaggy summer coat. He continued this intermittent retreat and disappeared over a ridge into the river valley below.

I hoped that this encounter with the wolverine was a good omen for our ursine quest, which had been unenlightening for the first two days. Our destination on the third day was the source of the McNeil, which was now only five miles southwest of our camp. As we set out across the slopes, the evidence of bear activity became increasingly apparent as tracks crisscrossed the snow in many places. The divides of several river drainages were in close proximity in this area and it was obvious that bears were traveling over these divides from one drainage to another.

Three miles from camp, while ascending the divide between the drainages of the McNeil and Little Kamishak rivers, we came upon a frozen alpine lake that was beginning to break up. From beneath its thawing ice a blue-green tributary of the McNeil surged to the surface and roared over smooth slabs of blue ice before disappearing again under a mantle of snow as it raced to the valley below. Despite the precarious state of the lake's ice, distinct tracks in the slushy snow revealed that a bear had just negotiated its way directly across the center of the lake. These fresh tracks continued up and over the divide ahead, so we hurried to reach the summit in hopes of spotting the bear.

The beautiful panorama from the crest of the ridge made us immediately forget about the tracks that vanished into the valley below. From this lofty outpost we looked at a succession of river valleys and mountain ranges that gradually faded toward the horizon in the hazy June air. Directly in front of us rose a group of 5,000-foot mountains that held numerous river-spawning glaciers on their steep slopes. The beginnings of the Little Kamishak River tumbled from a snow-covered glacier that extended up the mountain two miles to our right, and smaller glaciers filled the spaces between the many jagged peaks surrounding it. Our maps revealed that the McNeil originated from its glacial source on the opposite side of these peaks, so we skied toward a low saddle in the mountains ahead.

Big, black bumblebees zoomed overhead as we eagerly skied up the slushy melting snow to our final destination. Then, as we edged our way over the summit, the vertical slopes of the mountains plummeted before us and the deep, rugged drainage of the McNeil loomed suddenly into view. The head-waters of the river sprang from a large glacier that clung to the northwest face of a nameless mountain, and splashing white and lively the young McNeil fell from this lofty source and disappeared into the valley below.

We stopped and ate lunch on this spot and considered the bear sign we had observed that morning. It was obvious that many bears were moving from one river drainage to another through this high country. The snow-covered high country was used only for travel, as there was no sign of life, no reason that the brownies would be here. The bumblebees continued to zoom along overhead, but nothing else moved.

The mountains we sat on divided the Peninsula into two regions: the Aleutian Range on the east and the lowland lake country on the west. Most of the bears appeared to be traveling over these mountains from the lake country to the rugged river valleys that pour off the Aleutian Range on the eastern coast (for example, the McNeil). As the Aleutian Range covers only the extreme eastern coast of the Peninsula, it is quite probable that some McNeil bears seasonally occupy the more expansive lowland lake country in the spring and fall and move over the mountains into the McNeil drainage during July and August to fish the salmon run. The many migration routes in the mountains around us supported this hypothesis.

The wooded lake country is more protected from winter storms by the mountains, so the area receives less snow and spring arrives earlier than at McNeil. Some of our bears could find the early spring food prospects more rewarding along the many streams and lakes in that region. Such movements would be possible because the McNeil area and the lake country are only twenty-five miles apart, and travels of such length would be routine for the bears. Glenn in 1975, found that bears in the Black Lake region of the Peninsula 300 miles southwest of McNeil made similar migrations. In Glenn's study male bears ranged as far as 60 miles, though they averaged 26 miles between points of capture and recapture. Females traveled only half as far, but made movements as great as 36 miles. Considering this evidence, some bears could easily move from the lake country to McNeil and back again each year.

If many bears do make extensive migrations from other areas to fish McNeil Falls, then it is obvious why they do not forage on the new growth of sedge. That is, they have food sources in other areas that they utilize until moving to McNeil to fish. And in the meantime the valuable sedge goes untapped. The bears obviously cannot use every available food source, but they would be expected to go after the most nutritious items. What foods are better than sedge in early June is not known.

We awoke the next morning and found that the unusually nice weather had deteriorated. Low, wet clouds slid silently over the alpine slopes and reduced visibility to 30 feet or less. Two previous summers had left me very wary of McNeil weather, which can be stormy for a week or longer, so with this possibility in mind we cut the trip short and headed for home. Our old ski tracks were faintly visible at times in the wet snow, so we used these to decipher our original route through the dense fog that occasionally reduced visibility to near zero. The ceiling of low clouds rose above us as we skied down the slushy mountain slopes and reached the valley floor below.

We saw no bears during the two-day trip back to camp, and even the sedge flats were completely vacant when we arrived there in late evening. The sedge was growing rapidly, so the next day I collected samples for protein analysis from shoots that were eight to ten inches high. I checked around the flats and found no evidence of grazing by bears, which intensified my curiosity about their failure to utilize this food during its most nutritious stage of growth. They were ignoring a very valuable food source during a time of food scarcity. Even though we had just discovered evidence that bears had recently begun to move into the area from other regions, I was certain that some bears occupied the McNeil coastal area year round. Where were these permanent residents? What were they doing? What were they eating?

A number of bears prey on the red salmon during brief

runs that begin during mid-June on two or three tiny streams in the area. Red salmon are lake-system spawners, so they are not found on river systems, like the McNeil, that do not flow from lakes. Some McNeil bears fish for reds on Mikfik Creek, but this unreliable run does not represent a significant food source for them. The Mikfik salmon run is very small most years, and the sockeye often run the three miles of stream to Mikfik Lake quickly after each high tide. These sockeye salmon are at best morning and evening hors d'oeuvres for a few bears.

The Mikfik red salmon run approximately coincides with the early stages of sedge growth, but even though the bears fish on the creek within a mile of the sedge flats, most of them never utilize the sedge at this time. In contrast, they often alternate fishing with grazing after the fishing season begins at McNeil Falls, but for now the sedge grew and swayed silently with the sea breeze and was disturbed only by shore birds and an occasional flock of brant geese.

There were so few bears around that I suspected we could smoke some sockeyes without attracting any of them. Catching and filleting the reds on Mikfik Creek was the most dangerous part of the project for us, but the only close call came one afternoon when Patch Butt, a five-year-old male, came loping down the creek and surprised us just as we were preparing to leave. Moments before we were filleting reds beside the creek, but fortunately the fish were packed away in plastic bags by the time Patch Butt arrived, so we made a hasty exit and left him sniffing some bloody grass and rocks.

The next day we kept our reds on a string in the creek until we were ready to leave and then carried them intact onto the flats to fillet them. We watched carefully for bears until we left the brush and blind corners on the creek and reached the open flats, where we felt safe with our salmon. We dropped our guard a bit too soon, however, and it was just by chance that I happened to catch some movement behind us out of the corner of my eye and whirled around to see Blue, a seven-year-old female, about 20 feet away and approaching fast, with her eyes and nose glued to our salmon. Jim held her at bay by yelling and waving his arms at her, while I hurried across the creek and down the trail. Blue decided to circle around Jim, so he retreated across the creek and held her back by throwing big rocks in the water around her. Jim knew that warning shots would never even faze our old friend Blue, so he continued to yell and bombard her with rocks until I was halfway back to camp and out of sight.

Our salmon-smoking procedure was fairly simple. The fillets were cut into strips and soaked in a strong brine for about one hour. The strips were then skewered on green alder twigs and hung from wire racks in our smoker, which was nothing more than an old canvas wall tent with a wood stove inside. A few green alder twigs placed on top of the slow-burning stove provided just the right amount of delicate smoke flavor that we preferred.

The smoker was placed on a gravel spit away from camp so that if a bear got into the salmon it would not associate the fish with our camp proper. Fortunately we never had a single bear problem while smoking salmon four times in two summers. This vividly illustrates that few bears are in the McNeil area early in the summer, or we certainly would have lost some salmon.

When the flats continued to remain vacant, we decided to hike over to Akumwarvik Bay to see if any of our bears were grazing the massive sedge flats in that area. A trip to Akumwarvik along the beach was a difficult 15-mile hike that had to be timed between the tides, so we chose to survey the bay with spotting scopes from on top of McNeil Head. Jim had earlier discovered an easy route to the top of the mammoth headland that juts into the sea between McNeil Cove and Akumwarvik Bay. Alpine tundra covered about three square miles of this headland, which allowed us to move freely on top and use it as a fantastic observation post for surveying the McNeil landscape in every direction.

Our trail to McNeil Head left the gravel-covered beach north of camp and snaked up a deep, wide crack in the 50-foot conglomerate rock cliffs that border the sea and battle the breakers at every high tide. Wave erosion had sculpted smooth, rounded chambers and columns in the face of the cliffs. Above these vaulted formations and the breaking sea, our trail wound through waist-deep fields of fireweed, where the ground was soft and wet with the peaty remains of the lush vegetation. The tall fireweed and grasses gave way to varied tundra flora as we climbed higher along the cliffs, for here the Pacific gales roar up the sheer cliffs and cut across the headland to eliminate all but the low-growing tundra plants. Any stems protruding higher than one or two inches in winter are exposed above the thin crust of wind-swept snow, and ice crystals propelled by the wind quickly gnaw through these exposed stems and kill the plants. Wherever a slight ridge, slope, or depression provided a place for snow to drift or accumulate, scrubby, low-growing plants clung to the rocky soil with their stunted stems clipped and frayed evenly at snow level.

The trails that we followed were the well-defined bear trails that meander all over the McNeil landscape and form a network of travel routes that the bears maintain through habitual use. Most of the paths were worn down in a smooth continuous trail, but a few routes that were rarely used were nothing more than a series of round bald spots in the grass where bears had walked so uniformly that only their paw prints marked the trail. Even the trails on the barren tundra of McNeil Head were clearly visible through the lichens and crowberries.

While walking across the highest section of the headland, we spotted a large area of upturned earth where a bear had dug for roots or for an arctic ground squirrel. The ground was dug deeply on the slope in an area eight by three to four feet. Even though it was extensive, the dig had probably taken the bear only a short time to accomplish.

The most interesting observation about this dig was the fact that it was so rare. Jim and I had thoroughly explored this headland, and this was only the second dig by a bear that we had discovered during three summers of extensive hiking. This was a puzzling observation because brown bears supposedly utilize roots and ground squirrels when other foods are scarce, particularly in the spring. The paucity of bear activity was further demonstrated by the fact that we very rarely found a scat on this tundra and we never saw a bear grazing on the vast areas of crowberries found here. Later that summer, when we climbed the mountain that rises to the north of the falls, we were unable to locate any bear sign whatever on the berry fields found there. These observations emphasized the seasonal use of the area by McNeil bears and suggest that the permanent population may be small. The berries, sedge, ground squirrels, roots, and other vegetation appear to be significant food sources, but apparently they are not sufficient to maintain many bears the year around.

Farther down the trail we passed beside the 700-foot cliffs, where thousands of double-crested cormorants nest each summer. Hundreds of adults were soaring, landing, and then taking off again from their precariously placed nests on narrow ledges overlooking the sea. Hundreds of their broken eggs littered the top of the cliff—victims of predation by ravens and glaucous-winged gulls.

Behind us, camp was barely visible on the edge of McNeil Cove. The McNeil River drainage extended south and west into the Aleutian Range. We reached the southern edge of the headland and set up our spotting scopes on a smooth knoll overlooking the bays and flats 900 feet below. Akumwarvik Bay dominated the scene, and the several square miles of sedge flats around it were clearly visible past two smaller coves in the foreground. Ravines and ridges on the dark, barren mountains beyond the bay were still streaked with snowdrifts in surreal patterns of black and white that gave the country a most unusual appearance. Beyond this the glaciated mountains of Cape Douglas rose like great mounds of whipped cream through delicate wisps of ocean fog.

The sedge flats were three or more miles away, but our scopes zoomed in and clearly revealed that even these important flats had no grazing bears. We spent over an hour scanning the entire area, but we never located a single bear. Clearly our bears had not begun their annual pilgrimage to the McNeil area.

The initial arrival and congregation of bears on coastal sedge flats represents the first close contact of the year for most bears. In fact, the last time many of them rubbed shoulders may well have been on salmon streams the summer and fall of the previous year. They are so used to being alone after this long period of isolation that they must rehabituate to one another's presence. This annual process of habituation is dramatic. Initially, an older bear may not tolerate a younger bear on the flats, even though it may be as much as a half mile away. Later the same two bears may graze closely, and still later they may fish almost shoulder to shoulder at the falls.

One of the best examples of their early-season intolerance occurred on the flats in 1973 during my first summer at McNeil. A five-year-old female, named Red, had been grazing on sedge alone for two days when Big Ears, a five-year-old male, first arrived on the flats. Big Ears quickly spotted Red about one-half mile away and immediately began pursuit. The instant that Red started to retreat Big Ears broke into an easy lope, which spurred Red into a frightened run. The running pair traversed the flats and headed up the beach toward camp with Red about

100 yards in the lead. As soon as she passed camp, Red darted into the alders and disappeared. Big Ears then broke off his chase and ambled back toward the flats, apparently satisfied that he had driven Red away.

This was an extreme example, but it vividly illustrated how intolerant brownies are of one another early in the summer. Most bears will simply avoid others by spreading out widely on the flats to graze. However, at this early date Big Ears apparently preferred to have the entire area to himself and it was a simple task to remove Red.

Big Ears remained on the flats and grazed intently, so we decided to hike down and inspect him at close range. I suspected that he would be a bit spooky on his first day back at McNeil, so we edged our way along the beach and hid behind sandbars to keep out of sight. We managed to sneak within about one hundred fifty feet of Big Ears before we entered his field of vision. He was grazing with his back to us, so we crept along slowly, being careful to freeze if he lifted his head or looked around. It was not long before he smelled us and started to look around carefully. We crouched behind clumps of beach grass so that Big Ears could not quite spot us. We were visible but as long as we did not move a muscle he was unable to locate us. After a thorough look around, he hesitantly returned to grazing, but the next puff of breeze brought more evidence of our intrusion and once again he stopped to look around. It seemed incredible that he was unable to perceive our stationary forms, but he smelled us and knew we were there, so he cautiously began moving away. He continued to look around while retreating to the alders on the edge of the flats, but we remained still and he never succeeded in locating us that day. This was further evidence that the bears' vision does not discern stationary forms very well.

Big Ears' solitary reign of the flats was broken the next day by the arrival of a thirteen-year-old sow named Hardass, who had received her unusual name the year before when two successive immobilizing darts failed to penetrate her rump. Hardass was a power to be reckoned with, but Big Ears either was unaware of this or else was just confident enough to attempt to push the old girl around. At any rate, he grazed slowly in her direction and then approached in a menacing stalk with his head down and ears back. Hardass had initially ignored him, but when he came closer she suddenly stopped grazing and lunged after him in an outright charge. Big Ears' bullying act fell apart on the spot and he turned tail and ran for his life. Needless to say, Hardass had no further trouble with Big Ears.

The aggression and intolerance that the first bears on the flats displayed began to diminish as more bears arrived in the area. It generally took only a day or two for most bears to become accustomed to sharing their grazing area with others. This process of habituation proceeded more rapidly as the number of bears increased. When new arrivals found several other bears already grazing on the flats, they generally joined the others without an extended period of adjustment. By the third week of July, it was common to see ten bears grazing calmly on the flats. Most often they were widely spaced over the area, but occasionally they would all be clumped on a three- or four-acre section. Aggression was minimal because any bear that felt crowded could simply graze in the opposite direction and avoid the others.

All of the action was taking place on the flats, but soon the fishing season would begin at the falls. It was time for the earliest chum salmon to begin running the McNeil, so Jim and I set out one afternoon for a preliminary check of the falls. The trail to the falls crossed Mikfik Creek at its mouth in a spot where our hip boots were just high enough to keep us dry as we inched across on tiptoes. The well-used path then left the flats and crossed over an area of alpine tundra before descending to the falls.

An elevated vantage point on the bluff above the river provided us with a complete view of the falls. Just upstream from us the blue-green McNeil came racing and foaming around a bend and roared inexorably toward the falls. Just below us a series of conglomerate slabs breached the riverbed and turned a 100-yard stretch of the McNeil into a thundering white-water spectacle.

The upper falls was the main fishing area for the bears because here the water cascaded over and around large slabs of conglomerate rock and created numerous good fishing spots where the salmon were highly vulnerable. The number-one fishing spot, the best spot on the falls, was on the opposite bank directly across from us. The far side of the river was a restricted area that was off limits for people because the more wary bears, particularly the large boars, used the area exclusively. There was a large rock outcropping, which we called the center rock, in the center of the upper falls. There were numerous fishing spots around this rock and it was a center of fishing activities on the falls. The middle and lower falls areas had fewer fishing locations and the bears did not utilize these areas as intensely.

Our summer-long observation spot, or the "cave" as it is often called, was parallel to the upper falls about 30 feet from the riverbank. It was not a cave at all, but rather a small eroded spot on a knoll beside the falls. However, it did provide us with protection from the rear and shielded us from the stormy east wind and rain. We always moved on top of this knoll in good weather for a commanding view of the falls. Several bear trails passed beside the cave en route to the fishing spots on the upper falls. The more casual bears routinely sauntered past us on these trails at distances as close as ten feet and often did not even glance in our direction.

We pulled our small handmade chairs out of the cave and sat down on top of the knoll to watch the falls for a while. Careful observation of the white water and deep pools on the upper falls revealed that the chum salmon had not begun their summer run. Any bears that might stop by the falls to check for salmon would not stay long.

The sedge flats would continue to be the center of activity until the salmon run began. The adult bears always grazed alone in their asocial, solitary existence. However, some young bears preferred to be with other young bears at times. As I mentioned before, newly weaned siblings remain together, but other unrelated juveniles of similar age will also stay together in loose association on the flats. It was not unusual to see a group of three, four, or five of these young bears grazing and playing together for days at a time. Other juveniles, like Big Ears, were more aggressive and preferred a solitary existence. This difference in bears seems to be simply a matter of individual temperament.

Most of the young bears at McNeil who have grown up around people can be downright brazen when it comes to dealing with them. A classic example of this involved the three siblings that were appropriately named the Marx Brothers. This fearless, mischievous trio were together from 1974 to 1979. They were the three survivors of a group of five (yes, five) spring cubs that Red Collar, a large and dominant McNeil sow, had brought to the falls in 1974. Red Collar did not wean them on schedule and kept them with her for a third summer. This gave the cubs one extra summer of experience and confidence building with their powerful mother.

The Marx Brothers began their most outrageous exploits in June 1977, when they appeared on the flats as newly weaned three-year-olds in close sibling bonds. I was able to observe them that year during a two-week visit in June and again for a week in July. (As it turned out, my visits in the years following my study resulted in some of the best information on sibling relationships. This was simply the result of greater numbers of cubs being produced and weaned.) Red Collar had never brought her cubs close to camp, but the proximity of the latter to the flats combined with their sensitive noses and tremendous curiosity soon brought them around camp.

One evening they were grazing on the flats, so I hiked down with my camera and photographed them at close range in the soft evening light. They ignored me and grazed undisturbed until a distant splash in the river sent all three racing through the shallows after an early running salmon. It was a wild race that could not be won, as all three were soon fighting over the single fish in a chorus of roars, growls, and grunts. Moments later they were grazing and playing again with the setting sun casting halos around their fluffy coats. They seemed not to notice me.

Later that evening they came ambling up the beach toward camp in the midnight twilight, romping and playing as they walked. Then, instead of continuing down the beach, they walked right into camp and proceeded to engage in an extended play bout in our front yard. We tried to discourage any bear from entering camp, so I yelled and clapped my hands and waved my arms at them, only to have them stop and casually look up as though they were saying, "Yes, can we help you?" I summoned the assistance of five tourists in camp and together we managed to shoo them away and down the beach.

The next day was unusually warm and late morning found me sitting in front of the cabin basking in the sun and reading. Harpo, who was the most audacious of the brothers, had spotted or smelled my tent while sauntering by the edge of camp and decided it was worthy of further inspection. Fortunately,

I looked up just as he started tugging on the rain cover, and so my screaming charge down the path came in time to save my tent. However, the next day while I was out on a hike, another young bear chewed on my rain cover, broke two tent poles, tore out the mosquito netting, and dragged out my sleeping bag before a hesitant tourist could scare him away. They apparently singled out my tent as it was the first item they encountered on the edge of camp.

Harpo was capable of considerable finesse during his forays into camp. Not long after my tent had undergone alterations, Larry and Mo Aumiller, the Alaska Department of Fish and Game employees who administered the McNeil Sanctuary, arrived for the summer with lots of fresh food, which they stored in a *cool box* in the ground beside their cabin. Two days later, while everyone was out of camp, Harpo lifted the lid of their cool box and neatly dined on a five-pound brick of cheese, luncheon meats, and some moose burger. He did not eat or disturb the remaining food and even the plastic package of lunch meat had been opened, the meat removed, and the package left behind. When Larry and Mo returned after a brief absence, they unknowingly found Harpo licking his lips on the beach in front of camp.

I first met Larry and Mo Aumiller in July 1976. McNeil couldn't have been blessed with a nicer pair. Soon, I learned that Larry was a quiet, easygoing, and sensitive man of slight stature with curly black hair and beard and dark, expressive eyes that perhaps revealed some of his deep love for Alaska and the McNeil bears. Mo was a delightful, jovial, blond Australian lady with a no-nonsense practicality that seemed perfect in every situation. I remember asking her how in the world an Aussie found her way to McNeil River, Alaska. Without a moment's hesitation she said, "the same way a Texan does, I expect." (I was born and raised in Texas.) Whenever I planned a visit back to McNeil, I looked forward to seeing Larry and Mo as much as the bears and the country.

During my visits, I tried to help Larry and Mo learn the identities of the resident bears. It was their diligent observations and data collection beginning in 1976 that helped me to keep in touch with events at McNeil. The Marx Brothers were early favorites of Mo and Larry, and so they kept me updated on their activities.

Before I left in 1977, Groucho, Chico, and Harpo had their greatest adventure at the expense of some commercial fishermen. These innocent visitors anchored their boat in the river at high tide, then left it moored beside a sandbar at low tide and hiked upstream to go sport fishing. While they were gone, the Marx Brothers all boarded the boat and proceeded to rummage around, perusing every section of the vessel. The crew returned and caught them red-handed before any damage was done, but imagine their surprise on finding three fair-size brown bears on board!

Despite all this early summer mischief, the Marx Brothers did not become real problem bears that summer or in later years. Unfortunately this was not always the case with young bears around camp. White started out much the same way the Marx Brothers did, but she went on to become a problem.

Most bear problems begin with human error and White was no exception. Her trouble began when she was four. She found and ate some salami that a careless tourist had left just outside the door of his tent. This treat prompted her to return to camp, and it was not long before she was peering in cabin windows and trying to enter tents. Twice she was shot in the rump with bird shot from a 20-gauge shotgun, but this failed to deter her, so the third time larger shot was used. This caused some injury and she rested without moving for two days on a nearby bluff. Soon she was back to normal, but she never again entered camp.

White undoubtedly associated people with the pain of being shot, but this did not cause her to be afraid of us. In fact, she continued to be one of the more fearless McNeil bears. She had the unusual habit of following people around with a rather menacing look and approaching to very close range to inspect them. She stopped this curious behavior for three summers, but it appeared once again in 1976 when she had spring cubs at age eight. Unsuspecting tourists were particularly unnerved when she approached with her cubs to within forty feet and stood on hind legs to look at them.

Not all young McNeil bears were as fearless as White and the Marx Brothers. Each summer there would be two or three juveniles who would invariably run from us on the flats in the early summer. Every year these bears had to habituate themselves to the presence of people in the same way that they had to get used to other bears.

Richard Ellis

THE PREDATORS

Everything about a big sword-
fish is massive. It is a heavily built fish, especially in the "shoulders," which taper off to a narrow caudal peduncle before the
crescent-shaped tail flares vertically, bisecting the plane of the
lateral keels. The fins are heavy and fleshy, not thin and bony-
ribbed as they are in many other fishes. Like most other fishes,
a swordfish is able to regulate its density to the ambient water
pressure by the use of a swim bladder, a gas-filled organ that
the fish can inflate or deflate autonomically, and bring its overall
density close to that of the surrounding water. This means that
a fish that is stationary in the water is weightless; it displaces

exactly its own weight in water, and since it is submerged, it is integrated completely in its medium.

The broadbill swam slowly at fifty fathoms, sculling in the dark water with measured, steady sweeps of its crescent tail. It presented a majestic, solitary aspect, purposeful but unhurried. The great fish braked gently by imperceptibly increasing the angle of its pectoral fins. Its own momentum carried it forward for another ten yards before it glided to a stop. It hung suspended in the darkness, sensing movement from far above. With the limited light reaching this level, even its large eyes were of little use.

The swordfish again altered the angle of its pectorals, dropping the trailing edge, and thereby giving it a gradual angle of lift. With a twitch of its tail, it began to swim toward the surface, sensing with all its faculties the movement that was taking place above it, close to the surface. As it rose, its swim bladder increased in size with the corresponding decrease in pressure. The swim bladder is also a resonating or "hearing" organ, and the surface movement was being transmitted to the fish's brain by the conductivity of the gas in the bladder, and through the water itself, to the lateral line.

Now at a depth of twenty fathoms and rising, the swordfish saw the shadow of an immense school of mullet, silhouetted against the surface sunlight, as they wheeled and turned in a tight, flowing formation. As the swordfish got closer, the mullet school collectively became aware of its presence, and tightened ranks, a common practice among schooling fishes or herding animals; in the face of danger, they crowd even closer together. This may be because the predator cannot concentrate on any single individual in a milling, tightly packed throng, and will be unable to attack accurately. For a predator that would bite its prey as a means of capture, the closing of the ranks might work. For the swordfish, the packed school presented a perfect target.

   Now the swordfish closed with the school, driving directly
at the moving center of the mass of seething, twisting fishes.
In the sparkling water of the ocean's surface, the moving mullet
shone; a giant, multifaceted entity, glinting now silvery green,
now iridescent blue as the angle of the school shifted in the
sunlight. As its sword broke the perimeter of the school, the
swordfish began to throw its head violently from side to side.
The tightly packed fish were slashed and smashed, cut in half,
decapitated. There was no escape for the mullet that chance
had selected as the swordfish's victims, for even with death in
their very midst, they still resorted to the only defense a schooling fish has: the mass of the school.
   Pieces of mullet began to sink, and the hungry swordfish
gulped them down. The feeding was as easy and casual as the
killing had been violent, and many of the mutilated fish were
missed as they floated down through the bright waters, eventually to come to rest on the bottom, one thousand feet below.
There was now blood in the water, in addition to the broken
and dying fish.
   The combination of the blood and the dying, thrashing
fish did not go unnoticed. Some 500 yards from the scene of
the slaughter, another predator reacted to the signals of the
fluttering of dying mullet. This predator could smell blood in
the water from great distances, and it began to seek out the
source of the disturbance. Banking sharply, it turned in the
direction of the swordfish. Here too were strong, curved pectoral fins, but they were wider than those of the swordfish,
giving this animal more maneuverability. This hunger had the
same flattened caudal keels, and a tail of almost the same shape:
crescent-shaped and equally lobed. Despite their similarities,
these two predators were very different. Smooth skin and a
bony skeleton, an absence of teeth, and but a single weapon,
its sword, characterized the swordfish. The approaching animal,
while sharing the fusiform body shape and the horizontal keels,

was a very different creature indeed. No smooth skin, but rough, pointed scales covered its body, and in place of the swordfish's lateral line, it was covered with a network of sensory organs, making it perhaps the most sensitive and responsive animal in the world. It could sense and smell disturbances or potential prey at great distances and, contrary to popular belief, it could also see quite well. It had a large, sensitive eye, and would rely on vision to a great extent in its impending confrontation with the broadbill.

Still, the feeding swordfish was too far away to see, and even the best vision cannot penetrate four hundred yards of water. Crossing and recrossing the line of stimuli like a quartering hunting dog, the approaching fish received the messages with greater and greater clarity, and soon it was marking a straight course toward the swordfish. Its pace quickened as it neared the source of the disturbance, and the powerful strokes of its tail propelled it through the water with increasing speed. Within one hundred yards of the swordfish the animal approached steadily, its gaping, snaggletoothed mouth open, its lidless black eyes seeking the blood-smell and the flapping fish. Its sharply pointed snout, recurved teeth, and deep blue color identified it immediately: it was a mako shark, perhaps the most beautifully proportioned of all the sharks, and one of the fastest and most dangerous animals that swim.

It closed quickly, mindlessly picking up and swallowing the pieces of broken and torn mullet that had been carried by the same current that brought it the smell. This thousand-pound mako needed more than a few morsels of mullet to satisfy its hunger, and soon the swordfish was in visual range. The two great fish were about equal in weight, but the shark was more agile, and far more deadly at close quarters. Its appetite whetted by the mullet, the shark instantly launched its attack—no passes, no bumping, no preliminaries. Its upper teeth, already bared, were further protruded by the jaw mechanism that enables a shark to bite deeply despite the underslung position of its mouth. The pointed snout was forced upward by the emergence of the knifelike teeth, and the shark raked the flank of the swordfish as it flicked its sickle-shaped tail and dove downward in an attempt to escape. Parallel scars appeared on the side of the swordfish, and blood welled out of the rips in its sensitive skin. As the swordfish dove, the blood turned brown and then black, for red is the color with the shortest wavelength, and therefore the first color to go as the water cuts off the spectrum. The shark followed quickly, close on the tail of the diving swordfish.

The battle between these two giant fishes contained blind spots for both hunter and hunted. As long as the mouth of the shark was directly behind the vertical, flexing tail of the swordfish, the shark could get no purchase. For it to turn on its side would have meant to lose speed. The swordfish could not dive deeply enough to escape. At about one hundred fathoms, the shelf loomed ahead, forcing the swordfish to bank sharply. Surprised by this sudden maneuver, the shark checked by dropping its broad pectoral fins, and the swordfish sprinted along the drop-off wall, a black plume of blood trailing from its wounds. Suddenly a volcanic outcropping loomed ahead of the swordfish, and it was forced to change its course again. This time, it had to turn almost ninety degrees and, as it threw itself over, it presented its wide, vulnerable flank to its pursuer. Remaining broadside to the charging mako would have meant certain disaster for the swordfish, so it made an even tighter turn, to bring its only weapon into play. The shark tried to brake, but succeeded only in coming obliquely upon the four-foot sword. Slashing as it came about, the swordfish defended itself against the driving mako. The two terrible weapons, the mouth of the shark and the sword of the billfish, met in a silent clash. As the sword entered the mako's mouth, the shark closed its powerful jaws on it. Although strong enough to crush the sword, they did not do so before the point had pierced the soft inner skin of the shark's mouth. The two fish were attached now, the mako impaled on the sword, its teeth and jaws grinding as it writhed furiously in an attempt to pull away. Flexing its supple body into a full horseshoe curve, the mako somersaulted backward in the water. The swordfish, its only weapon splintered and useless, turned and headed for the surface.

Recovering, the mako saw the bulky shape of the swordfish as it shot toward the light. The open ocean gave no escape, no place to hide. In order for this battle to end, one of these fish would die. They were both wounded now—the swordfish raked by the teeth of the shark and its sword broken, the shark with a gaping, bleeding wound below its black, expressionless eye.

With astonishing speed, the two fish rose in tandem from the depths, through the greenish blue of the shallower waters and up into the clear green water of the surface. The distance between them remained the same as the swordfish burst out of the water, shimmering purple-bronze in the hot sunlight. Throwing off a spray of white water and bright red blood, the swordfish arched and twisted as if it were trying to leave the ocean forever—before it hit the water again. The forces of nature and gravity conspired to draw the fish back to the water with a gigantic splash, but the tail of the fish had not disappeared below the roiled surface when the mako appeared in airborne pursuit. Iridescent ultramarine above and snowy white below, sharp contrast to the vivid trail of blood that flowed from its injured mouth, the flying shark presented an awesome spectacle, clearing the water by at least twenty feet before it reentered in a crashing shower of spray. Diving again, the swordfish twisted, trying to shake its dogged pursuer. As the swordfish circled, the mako cut across the diameter of the circle, again coming broadside upon the fish's vulnerable flank. With a great thrust of the powerful muscles that move its tail, the swordfish lunged forward. The shark's reaction was instinctive and immediate: it did the same thing, and this brought its gaping mouth within inches of the flattened base of the broadbill's tail. In a second, they both lunged again, but this time the flick of the swordfish's tail smashed into the open mouth of the shark. The teeth of a shark are not fixed in the gums or in the cartilage of the jaw itself, but attached to a membrane inside the jaw. Some of the two-inch teeth were knocked out by this desperate swipe of the swordfish's tail, but the tail stock was driven deep into the shark's mouth. Reflex action closed the mouth of the mako, and both fish began to squirm and twist—the swordfish to escape the terrible tearing jaws, the shark to exert maximum pressure on its victim.

The water churned red with blood and froth, and with one frantic lunge, the swordfish wrenched free. The mako had triumphed, however, for it had bitten through the flesh and bone of the caudal peduncle of the swordfish, and severed its tail. Streaming blood, the once-mighty fish tried again to escape, but without its tail it had no power of propulsion. It fluttered its pectoral fins weakly and rolled its great black eye, trying to keep the shark in view as it came up beneath the broadbill's belly. Stretching its mouth open to a remarkable degree, the mako bit deeply into the soft flesh of the swordfish. Again the shark convulsed its entire body and the hard muscles rippled as it tore a great mouthful of meat from the body of the swordfish. The shark swallowed this twenty-pound gobbet whole, circled again, and hit the fish from the other side. Blood spread through the water as the mako fed on the torn, living carcass of the swordfish.

Drawn by the blood in the water and the sounds of the battle, other sharks arrived—oceanic whitetips, accompanied by their attendant pilot fish. The sharks, smaller and lighter than the mako, and a dirty ocher color, with white-tipped dorsal, pectoral, and tail fins, are numerous and ever present at the scene of a disaster in the sea. They too tore great bites out of the ragged body of the swordfish, while the pilot fish picked up the shreds. The head with the splintered sword sank to the sand at the base of the drop-off, there to be picked clean by crabs and other scavengers.

Edward Abbey

THE SERPENTS
OF PARADISE

The April mornings are bright, clear and calm. Not until the afternoon does the wind begin to blow, raising dust and sand in funnelshaped twisters that spin across the desert briefly, like dancers, and then collapse —whirlwinds from which issue no voice or word except the forlorn moan of the elements under stress. After the reconnoitering dust-devils comes the real, the serious wind, the voice of the desert rising to a demented howl and blotting out sky and sun behind yellow clouds of dust, sand, confusion, embattled birds, last year's scrub-oak leaves, pollen, the husks of locusts, bark of juniper. . . .

Time of the red eye, the sore and bloody nostril, the sand-pitted windshield, if one is foolish enough to drive his car into such a storm. Time to sit indoors and continue that letter which is never finished—while the fine dust forms neat little windows under the edge of the door and on the windowsills. Yet the springtime winds are as much a part of the canyon country as the silence and the glamorous distances; you learn, after a number of years, to love them also.

The mornings therefore, as I started to say and meant to say, are all the sweeter in the knowledge of what the afternoon is likely to bring. Before beginning the morning chores I like to sit on the sill of my doorway, bare feet planted on the bare ground and a mug of hot coffee in hand, facing the sunrise. The air is gelid, not far above freezing, but the butane heater inside the trailer keeps my back warm, the rising sun warms the front, and the coffee warms the interior.

Perhaps this is the loveliest hour of the day, though it's hard to choose. Much depends on the season. In midsummer the sweetest hour begins at sundown, after the awful heat of the afternoon. But now, in April, we'll take the opposite, that hour beginning with the sunrise. The birds, returning from wherever they go in winter, seem inclined to agree. The pinyon jays are whirling in garrulous, gregarious flocks from one stunted tree to the next and back again, erratic exuberant games without any apparent practical function. A few big ravens hang around and croak harsh clanking statements of smug satisfaction from the rimrock, lifting their greasy wings now and then to probe for lice. I can hear but seldom see the canyon wrens singing their distinctive song from somewhere up on the cliffs: a flutelike descent—never ascent—of the whole-tone scale. Staking out new nesting claims, I understand. Also invisible but invariably present at some indefinable distance are the mourning doves whose plaintive call suggests irresistibly a kind of seeking-out, the attempt by separated souls to restore a lost communion:

"*Hello* . . . they seem to cry, *who . . . are . . . you?*

And the reply from a different quarter. *Hello* . . . (pause) *where . . . are . . . you?*

No doubt this line of analogy must be rejected. It's foolish and unfair to impute to the doves, with serious concerns of their own, an interest in questions more appropriate to their human kin. Yet their song, if not a mating call or a warning, must be what it sounds like, a brooding meditation on space, on solitude. The game.

Other birds, silent, which I have not yet learned to identify, are also lurking in the vicinity, watching me. What the ornithologist terms l.g.b.'s—little gray birds—they flit about from point to point on noiseless wings, their origins obscure.

As mentioned before, I share the housetrailer with a number of mice. I don't know how many but apparently only a few, perhaps a single family. They don't disturb me and are welcome to my crumbs and leavings. Where they came from, how they got into the trailer, how they survived before my arrival (for the trailer had been locked up for six months), these are puzzling matters I am not prepared to resolve. My only reservation concerning the mice is that they do attract rattlesnakes.

I'm sitting on my doorstep early one morning, facing the sun as usual, drinking coffee, when I happen to look down and see almost between my bare feet, only a couple of inches to the rear of my heels, the very thing I had in mind. No mistaking that wedgelike head, that tip of horny segmented tail peeping out of the coils. He's under the doorstep and in the shade where the ground and air remain very cold. In his sluggish condition he's not likely to strike unless I rouse him by some careless move of my own.

There's a revolver inside the trailer, a huge British Webley

.45, loaded, but it's out of reach. Even if I had it in my hands I'd hesitate to blast a fellow creature at such close range, shooting between my own legs at a living target flat on solid rock thirty inches away. It would be like murder; and where would I set my coffee? My cherrywood walking stick leans against the trailerhouse wall only a few feet away but I'm afraid that in leaning over for it I might stir up the rattler or spill some hot coffee on his scales.

Other considerations come to mind. Arches National Monument is meant to be among other things a sanctuary for wildlife—for all forms of wildlife. It is my duty as a park ranger to protect, preserve and defend all living things within the park boundaries, making no exceptions. Even if this were not the case I have personal convictions to uphold. Ideals, you might say. I prefer not to kill animals. I'm a humanist; I'd rather kill a *man* than a snake.

What to do. I drink some more coffee and study the dormant reptile at my heels. It is not after all the mighty diamondback, *Crotalus atrox*, I'm confronted with but a smaller species known locally as the horny rattler or more precisely as the Faded Midget. An insulting name for a rattlesnake, which may explain the Faded Midget's alleged bad temper. But the name is apt: he is small and dusty-looking, with a little knob above each eye—the horns. His bite though temporarily disabling would not likely kill a full-grown man in normal health. Even so I don't really want him around. Am I to be compelled to put on boots or shoes every time I wish to step outside? The scorpions, tarantulas, centipedes, and black widows are nuisance enough.

I finish my coffee, lean back and swing my feet up and inside the doorway of the trailer. At once there is a buzzing sound from below and the rattler lifts his head from his coils, eyes brightening, and extends his narrow black tongue to test the air.

After thawing out my boots over the gas flame I pull them on and come back to the doorway. My visitor is still waiting beneath the doorstep, basking in the sun, fully alert. The trailerhouse has two doors. I leave by the other and get a long-handled spade out of the bed of the government pickup. With this tool I scoop the snake into the open. He strikes; I can hear the click of the fangs against steel, see the strain of venom. He wants to stand and fight, but I am patient; I insist on herding him well away from the trailer. On guard, head aloft—that evil slit-eyed weaving head shaped like the ace of spades—tail whirring, the rattler slithers sideways, retreating slowly before me until he reaches the shelter of a sandstone slab. He backs under it.

You better stay there, cousin, I warn him; if I catch you around the trailer again I'll chop your head off.

A week later he comes back. If not him, his twin brother. I spot him one morning under the trailer near the kitchen drain, waiting for a mouse. I have to keep my promise.

This won't do. If there are midget rattlers in the area there may be diamondbacks too—five, six or seven feet long, thick as a man's wrist, dangerous. I don't want *them* camping under my home. It looks as though I'll have to trap the mice.

However, before being forced to take that step I am lucky enough to capture a gopher snake. Burning garbage one morning at the park dump, I see a long slender yellow-brown snake emerge from a mound of old tin cans and plastic picnic plates and take off down the sandy bed of a gulch. There is a burlap sack in the cab of the truck which I carry when plucking Kleenex flowers from the brush and cactus along the road; I grab that and my stick, run after the snake and corner it beneath the exposed roots of a bush. Making sure it's a gopher snake and not something less useful, I open the neck of the sack and with a great deal of coaxing and prodding get the snake into it. The gopher snake, *Drymarchon corais couperi*, or bull snake, has a reputation as the enemy of rattlesnakes, destroying or driving them away whenever encountered.

Hoping to domesticate this sleek, handsome and docile reptile, I release him inside the trailerhouse and keep him there for several days. Should I attempt to feed him? I decide against it—let him eat mice. What little water he may need can also be extracted from the flesh of his prey.

The gopher snake and I get along nicely. During the day he curls up like a cat in the warm corner behind the heater and at night he goes about his business. The mice, singularly quiet for a change, make themselves scarce. The snake is passive, apparently contented, and makes no resistance when I pick him up with my hands and drape him over an arm or around my neck. When I take him outside into the wind and sunshine his favorite place seems to be inside my shirt, where he wraps himself around my waist and rests on my belt. In this position he sometimes sticks his head out between shirt buttons for a survey of the weather, astonishing and delighting any tourists who may happen to be with me at the time. The scales of a snake are dry and smooth, quite pleasant to the touch. Being a cold-blooded creature, of course, he takes his temperature from that of the immediate environment—in this case my body.

We are compatible. From my point of view, friends. After a week of close association I turn him loose on the warm sandstone at my doorstep and leave for patrol of the park. At noon when I return he is gone. I search everywhere beneath, nearby and inside the trailerhouse, but my companion has disappeared. Has he left the area entirely or is he hiding somewhere close by? At any rate I am troubled no more by rattlesnakes under the door.

The snake story is not yet ended.

In the middle of May, about a month after the gopher snake's disappearance, in the evening of a very hot day, with all the rosy desert cooling like a griddle with the fire turned off, he reappears. This time with a mate.

I'm in the stifling heat of the trailer opening a can of beer, barefooted, about to go outside and relax after a hard day watching cloud formations. I happen to glance out the little window near the refrigerator and see two gopher snakes on my verandah engaged in what seems to be a kind of ritual dance. Like a living caduceus they wind and unwind about each other in undulant, graceful, perpetual motion, moving slowly across a dome of sandstone. Invisible but tangible as music is the passion which joins them—sexual? combative? both? A shameless *voyeur*, I stare at the lovers, and then to get a closer view run outside and around the trailer to the back. There I get down on hands and knees and creep toward the dancing snakes, not wanting to frighten or disturb them. I crawl to within six feet of them and stop, flat on my belly, watching from the snakes'-eye level. Obsessed with their ballet, the serpents seem unaware of my presence.

The two gopher snakes are nearly identical in length and coloring; I cannot be certain that either is actually my former household pet. I cannot even be sure that they are male and female, though their performance resembles so strongly a *pas de deux* by formal lovers. They intertwine and separate, glide side by side in perfect congruence, turn like mirror images of each other and glide back again, wind and unwind again. This is the basic pattern but there is a variation: at regular intervals the snakes elevate their heads, facing one another, as high as they can go, as if each is trying to outreach or overawe the other. Their heads and bodies rise, higher and higher, then topple together and the rite goes on.

I crawl after them, determined to see the whole thing. Suddenly and simultaneously they discover me, prone on my belly a few feet away. The dance stops. After a moment's pause the two snakes come straight toward me, still in flawless unison, straight toward my face, the forked tongues flickering, their intense wild yellow eyes staring directly into my eyes. For an instant I am paralyzed by wonder; then, stung by a fear too ancient and powerful to overcome I scramble back, rising to my knees. The snakes veer and turn and race away from me in parallel motion, their lean elegant bodies making a soft hissing noise as they slide over the sand and stone. I follow them for a short distance, still plagued by curiosity, before remembering my place and the requirements of common courtesy. For godsake let them go in peace, I tell myself. Wish them luck and (if lovers) innumerable offspring, a life of happily ever after. Not for their sake alone but for your own.

In the long hot days and cool evenings to come I will not see the gopher snakes again. Nevertheless I will feel their presence watching over me like totemic deities, keeping the rattlesnakes far back in the brush where I like them best, cropping off the surplus mouse population, maintaining useful connections with the primeval. Sympathy, mutual aid, symbiosis, continuity.

How can I descend to such anthropomorphism? Easily—but is it in this case entirely false? Perhaps not. I am not attributing human motives to my snake and bird acquaintances. I recognize that when and where they serve purposes of mine they do so for beautifully selfish reasons of their own. Which is exactly the way it should be. I suggest, however, that it's a foolish, simple-minded rationalism which denies any form of emotion to all animals but man and his dog. This is no more justified than the Moslems are in denying souls to women. It seems to me possible, even probable, that many of the non-human undomesticated animals experience emotions unknown to us. What do the coyotes mean when they yodel at the moon? What are the dolphins trying so patiently to tell us? Precisely what did those two enraptured gopher snakes have in mind when they came gliding toward my eyes over the naked sandstone? If I had been as capable of trust as I am susceptible to fear I might have learned something new or some truth so very old we have all forgotten it.

They do not sweat and whine about their condition,
They do not lie awake in the dark and weep for their
    sins. . . .

All men are brothers, we like to say, half-wishing sometimes in secret it were not true. But perhaps it is true. And is the evolutionary line from protozoan to Spinoza any less certain? That also may be true. We are obliged, therefore, to spread the news, painful and bitter though it may be for some to hear, that all living things on earth are kindred.

Stephen J. Gould

THE MISNAMED, MISTREATED, AND MISUNDERSTOOD IRISH ELK

Nature herself seems by the vast magnitude and stately horns, she has given this creature, to have singled it out as it were, and showed it such regard, with a design to distinguish it remarkably from the common herd of all other smaller quadrupeds.

THOMAS MOLYNEUX, 1697

The Irish Elk, the Holy Roman Empire, and the English Horn form a strange ensemble indeed. But they share the common distinction of their completely inappropriate names. The Holy Roman Empire, Voltaire tells us, was neither holy, nor Roman, nor an empire. The English horn is a continental oboe; the original versions were curved, hence "angular" (corrupted to English) horn. The Irish Elk was neither exclusively Irish, nor an elk. It was the largest deer that ever lived. Its enormous antlers were even more impressive. Dr. Molyneux marveled at "these spacious horns" in the first published description of 1697. In 1842, Rathke described them in a language unexcelled for the expression of enormity as *bewunderungswuerdig*. Although the Guinness book of world records ignores fossils and honors the American moose, the antlers of the Irish Elk have never been exceeded, or even approached, in the history of life. Reliable estimates of their total span range up to 12 feet. This figure seems all the more impressive when we recognize that the antlers were probably shed and regrown annually, as in all other true deer.

Fossil antlers of the giant deer have long been known in Ireland, where they occur in lake sediments underneath peat deposits. Before attracting the attention of scientists, they had been used as gateposts, and even as a temporary bridge to span a rivulet in County Tyrone. One story, probably apocryphal, tells of a huge bonfire made of their bones and antlers in County Antrim to celebrate the victory over Napoleon at Waterloo. They were called elk because the European moose (an "elk" to Englishmen) was the only familiar animal with antlers that even approached those of the giant deer in size.

The first known drawing of giant deer antlers dates from 1588. Nearly a century later, Charles II received a pair of antlers and (according to Dr. Molyneux) "valued them so highly for their prodigious largeness" that he set them up in the horn gallery of Hampton Court, where they "so vastly exceed" all others in size "that the rest appear to lose much of their curiosity."

Ireland's exclusive claim vanished in 1746 (although the name stuck) when a skull and antlers were unearthed in Yorkshire, England. The first continental discovery followed in 1781 from Germany, while the first complete skeleton (still standing in the museum of Edinburgh University) was exhumed from the Isle of Man in the 1820s.

We now know that the giant deer ranged as far east as Siberia and China and as far south as northern Africa. Specimens from England and Eurasia are almost always fragmentary, and nearly all the fine specimens that adorn so many museums throughout the world come from Ireland. The giant deer evolved during the glacial period of the last few million years and may have survived to historic times in continental Europe, but it became extinct in Ireland about 11,000 years ago.

"Among the fossils of the British empire," wrote James Parkinson in 1811, "none are more calculated to excite astonishment." And so it has been throughout the history of paleontology. Putting aside both the curious anecdotes and the sheer wonder that immensity always inspires, the importance of the giant deer lies in its contribution to debates about evolutionary theory. Every great evolutionist has used the giant deer to defend his favored views. The controversy has centered on two main issues: (1) Could antlers of such bulk be of any use? and (2) Why did the giant deer become extinct?

Since debate on the Irish Elk has long centered on the reasons for its extinction, it is ironic that the primary purpose of Molyneux's original article was to argue that it must still be alive. Many seventeenth-century scientists maintained that the extinction of any species would be inconsistent with God's goodness and perfection. Dr. Molyneux's article of 1697 begins:

> That no real species of living creatures is so utterly extinct, as to be lost entirely out of the World, since it was first created, is the opinion of many naturalists; and 'tis grounded on so good a principle of Providence taking care in general of all its animal productions, that it deserves our assent.

Yet the giant deer no longer inhabited Ireland, and Molyneux was forced to search elsewhere. After reading travelers' reports of antler size in the American moose, he concluded that the Irish Elk must be the same animal; the tendency toward exaggeration in such accounts is apparently universal and timeless. Since he could find neither figure nor an accurate description of the moose, his conclusions are not as absurd as modern knowledge would indicate. Molyneux attributed the giant deer's demise in Ireland to an "epidemick distemper," caused by "a certain ill constitution of air."

For the next century arguments raged along Molyneux's line—to which modern species did the giant deer belong? Opinion was equally divided between the moose and the reindeer.

As eighteenth-century geologists unraveled the fossil record of ancient life, it became more and more difficult to argue that the odd and unknown creatures revealed by fossils were all still living in some remote portion of the globe. Perhaps God had not created just once and for all time; perhaps He had experimented continually in both creation and destruction. If so, the world was surely older than the six thousand years that literalists allowed.

The question of extinction was the first great battleground of modern paleontology. In America, Thomas Jefferson maintained the old view, while Georges Cuvier, the great French paleontologist, was using the Irish Elk to prove that extinction did occur. By 1812 Cuvier had resolved two pressing issues: by minute anatomical description, he proved that the Irish Elk was not like any modern animal; and by placing it among many fossil mammals with no modern counterparts, he established the fact of extinction and set the basis for a geologic time scale.

Once the fact of extinction had been settled, debate moved to the time of the event: in particular, had the Irish Elk survived the flood? This was no idle matter, for if the flood or some previous catastrophe had wiped out the giant deer, then its demise had natural (or supernatural) causes. Archdeacon Maunsell, a dedicated amateur, wrote in 1825: "I apprehended they must have been destroyed by some overwhelming deluge." A certain Dr. MacCulloch even believed that the fossils were found standing erect, noses elevated—a final gesture to the rising flood, as well as a final plea: don't make waves.

If, however, they had survived the flood, then their exterminating angel could only have been the naked ape himself. Gideon Mantell, writing in 1851, blamed Celtic tribes; in 1830, Hibbert implicated the Romans and the extravagant slaughters of their public games. Lest we assume that our destructive potential was recognized only recently, Hibbert wrote in 1830: "Sir Thomas Molyneux conceived that a sort of distemper, or pestilential murrain, might have cut off the Irish Elks. . . . It is, however, questionable, if the human race has not occasionally proved as formidable as a pestilence in exterminating from various districts, whole races of wild animals."

In 1846, Britain's greatest paleontologist, Sir Richard Owen, reviewed the evidence and concluded that in Ireland at least, the giant deer had perished before man's arrival. By this time, Noah's flood as a serious geologic proposition had passed from the scene. What then had wiped out the giant deer?

Charles Darwin published the *Origin of Species* in 1859. Within ten years virtually all scientists had accepted the *fact* of evolution. But the debate about causes and mechanisms was not resolved (in Darwin's favor) until the 1940s. Darwin's theory of natural selection requires that evolutionary changes be adaptive—that is, that they be useful to the organism. Therefore, anti-Darwinians searched the fossil record for cases of evolution that could not have benefited the animals involved.

The theory of orthogenesis became a touchstone for anti-Darwinian paleontologists, for it claimed that evolution proceeded in straight lines that natural selection could not regulate.

Certain trends, once started, could not be stopped even if they led to extinction. Thus certain oysters, it was said, coiled their valves upon each other until they sealed the animal permanently within; saber-toothed "tigers" could not stop growing their teeth or mammoths their tusks.

But by far the most famous example of orthogenesis was the Irish Elk itself. The giant deer had evolved from small forms with even smaller antlers. Although the antlers were useful at first, their growth could not be contained and, like the sorcerer's apprentice, the giant deer discovered only too late that even good things have their limits. Bowed by the weight of their cranial excrescences, caught in the trees or mired in the ponds, they died. What wiped out the Irish Elk? They themselves or, rather, their own antlers did.

In 1925, the American paleontologist R. S. Lull invoked the giant deer to attack Darwinism: "Natural selection will not account for overspecialization, for it is manifest that, while an organ can be brought to the point of perfection by selection, it would never be carried to a condition where it is an actual menace to survival . . . [as in] the great branching antlers of the extinct Irish deer."

Darwinians, led by Julian Huxley, launched a counterattack in the 1930s. Huxley noted that as deer get larger—either during their own growth or in the comparison of related adults of different sizes—the antlers do not increase in the same proportion as body size; they increase faster, so that the antlers of large deer are not only absolutely larger but also relatively larger than those of small deer. For such regular and orderly change of shape with increasing size, Huxley used the term allometry.

Allometry provided a comfortable explanation for the giant deer's antlers. Since the Irish Elk had the largest body size of any deer, its relatively enormous antlers could have been a simple result of the allometric relationship present among all deer. We need only assume that increased body size was favored by natural selection; the large antlers might have been an automatic consequence. They might even have been slightly harmful in themselves, but this disadvantage was more than compensated by the benefits of larger size, and the trend continued. Of course, when problems of larger antlers outweighed the advantages of larger bodies, the trend would cease since it could no longer be favored by natural selection.

Almost every modern textbook on evolution presents the Irish Elk in this light, citing the allometric explanation to counter orthogenetic theories. As a trusting student, I had assumed that such constant repetition must be firmly based on copious data. Later I discovered that textbook dogma is self-perpetuating; therefore, three years ago I was disappointed, but not really surprised, to discover that this widely touted explanation was based on no data whatsoever. Aside from a few desultory attempts to find the largest set of antlers, no one had ever measured an Irish Elk. Yardstick in hand, I resolved to rectify this situation.

The National Museum of Ireland in Dublin has seventeen specimens on display and many more, piled antler upon antler, in a nearby warehouse. Most large museums in western Europe and America own an Irish Elk, and the giant deer adorns many trophy rooms of English and Irish gentry. The largest antlers grace the entranceway to Adare Manor, home of the Earl of Dunraven. The sorriest skeleton sits in the cellar of Bunratty Castle, where many merry and slightly inebriated tourists repair for coffee each evening after a medieval banquet. This poor fellow, when I met him early the morning after, was smoking a cigar, missing two teeth, and carrying three coffee cups on the tines of his antlers. For those who enjoy invidious comparisons, the largest antlers in America are at Yale; the smallest in the world at Harvard.

To determine if the giant deer's antlers increased allometrically, I compared antler and body size. For antler size, I used a compounded measure of antler length, antler width, and the lengths of major tines. Body length, or the length and width of major bones, might be the most appropriate measure of body size, but I could not use it because the vast majority of specimens consist only of a skull and its attached antlers. Moreover, the few complete skeletons are invariably made up of several animals, much plaster, and an occasional ersatz (the first skeleton in Edinburgh once sported a horse's pelvis). Skull length therefore served as my measure of overall size. The skull reaches its final length at a very early age (all my specimens are older) and does not vary thereafter; it is, therefore, a good indicator of body size. My sample included seventy-nine skulls and antlers from museums and homes in Ireland, Britain, continental Europe, and the United States.

My measurements showed a strong positive correlation between antler size and body size, with the antlers increasing in size two and one-half times faster than body size from small to large males. This is not a plot of individual growth; it is a relationship among adults of different body size. Thus, the allometric hypothesis is affirmed. If natural selection favored large deer, then relatively larger antlers would appear as a correlated result of no necessary significance in itself.

Yet, even as I affirmed the allometric relationship, I began to doubt the traditional explanation—for it contained a curious remnant of the older, orthogenetic view. It assumed that the antlers are not adaptive in themselves and were tolerated only because the advantages of increased body size were so great. But why must we assume that the immense antlers had no primary function? The opposite interpretation is equally possible: that selection operated primarily to increase antler size, thus yielding increased body size as a secondary consequence. The case for inadaptive antlers has never rested on more than subjective wonderment born of their immensity.

Views long abandoned often continue to exert their influence in subtle ways. The orthogenetic argument lived on in the allometric context proposed to replace it. I believe that the supposed problem of "unwieldy" or "cumbersome" antlers is an illusion rooted in a notion now abandoned by students of animal behavior.

To nineteenth-century Darwinians, the natural world was a cruel place. Evolutionary success was measured in terms of battles won and enemies destroyed. In this context, antlers were viewed as formidable weapons to be used against predators and rival males. In his *Descent of Man* (1871), Darwin toyed with another idea: that antlers might have evolved as ornaments to attract females. "If, then, the horns, like the splendid accouterments of the knights of old, add to the noble appearance of stags and antelopes, they may have been modified partly for this purpose." Yet he quickly added that he had "no evidence in favor of this belief," and went on to interpret antlers according to the "law of battle" and their advantages in "reiterated deadly contests." All early writers assumed that the Irish Elk used its antlers to kill wolves and drive off rival males in fierce battle. To my knowledge this view has been challenged only by the Russian paleontologist L. S. Davitashvili, who asserted in 1961 that the antlers functioned primarily as courtship signals to females.

Now, if antlers are weapons, the orthogenetic argument is appealing, for I must admit that ninety pounds of broad-palmed antler, regrown annually and spanning twelve feet from tip to tip, seems even more inflated than our current military budget. Therefore, to preserve a Darwinian explanation, we must invoke the allometric hypothesis in its original form.

But what if antlers do not function primarily as weapons? Modern studies of animal behavior have generated an exciting concept of great importance to evolutionary biology: many structures previously judged as actual weapons or devices for display to females are actually used for ritualized combat among males. Their function is to prevent actual battle (with consequent injuries and loss of life) by establishing hierarchies of dominance that males can easily recognize and obey.

Antlers and horns are a primary example of structures used for ritualized behavior. They serve, according to Valerius Geist, as "visual dominance-rank symbols." Large antlers confer high status and access to females. Since there can be no evolutionary advantage more potent than a guarantee of successful reproduction, selective pressures for larger antlers must often be intense. As more and more horned animals are observed in their natural environment, older ideas of deadly battle are yielding to evidence of purely ritualized display without body contact, or fighting in ways clearly designed to prevent bodily injury. This has been observed in red deer by Beninde and Darling, caribou by Kelsall, and in mountain sheep by Geist.

As devices for display among males, the enormous antlers of the Irish Elk finally make sense as structures adaptive in themselves. Moreover, as R. Coope of Birmingham University pointed out to me, the detailed morphology of the antlers can be explained, for the first time, in this context. Deer with broad-palmed antlers tend to show the full width of their antlers in display. The modern fallow deer (considered by many as the Irish Elk's nearest living relative) must rotate its head from side to side in order to show its palm. This would have created great problems for giant deer, since the torque produced by swinging ninety-pound antlers would have been immense. But the antlers of the Irish Elk were arranged to display the palm fully when the animal looked straight ahead. Both the unusual configuration and the enormous size of the antlers can be explained by postulating that they were used for display rather than for combat.

If the antlers were adaptive, why did the Irish Elk become extinct (at least in Ireland)? The probable answer to this old dilemma is, I am afraid, rather commonplace. The giant deer flourished in Ireland for only the briefest of times—during the so-called Alleröd interstadial phase at the end of the last glaciation. This period, a minor warm phase between two colder epochs, lasted for about 1,000 years, from 12,000 to 11,000 years before the present. (The Irish Elk had migrated to Ireland during the previous glacial phase when lower sea levels established a connection between Ireland and continental Europe.) Although it was well adapted to the grassy, sparsely wooded, open country of Alleröd times, it apparently could not adapt either to the subarctic tundra that followed in the next cold epoch or to the heavy forestation that developed after the final retreat of the ice sheet.

Extinction is the fate of most species, usually because they fail to adapt rapidly enough to changing conditions of climate or competition. Darwinian evolution decrees that no animal shall actively develop a harmful structure, but it offers no guarantee that useful structures will continue to be adaptive in changed circumstances. The Irish Elk was probably a victim of its own previous success. *Sic transit gloria mundi.*

Charles D. Stewart

CHICAGO SPIDERS

Being a spider in Chicago is a very satisfactory vocation. In the evening, when it is time to take down the old web and put up the new, a spider will gather a section into a ball or skein that is positively black, and kick it out behind him into the street below as if he were disgusted with such a grimy mess. It is so bulky with dirt that a small piece of web makes a large armful for him. And after the new one has been spread for an hour or two, its sticky filaments are so coated with particles of atmosphere that it will hardly catch anything else. Only by going through a sort of jumping-jack performance can a Chicago spider manage to make a fly stick.

Whether a country spider, with a whole garden fence at his disposal, takes down his old web, I do not know, though it would seem that there he could, by merely moving a foot or two, save himself all the work; but in Chicago, where corner locations are the most valuable—especially the corners of windows where houseflies long to enter—and where each corner is pre-empted by a particular spider, the taking down of the old web is necessary to the greatest daily profit. It pays better than to move.

A Chicago spider can take down a web and put up another in about twenty minutes—and from this I am anxious to have the reader infer that the daily presence of a great number of them does not mean a neglected window. If anyone thinks his household guiltless in this regard, let him observe his own window closely. I dare say he will find this story sumptuously illustrated.

Before I was laid on a bed by a window and tied down as firmly as any Gulliver by Chicago pygmies, most of whom belonged to the tribe of Typhus, I would have considered it poor employment for any man to enter into the affairs of creatures so much smaller than himself. But they did shrewd things before my eyes every day, and when I began to understand, I became interested; and thus, for three weeks, I found myself bound out to the trade.

It was the jumping-jack trick that I first discovered and appreciated. The spider, sitting patiently at the focus of his elastic wheel with all legs on the lines, is in telegraphic communication with every part of it; now let a fly so much as flutter a filament, and the spider jumps up and down as if he were trying to shake the whole structure from its moorings. This bounces the fly till he has his feet solidly on the line, and perhaps tangled in other lines. After taking this precaution, the spider, if he has been lucky, runs out and ties up his victim in the usual bundle, ready to carry. He does up a fly like a turkey trussed and ready for the table.

To one who has had a motionless and half-forgotten spider in his eye for an hour or so, this sudden exhibition of vigor in jumping up and down is startling. He does it as if he were in a great fit of temper. From this practice it is evident that he cannot depend upon the web alone to catch the prey, and hold it long enough for him to get out to it. The web is not merely a stationary snare, like a tree with birdlime on it, but a contrivance that may be operated personally by the spider as a trap. The structure, being elastic, works up and down when he jumps, so that each row of lines traverses at least the distance between it and the next row of lines. Thus, despite the open spaces between them, he is virtually in possession of the whole plane of space, for anything with air-disturbing wings can hardly pass through it without sending in an alarm and being caught. All spiders, I suppose, know this trick of the trade; but a Chicago spider must stick to his post and practice it in every case. If he did not, his daily catch would be all soot and no flies.

The same spiders did not occupy the window throughout the three weeks; but with the exception of one red spider who came along and seemed very doubtful about setting to work, they were all of one kind, big and little. This auburn-hued spider was more slender and shapely—not so fat and commercial-looking as the others. There were little spiders who spun little webs of such fineness that they were visible only when the sun fell just right on the glinting new gossamer; and for over a week a very big fellow, with a yellow hieroglyph on him like gold bullion on the back of a priest, held sway in webs a foot across. He sat with his back toward the room, whereas most of them made a practice of keeping their under sides toward the window. In this, there seems to be a difference in practice; but all of them sit upside-down, head-downwards, invariably.

I discovered, to my own satisfaction at least, why a spider sits in his web upside-down. A spider has eight legs, besides a very short pair in front which are more like arms; but in truth a spider's legs are all fingers, and he needs as many as possible to handle his prey. Were he to support himself right side up in grappling with a victim, it would require four of the legs merely to hold him in that position, for he would have to grasp more than one thread; but he can hang head-downwards with only the one hind pair of legs, and have all the rest free to handle the prey before him. His hind pair of legs extend almost straight behind him for the purpose of being his sole support in such cases; and because he is built in this way, in order to cope successfully with other insects, the upside-down attitude is his easiest way of staying on watch. It is his most restful position.

One of the big spiders was one day surprised by a chrysalis that fell down from some place into his web. It turned out to be a very windfall of fortune, for the luscious larva was quite to his taste. At least, he examined it thoroughly, and kept it, as if he were satisfied with what he found inside of the cocoon. It was almost as long as himself, and he showed great dexterity in turning it about and examining it in all positions with his six free legs, holding it before him as he hung head-downwards. A spider can handle himself in all positions with equal facility, and when he is surprised he will suddenly turn head-upward as he surveys the web, and keep that position for a while. But when all is quiet on the Potomac, he turns upside-down again and takes his ease.

I read in a book review that the male spider is said to dance in order to please his inamorata. I have seen such a performance, and would describe it as follows. One of the spiders retreats backward an inch or two from the other; he pauses there a moment and advances; and when the two are face to face, they go through certain antics, both of them, with their front legs. It is exactly as if one were to interlock his fingers loosely and then twiddle them. After this twiddling of legs, the visitor backs up, pauses, and comes forward again; and they will keep up this performance for quite a while. Whether this is flirtation I do not know; much less do I understand the code. And whether it is dancing or not depends upon—the figure of speech.

These spiders, according to the dictionary, are geometrical or garden spiders; but the ones with whom I was personally acquainted saw nothing more verdant than a rubber plant and one smoke-blasted tree. This ailing tree was the only survivor in those parts, and so its twiggery had to accommodate the sparrows of a large territory every evening; it was little more than a community perch or convention tree, and it had more sparrows on it than leaves. Regularly they would come home to Bedlam at night, and they would seem much excited over the return to nature. As to the spiders, they were garden spiders in the sense that Chicago is the Garden City.

Before proceeding further, I must explain that this comment on the secrets of the craft is merely by way of introducing the reader to a particular spider, who had an admirable adventure. I shall come to him later on. I should confess that I do not know spiders anatomically or microscopically, but only personally; I know only that about a spider which he knows himself—namely, his trade. This, I think, is worth describing, step by step.

It will be best to take a Chicago spider who is building in the upper corner of a window, for here is a set of conditions which are uniform throughout the country, and which everyone is familiar with. The spider, having found this unoccupied place, walks on the window frame away from the corner and stops at the right distance for the size of the web, which depends upon the size of the spider. The corner of the window frame offers the foundation, or outline, for two sides of his web; but he must himself complete the circumference within which to spread his work. Now, a line stretched from where he stands, on the top frame, to a point on the side frame will give him a triangle; and he must project this line transversely through the air.

This is easily done. Pressing the end of the line to the window frame, he takes hold of it with one hind leg and runs along with it to the corner, spinning it out as he goes; and he holds the line out with his hind leg like a boy flying a kite. He must hold it well out and keep it taut, for it must not touch the wood anywhere along its length. Having reached the corner, he turns and runs down the side frame; and now it is as if the kite were going up in the air. As he runs downward from the corner, paying out the line, it opens, fanwise, from the upper frame; and when it has formed the triangle he stops and fastens that end.

This is to be his main cable, which must, on that side, support the ends of all the lines. And these inner lines are to be stretched with considerable tension. For such a heavy strain the single strand is not enough, so he now runs back and forth along its length and keeps paying out till he has augmented it with several plies of filament—a cable. It is now strong enough, but as the tension on it is to be sidewise it is not rigid enough; it would bow inwards as he stretched the web from it, and so it needs a few small guy lines, or stays, to brace it. These stays he fastens farther out on the wood, or to points on the glass itself. He could, in fact, as far as his abilities are concerned, fasten every line of his web to the glass; but the wind would blow it against the pane and interfere with its workings. Therefore he makes the cable to stretch it to, a little distance from the window.

The outline or foundation is now done. Inside this triangular circumference he has now to make the spokes of his wheel before stretching upon them the circular lines. In like manner as he put up the main cable, he runs a single line across this triangular space, about the middle of it. Having this line stretched, he climbs to the middle of it and there stops, for this is to be the center of his wheel. In stretching this diametrical line he has really made two spokes at one operation; but now he must pursue a different method, making one spoke at a time. If he were to try to keep up this way of making two spokes at a time, fastening a line at one side and running around the circumference to the opposite side to fasten it there, his line would become entangled with the one stretched before; it would stick, and he could not raise the new line to the middle of the other where it ought to cross. Therefore he must now work from the middle outwards, stretching one spoke at a time. He fastens the end of the spoke he is about to spin to the middle of this diametrical line, takes this new line in his hind leg in order to hold it free of the other as he climbs it, and thus he gets the spoke to the window frame. Then he proceeds with it along the window frame a short distance, the second line opening out, fanwise, from the first; and when it has opened to the proper angle he fastens it down to the wood. He then descends the new one and repeats the operation; and so he keeps on, always using the one he stretched last to return upon and bring out another, and always holding the new line clear and taut as he pays it out, exactly like a boy flying a kite. It must not touch and tangle. And, like the boy, he runs along at a good gait as if he had no time to lose.

By this simple method, the spokes are all put in; and it is very easy according to his system. It is worth considering, however, that he is always very fortunate in coming out so nearly uniform in the spacing of his spokes—and this is an irregular triangle upon which the spokes must fall at all sorts of distances in order to be equally spaced. He seems to be an expert in division. But it is not the *outside* of his space that he can measure off in an automatic way, for there the distances are not uniform. I think he must accomplish it all by watching the new line open fanwise from the middle, and so I regard him as a sort of surveyor with a good eye for angles. The wheel part is now done, and he has to weave on it the circling strands.

He takes his place at the middle of the wheel, and keeping his head always toward the center, he steps sidewise from spoke to spoke, fastening the thread to a spoke, drawing it across to the next one at the right tension, dabbing it down to fasten it, and so on, round and round. And he works with considerable speed.

But this mode of operation cannot be kept up to the end. When he has worked out a short distance from the center, the radiating spokes are too far apart for him to straddle across. Here he changes the method. Instead of straddling across, he goes out on a single spoke, fastens his thread to it, comes in and crosses to the next spoke by means of the line that he stretched on his last trip around. He then goes out on the next spoke, carrying the line in his hind leg, and fastens it—and he always handles it with his leg, so that there is no surplus spun out, and it has the right tension. Thus he continues till his wheel is big enough, always using his last circle as a bridge from spoke to spoke as he adds the next surrounding circle. This part, when done, is really a spiral.

The garden spider, in making a web that fulfills the ideal, puts in this spiral I have just described with the lines very far apart—very open. He then starts at the circumference and fills it in finer, working round and round toward the middle. This first spiral may be considered his scaffold. As we see, it was constructed under certain drawbacks; but now that he has so much put in coarsely, he can walk round and round with more footing, and work with less trouble.

When the web seems finished, one thing yet remains to be done. Where the spokes have each been fastened to the center, there is a mass of fiber, the tag-ends of the whole job, which would be in his way as he sat in the middle of the web. He takes this out neatly, leaving a hole. Had he taken this out before the spiral was put on, the whole wheel would, of course, have collapsed. He throws the fiber into the street below, and takes his place over the hole with his legs holding the lines around him; and now it is time for Providence to send a fly.

The spider does his work behind his back, as it were; he cannot see what he is doing; and yet in certain of his operations he must make strokes that are instantly accurate and "to the point." This would call for some miraculous knowledge of location—which he has not; and his way of meeting the problem is interesting. In that division of his work which consists in stretching the cable and spokes, his problem is simple; it is merely the fastening of sticky threads to the window frame, a surface which is firm and flat. As it is flat, he does not need to strike a fine particular point on it; and as it is perfectly stable, he simply presses the line down firmly behind him as it comes from his spinneret. But in stretching the spiral from spoke to spoke of the web itself, he must strike a certain point on his line against a particular point on the web, in order to have the right tension; he must unite them firmly at that point and do it at a dab. It is a fine point to find; and to do such work behind him, against a yielding, air-blown filament, is quite a different matter from pressing his line to a flat, firm surface. He proceeds, accordingly, on the same principle, but takes it another way about. Instead of merely dabbing down the line he is spinning, he seizes with a hind leg the line *to which* he wishes to make a fastening and presses that against a particular part of *himself*; that is, he raises the spoke and touches it firmly to the point where the new line is spinning out. Thus the spiral is put in. The whole extraneous difficulty is transmuted into a mere matter of self-knowledge—like finding one's mouth in the dark.

During this part of the work he does not need to use one

leg to prevent entanglement, the parallel spans being shorter and more widely separate from the beginning; and it is lucky for him that he can now spare that member, for in the operations of putting in the spiral his multitude of legs are busy indeed. One is seizing the spoke and dabbing it to his spinneret; one is pressing on the new-spun line, as if to regulate the tension; the others are stepping about lively in order to accommodate his body to the advancing work—and altogether it is as rapid and unobservable as the flight of knitting needles. But once it is caught by the eye, the mystery of his accuracy is small, and its ingenuity is great. But the very fact that he has to descend to mere ingenuity, in lieu of instinct, which can perform miracles, presents him to us as a humble spinner, and human. I think it is a person of little promise who can look through his web and not find that this display of window work, spread out between us and the universe, is a sort of trap for the mind, tending to keep it within bounds.

The large spiders, so far as I have observed, are the most careless workmen. In some of their webs the geometrical design could hardly be perceived were it not for the radiating spokes; and these are not straight, but drawn to this side and that by the connecting lines. And these lines, that ought to be the spiral, have been put in any way at all, as if one at a time, here and there; and moreover they have been put in loosely and then tightened to the spoke with other little guy lines, so that they have the shape of a Y. The web seems to be not only patched, but all patchwork from the start. It has the wheel shape in it, however, and the same principles are employed throughout; in fact, there is more individuality and a greater display of mechanical science in such a web than in one that conforms to the ideal. It takes a better mechanic to patch a job than to follow specifications to a successful conclusion. The little spiders do the most perfect work, strikingly geometrical, with the lines of the spiral exactly parallel. I once picked from a bush a withered leaf that had curled up at the end, and in this space, smaller in extent than a quarter of a dollar, was a spider's web perfect in every detail.

Other webs would differ from this window web; but the difference would not be in the web proper so much as in the out-rigging or foundation for it. In truth, the most interesting part of a spider's work is not in the geometrical part that excites our first wonder, but in his ways of devising the irregular circumference, the making use of vantage points, the solving of problems peculiar to each set of surroundings. Here is individual work, separate planning to suit each case, the application of principles rather than automatic and uniform procedure— the work of a mechanic.

The opportunities for studying nature in a city apartment are growing every day. The renaissance of Colonial architecture, with the small windowpanes, allows the spiders to cultivate the whole field of glass. A spider soon learns all about glass; a fly never. The spider works with it familiarly; he even uses its surface to moor the stays of his cable; but the fly buzzes and butts his head against it, utterly unable to learn that the invisible can have existence. The invention of glass was a godsend to spiders, and a sorry thing for flies.

There is much more to the trade of building a web, but so technical in detail that it would have to be considered at much length in order to arrive at the ultimate mechanical reasons. A thing superficially perceived or half explained might as well not be explained at all. Much "nature study" consists in these mere semblances of explanations—incomplete perceptions. The most profitable work in this line, I think, would be the work of the skilled mechanic, rather than the poetic "nature student" or the mere microscopic observer; for this shrewd stealing of secrets, by both observation and basic reasoning, has been his lifelong attitude in filching his own trade from others, as well as from nature. And as to the writing of it, the simple and luminous expression of such things calls for the very highest and completest set of mental faculties. Contrary to the popular notion, the creation of so-called "atmospheric" impression in literature is much easier, and of a lower order of intellect, than to convey in familiar words exactly what was done, and why. This also takes imagination.

But, as I have said, it was not my intention in writing this to record all that I learned of the trade so far as I advanced, but rather to make public a tragi-comedy that was enacted in spider life. To recount all that I observed would be robbing the reader of his privilege of discovering things for himself— even denying him the right to look out of his own window— which is one of the things I protest against. I have told this much because it was necessary thus to introduce, in their proper persons, the two characters of the play.

It was drawing on toward evening. The day had been— simply another day; a wilderness of roofs in a soft-coal mist, a turbid patch of sky, and the people below moving monotonously past like cattle in a canyon. The street nearby became darker with the stream of people hurrying home from store and factory; Chicago had let out. The worn-out tree was receiving back the sparrows, and every twig was a perch. I was tired of all this; there was nothing interesting about it; and so from trying to see something *out* of the window I turned again to look *at* it, for it was time for the spiders to go to work.

The corner nearest me, which had to be renovated of its dusty and damaged web, belonged to a medium-sized spider; and promptly he came forth to the work. Another corner was held—I cannot say occupied—by a set of legs on a very old web. A spider, with all his skill in taking down a web, moves away and leaves his dirt behind him. Not only this, but he has a habit, when he has his new set of legs, of leaving the old ones on the web; and there they remain, occupying the position that he last held. They do not come off him singly, but in a complete set, like a truck that has been removed from a car. And it is wonderful how long a web will withstand the weather and bear this grisly semblance of a spider with each leg set on a line. This particular set of sere and yellowish legs danced in every breeze, and seemed even more active than when they had a spider to operate them. I often wished that some enterprising spider would come along and take it all down; but none ever did. From watching to see whether this would happen, I turned my attention to the medium-sized spider as he cleared his space. Finally, he had his old web all down and disposed of; and the new one was put up with "neatness and dispatch."

When the web was seemingly done, the spider spent a little while on the window frame among his guy lines—possibly making things still more taut. There now appeared suddenly on the top of the frame, at the opposite corner of the web, a big able-bodied spider. He was much larger than the other—let us call them David and Goliath. He stopped short at the edge of the web as if pausing to look across at the owner and make up his mind. The other spider stopped work suddenly, as if looking back at him. I immediately suspected that here was a situation, and so I watched closely; there seemed to be spider thinking going on. The big spider stepped deliberately on the web, and then, with a sudden dash, went out on it. He had no more than reached the middle when he was snapped back to where he came from, and thrown against the upper frame of the window as if he had been shot from a rubber sling—and the web was gone. In that instant, the smaller spider had cut the main cable. David's elastic sling had not only thrown Goliath back where he belonged, but had knocked him against the frame and slapped him in the face for his impudence.

The big spider, we can only conclude, meant harm—either robbery or bodily injury—and the other spider knew it. But this does not explain what we like always to see in nature—an object in everything. What was the beneficent object? It was not a provision on the part of instinct to enable the spider to save its web from the robber, for the web was utterly sacrificed. As to the loss of property, the little spider might just as well have run away and let the big one have it. And as to the little spider saving its life, it might as well have run at once, for a spider can pursue another anywhere, even if there is no web. To me it seemed to be a pure case of "You won't get the best of me." Does Nature, in her wise regard for the needs of all her creatures, make provision for the satisfaction of transcendental justice?

It looked like an original act of thought—the presence of mind of a good mechanic who understands his machine. I have often wondered, on the theory that it might have been a way of saving the smaller spider's life, whether the big spider was injured; and if the smaller spider had simply run away and left his web, would not the other have been satisfied with it, and not bothered to pursue him? Why this provision of instinct—if it was mere instinct?

I am sorry to say that I was not myself in a condition to look into the physical state of Goliath and see whether he was disabled. I was so taken up with the tragi-comic view, the human phase of it, that I did not even think of these other things. In fact I was so delighted over the victory that, weak as I was, and bound down as by cords made of my own tendons, I raised myself up and inwardly exclaimed—*Foiled*!

Spiders are interesting companions—under conditions. And the outcome of all one's observations is finally a question—Is it God that is doing these things, or is it a spider?

G. Murray Levick

ANTARCTIC PENGUINS

The penguins of the Antarctic regions very rightly have been termed the true inhabitants of that country. The species is of great antiquity, fossil remains of their ancestors having been found which showed that they flourished as far back as the eocene epoch. To a degree far in advance of any other bird, the penguin has adapted itself to the sea as a means of livelihood, so that it rivals the very fishes. The proficiency in the water has been gained at the expense of its power of flight, but this is a matter of small moment, as it happens.

In few other regions could such an animal as the penguin

rear its young, for when on land its short legs offer small advantage as a means of getting about, and as it cannot fly, it would become an easy prey to any of the carnivora which abound in other parts of the globe. Here, however, there are none of the bears and foxes which inhabit the North Polar regions, and once ashore the penguin is safe.

The reason for this state of things is that there is no food of any description to be had inland. Ages back, a different state of things existed: tropical forests abounded, and at one time, the seals ran about on shore like dogs. As conditions changed, these latter had to take to the sea for food, with the result that their four legs, in course of time, gave place to wide paddles or "flippers," as the penguins' wings have done, so that at length they became true inhabitants of the sea.

When seen for the first time, the Adelie penguin gives you the impression of a very smart little man in an evening dress suit, so absolutely immaculate is he, with his shimmering white front and black back and shoulders. He stands about two feet five inches in height, walking very upright on his little legs.

His carriage is confident as he approaches you over the snow, curiosity in his every movement. When within a yard or two of you, as you stand silently watching him, he halts, poking his head forward with little jerky movements, first to one side, then to the other, using his right and left eye alternately during his inspection. He seems to prefer using one eye at a time when viewing any near object, but when looking far ahead, or walking along, he looks straight ahead of him, using both eyes. He does this, too, when his anger is aroused, holding his head very high, and appearing to squint at you along his beak.

After a careful inspection, he may suddenly lose all interest in you, and ruffling up his feathers, sink into a doze. Stand still for a minute till he has settled himself to sleep, then make sound enough to wake him without startling him; and he opens his eyes, stretching himself, yawns, then finally walks off, caring no more about you.

The Adelie penguin is excessively curious, taking great pains to inspect any strange object he may see. When we were waiting for the ship to fetch us home, some of us lived in little tents which we pitched on the snow about fifty yards from the edge of the sea. Parties of penguins from Cape Royds rookery frequently landed here, and almost invariably the first thing they did on seeing our tents was at once to walk up the slope and inspect these, walking all round them, and often staying to doze by them for hours. Some of them, indeed, seemed to enjoy our companionship. When you pass on the sea-ice anywhere near a party of penguins, these generally come up to look at you; and we had great trouble to keep them away from the sledge dogs when these were tethered in rows near the hut at Cape Evans. The dogs killed large numbers of them in consequence, in spite of all we could do to prevent this.

The Adelies are extremely brave, and though panic occasionally overtakes them, I have seen a bird return time after time to attack a seaman who was brutally sending it flying by kicks from his sea-boot, before I arrived to interfere.

The Adelie penguins spend their summer and bring forth their young in the far South. Nesting on the shores of the Antarctic continent, and on the islands of the Antarctic seas, they are always close to the water, being dependent on the sea for their food, as are all Antarctic fauna; the frozen regions inland, for all practical purposes, being barren of both animal and vegetable life.

Their requirements are few: they seek no shelter from the terrible Antarctic gales, their rookeries in most cases being open wind-swept spots. In fact, three of the four rookeries I visited were possibly in the three most windy regions of the Antarctic. The reason for this is that only wind-swept places are so kept bare of snow that solid ground and pebbles for making nests are to be found.

When the chicks are hatched and fully fledged, they are taught to swim; and when this is accomplished and they can catch food for themselves, both young and old leave the Southern limits of the sea, and make their way to the pack-ice out to the northward; thus escaping the rigors and darkness of the Antarctic winter, and keeping where they will find the open water which they need. For in the winter the seas where they nest are completely covered by a thick sheet of ice which does not break out until early in the following summer. Much of this ice is then borne northward by tide and wind, and accumulates to form the vast rafts of what is called "pack-ice," many hundreds of miles in extent, which lie upon the surface of the Antarctic seas.

When young and old leave the rookery at the end of the breeding season, the new ice has not yet been formed, and their long journey to the pack has to be made by water; but they are wonderful swimmers and seem to cover the hundreds of miles quite easily.

Arrived on the pack, the first year's birds remain there for two winters. It is not until after their first molt, the autumn following their departure from the rookery, that they grow the distinguishing mark of the adult, black feathers replacing the white plumage which has hitherto covered the throat.

The spring following this, and probably every spring for the rest of their lives, they return South to breed, performing their journey, very often, not only by water, but on foot across many miles of frozen sea.

For those birds who nest in the southernmost rookeries, such as Cape Crozier, this journey must mean for them a journey of at least four hundred miles by water, and an unknown but considerable distance on foot over ice.

The first Adelie penguins arrived at the Ridley Beach rookery, Cape Adare, on October 13. A blizzard came on then, with thick drift which prevented any observations being made. The next day, when this subsided, there were no penguins to be seen.

By the morning of October 19 there had been a good many more arrivals, but the rookery was not yet more than one-twentieth part full. All the birds were fasting absolutely. Nest building was now in full swing, and the whole place waking up to activity. Most of the pebbles for the new nests were being taken from old nests, but a great deal of robbery went on nevertheless. Depredators when caught were driven furiously away, and occasionally chased for some distance, and it was curious to see the difference in the appearance between the fleeing thief and his pursuer. As the former raced and ducked about among the nests, doubling on his tracks, and trying by every means to get lost in the crowd and so rid himself of his pursuer, his feathers lay close back on his skin, giving him a sleek look which made him appear half the size of the irate nest-holder who sought to catch him, with feathers ruffled in indignation. This at first led me to think that the hens were larger than the cocks, as it was generally the hen who was at home, and the cock who was after the stones, but later I found that sex makes absolutely no difference in the size of the birds, or indeed in their appearance at all, as seen by the human eye. After mating, their behavior as well as various outward signs serve to distinguish male from female. Besides this, certain differences in their habits are to be noted.

The consciousness of guilt, however, always makes a penguin smooth his feathers and look small; while indignation has the opposite effect. Often when observing a knoll crowded with nesting penguins, I have seen an apparently undersized individual slipping quietly along among the nests, and always by his subsequent proceedings he has turned out to be a robber on the hunt for his neighbors' stones. The others, too, seemed to know it, and would have a peck at him as he passed them.

At last he would find a hen seated unwarily on her nest, slide up behind her, deftly and silently grab a stone, and run off triumphantly with it to his mate who was busily arranging her own home. Time after time he would return to the same spot, the poor depredated nest-holder being quite oblivious of the fact that the side of her nest which lay behind her was slowly but surely vanishing stone by stone.

Here could be seen how much individual character makes for success or failure in the efforts of the penguins to produce and rear their offspring. There are vigilant birds, always alert, who seem never to get robbed or molested in any way: these have big high nests, made with piles of stones. Others are unwary and get huffed as a result. There are a few even who, from weakness of character, actually allow stronger-natured and more aggressive neighbors to rob them under their very eyes.

In speaking of the robbery which is such a feature of the rookery during nest building, special note must be made of the fact that violence is never under any circumstances resorted to by the thieves. When detected, these invariably beat a retreat, and offer not the least resistance to the drastic punishment they receive if they are caught by their indignant pursuers. The only disputes that ever take place over the question of property are on the rare occasions when a bona-fide misunderstanding arises over the possession of a nest. These must be very rare indeed, as only on one occasion have I seen such a quarrel take place. The original nesting sites being, as I will show, chosen by the hens, it is the lady, in every case, who is the cause of the battle; and when she is won her scoop goes with her to the victor.

As I grew to know these birds from continued observation, it was surprising and interesting to note how much they differed in character; though the weaker-minded who would actually allow themselves to be robbed were few and far between, as might be expected. Few, if any, of these ever could succeed in hatching their young and winning them through to the feathered stage.

When starting to make her nest, the usual procedure is for the hen to squat on the ground for some time, probably to thaw it, then working with her claws to scratch away at the material beneath her, shooting out the rubble behind her. As she does this she shifts her position in a circular direction until she has scraped out a round hollow. Then the cock brings stones, performing journey after journey, returning each time with one pebble in his beak which he deposits in front of the hen, who places it in position.

Sometimes the hollow is lined with a nest pavement of stones placed side by side, one layer deep, on which the hen squats, afterwards building up the sides about her. At other times the scoop would be filled up indiscriminately by a heap of pebbles, on which the hen then sat, working herself down into a hollow in the middle.

Individuals differ, not only in their building methods, but also in the size of the stones they select. Side by side may be seen a nest composed wholly of very big stones, so large that it is a matter of wonder how the birds can carry them, and another nest of quite small stones.

Different couples seem to vary much in character or mood. Some can be seen quarreling violently, whilst others appear most affectionate, and the tender politeness of some of these latter toward one another is very pretty to see.

I may here mention that the temperatures were rising considerably by October 19, ranging about zero F.

During October 20 the stream of arrivals was incessant. Some mingled at once with the crowd; others lay in batches on the sea-ice a few yards short of the rookery, content to have got so far, and evidently feeling the need for rest after their long journey from the pack. The greater part of this journey was doubtless performed by swimming, as they crossed open water, but I think that much of it must have been done on foot over many miles of sea-ice, to account for the fatigue of many of them.

On the ice they have two modes of progression. The first is simple walking. Their legs being very short, their stride amounts at most to four inches. Their rate of stepping averages about one hundred and twenty steps per minute when on the march.

Their second mode of progression is "tobogganing." When wearied by walking or when the surface is particularly suitable, they fall forward on to their white breasts, smooth and shimmering with a beautiful metallic luster in the sunlight, and push themselves along by alternate powerful little strokes of their legs behind them.

When quietly on the march, both walking and tobogganing produce the same rate of progression, so that the string of arriving birds, tailing out in a long line as far as the horizon, appears as a well-ordered procession. I walked out a mile or so along this line, standing for some time watching it tail past me and taking photographs. Most of the little creatures seemed much out of breath, their wheezy respiration being distinctly heard.

First would pass a string of them walking, then a dozen or so tobogganing. Suddenly those that walked would flop on their breasts and start tobogganing; and conversely strings of tobogganers would as suddenly pop up on to their feet and start walking. In this way they relieved the monotony of their march, and gave periodical rest to different groups of muscles and nerve-centers.

The surface of the snow on the sea-ice varied continually, and over any very smooth patches the pedestrians almost invariably started to toboggan, whilst over "bad going" they all had perforce to walk.

On October 21 many thousands of penguins arrived from

the northerly direction and poured on to the beach in a continuous stream, the snaky line of arrivals extending unbroken across the sea-ice as far as the eye could see.

Although squabbles and encounters had been frequent since their arrival in any numbers, it now became manifest that there were two very different types of battles; first, the ordinary quarreling consequent on disputes over nests and the robbery of stones for these, and secondly, the battles between cocks who fought for the hens. These last were more earnest and severe, and were carried to a finish, whereas the first-named rarely proceeded to extremes.

In regard to the mating of the birds, the following most interesting customs seemed to be prevalent.

The hen would establish herself on an old nest, or in some cases scoop out a hollow in the ground and sit in or by this, waiting for a mate to propose himself. She would not attempt to build while she remained unmated. During the first week of the nesting season, when plenty of fresh arrivals were continually pouring into the rookery, she did not have long to wait as a rule. Later, when the rookery was getting filled up, and only a few birds remained unmated in that vast crowd of some three-quarters of a million, her chances were not so good.

For example, on November 16 on a knoll thickly populated by mated birds, many of which already had eggs, a hen was observed to have scooped a little hollow in the ground and to be sitting in this. Day after day she sat on looking thinner and sadder as time passed and making no attempt to build her nest. At last, on November 27, she had her reward, for I found that a cock had joined her, and she was busily building her nest in the little scoop she had made so long before, her husband steadily working away to provide her with the necessary pebbles. Her forlorn appearance of the past ten days had entirely given place to an air of occupation and happiness.

As time went on I became certain that invariably pairing

took place after arrival at the rookery. On October 23 I went to the place where the stream of arrivals was coming up the beach, and presently followed a single bird, which I afterwards found to be a cock, to see what it was going to do. He threaded his way through nearly the whole length of the rookery by himself, avoiding the tenanted knolls where the nests were, by keeping to the emptier hollows. About every hundred yards or so he stopped, ruffled up his feathers, closed his eyes for a moment, then "smoothed himself out" and went on again, thus evidently struggling against desire for sleep after his journey. As he progressed he frequently poked his little head forward and from side to side, peering up at the knolls, evidently in search of something.

Arrived at length at the south end of the rookery, he appeared suddenly to make up his mind, and boldly ascending a knoll which was well tenanted and covered with nests, walked straight up to one of these on which a hen sat. There was a cock standing at her side, but my little friend either did not see him or wished to ignore him altogether. He stuck his beak into the frozen ground in front of the nest, lifted up his head and made as if to place an imaginary stone in front of the hen, a most obvious piece of dumb show. The hen took not the slightest notice nor did her mate.

My friend then turned and walked up to another nest, a yard or so off, where another cock and hen were. The cock flew at him immediately, and after a short fight, in which each used his flippers savagely, he was driven clean down the side of the knoll away from the nests, the victorious cock returning to his hen. The newcomer, with the persistence which characterizes his kind, came straight back to the same nest and stood close by it, soon ruffling his feathers and evidently settling himself for a doze; but, I suppose because he made no further overtures, the others took no notice of him at all, as, overcome by sheer weariness, he went to sleep and remained so until I was too cold to await further developments. On my way back to our hut I followed another cock for about thirty yards, when he walked up to another couple at a nest and gave battle to the cock. He, too, was driven off after a short and decisive fight. Soon there were many cocks on the war-path. Little knots of them were to be seen about the rookery, the lust of battle in them, watching and fighting each other with desperate jealousy, and the later the season advanced the more "bersac" they became. The roar of battle and thuds of blows could be heard continuously, and of the hundreds of such fights, all plainly had their cause in rivalry for the hens.

When starting to fight, the cocks sometimes peck at each other with their beaks, but always they very soon start to use their flippers, standing up to one another and raining in the blows with such rapidity as to make a sound which, in the words of Dr. Wilson, resembles that of a boy running and dragging his hoop-stick along an iron paling. Soon they start "in-fighting," in which position one bird fights right-handed, the other left-handed; that is to say, one leans his left breast against his opponent, swinging in his blows with his right flipper, the other presenting his right breast and using his left flipper. My photographs of cocks fighting all show this plainly. It is interesting to note that these birds, though fighting with one flipper only, are ambidextrous. Whilst battering one another with might and main they use their weight at the same time, and as one outlasts the other, he drives his vanquished opponent before him over the ground, as a trained boxing man when "in-fighting" drives his exhausted opponent round the ring.

Desperate as these encounters are, I don't think one penguin ever kills another. In many cases blood is drawn. I saw one with an eye put out, and that side of its beak (the right side) clotted with blood, while the crimson print of a blood-stained flipper across a white breast was no uncommon sight.

Hard as they can hit with their flippers, however, they are also well protected by their feathers, and being marvelously tough and enduring, the end of a hard fight merely finds the vanquished bird prostrated with exhaustion and with most of the breath beaten out of his little body. The victor is invariably satisfied with this, and does not seek to dispatch him with his beak.

It was very usual to see a little group of cocks gathered together in the middle of one of the knolls squabbling noisily. Sometimes half a dozen would be lifting their raucous voices at one particular bird, then they would separate into pairs, squaring up to one another and emphasizing their remarks from time to time by a few quick blows from their flippers. It seemed that each was indignant with the others for coming and spoiling his chances with a coveted hen, and trying to get them to depart before he went to her.

It was useless for either to attempt overtures whilst the others were there, for the instant he did so, he would be set upon and a desperate fight would begin. Usually, as in the case I described above, one of the little crowd would suddenly "see red" and sail into an opponent with desperate energy, invariably driving him in the first rush down the side of the knoll to the open space surrounding it, where the fight would be fought out, the victor returning to the others, until by his prowess and force of character, he would rid himself of them all. Then came his overtures to the hen. He would, as a rule, pick up a stone and lay it in front of her if she were sitting in her "scoop," or if she were standing by it he might himself squat in it. She might take to him kindly, or, as often happened, peck him furiously. To this he would submit tamely, hunching up his feathers and shutting his eyes while she pecked him cruelly. Generally after a little of this she would become appeased. He would rise to his feet, and in the prettiest manner edge up to her, gracefully arch his neck, and with soft guttural sounds pacify her and make love to her.

Both perhaps would then assume the "ecstatic" attitude, rocking their necks from side to side as they faced one another, and after this a perfect understanding would seem to grow up between them, and the solemn compact was made.

It is difficult to convey in words the daintiness of this pretty little scene. I saw it enacted many dozens of times, and it was wonderful to watch one of these hardy little cocks pacifying a fractious hen by the perfect grace of his manners.

This antic is gone through by both sexes and at various times, though much more frequently during the actual breeding season. The bird rears its body upward and stretching up its neck in a perpendicular line, discharges a volley of guttural sounds straight at the unresponding heavens. At the same time the clonic movements of its syrinx or "sound box" distinctly can be seen going on in its throat. Why it does this I have never been able to make out, but it appears to be thrown into this ecstasy when it is pleased; in fact, the zoologist of the "*Pourquoi Pas*" expedition termed it the "*Chant de Satisfaction.*" I suppose it may be likened to the crowing of a cock or the braying of an ass. When one bird of a pair starts to perform in this way, the other usually starts at once to pacify it.

On November 3 several eggs were found, and on the 4th these were beginning to be plentiful in places, though many of the colonies had not yet started to lay.

Let me here call attention to the fact that up to now not a single bird out of all those thousands had left the rookery once it had entered it. Consequently not a single bird had taken food of any description during all the most strenuous part of the breeding season, and as they did not start to feed till November 8, thousands had to my knowledge fasted for no fewer than twenty-seven days. Now of all the days of the year these twenty-seven are certainly the most trying during the life of the Adelie.

With the exception, in some cases, of a few hours immediately after arrival (and I believe the later arrivals could not afford themselves even this short respite), constant vigilance had been maintained; battle after battle had been fought; some had been nearly killed in savage encounters, recovered, fought again and again with varying fortune. They had mated at last, built their nests, procreated their species, and, in short, met the severest trials that Nature can inflict upon mind and body; and at the end of it, though in many cases blood-stained and in all caked and bedraggled with mire, they were as active and as brave as ever.

By November 7, though many nests were still without eggs, a large number now contained two, and their owners started, turn and turn about, to go to the open water leads about a third of a mile distant, to feed, and as a result of this a change began gradually to come over the face of the rookery. Hitherto the whole ground in the neighborhood of the nests had been stained a bright green. This was due to the fasting birds continually dropping their watery, bile-stained excreta upon it. (The gall of penguins is bright green.) These excreta practically contained no solid matter except epithelial cells and salts.

The nests themselves are never fouled, the excreta being squirted clear of them for a distance of a foot or more, so that each nest has the appearance of a flower with bright green petals radiating from its center. Even when the chicks have come and are being sat upon by the parents, this still holds good, because they lie with their heads under the old bird's belly and their hindquarters just presenting themselves, so that they may add their little decorative offerings, petal by petal! Now that the birds were going to feed, the watery-green stains upon the ground gave place to the characteristic bright brick-

red guano, resulting from their feeding on the shrimp-like euphausia in the sea; and the color of the whole rookery was
changed in a few days, though this was first noticeable, of
course, in the region of those knolls which had been occupied
first, and which were now settled down to the peaceable and
regular family life which was to last until the chicks had grown.

During the fasting season, as none of the penguins had
entered the water, they all became very dirty and disreputable
in appearance, as well may be imagined considering the life
they led; but now that they went regularly to swim, they immediately got back their sleek and spotless state.

From the ice-foot to the open water, the half mile or so
of sea-ice presented a lively scene as the thousands of birds
passed to and fro over it, outward bound parties of dirty birds
from the rookery passing the spruce bathers, homeward bound
after their banquet and frolic in the sea. So interesting and
instructive was it to watch the bathing parties, that we spent
whole days in this way.

As I have said before, the couples took turn and turn about
on the nest, one remaining to guard and incubate while the
other went off to the water.

On leaving their nests, the birds made their way down the
ice-foot on to the sea-ice. Here they would generally wait about
and join up with others until enough had gathered together to
make up a decent little party, which would then set off gaily
for the water. They were now in the greatest possible spirits,
chattering loudly and frolicking with one another, and playfully
chasing each other about, occasionally indulging in a little
friendly sparring with their flippers.

Arrived at length at the water's edge, almost always the
same procedure was gone through. The object of every bird
in the party seemed to be to get one of the others to enter the
water first. They would crowd up to the very edge of the ice,
dodging about and trying to push one another in. Sometimes

those behind nearly would succeed in pushing the front rank in, who then would just recover themselves in time, and rushing round to the rear, endeavor to turn the tables on the others. Occasionally one actually would get pushed in, only to turn quickly under water and bound out again onto the ice like a cork shot out of a bottle. Then for some time they would chase one another about, seemingly bent on having a good game, each bird intent on finding any excuse from being the first in. Sometimes this would last a few minutes, sometimes for the better part of an hour, until suddenly the whole band would change its tactics, and one of the number start to run at full tilt along the edge of the ice, the rest following closely on his heels, until at last he would take a clean header into the water. One after another the rest of the party followed him, all taking off exactly from the spot where he had entered, and following one another so quickly as to have the appearance of a lot of shot poured out of a bottle into the water.

A dead silence would ensue till a few seconds later, when they would all come to the surface some twenty or thirty yards out, and start rolling about and splashing in the water, cleaning themselves and making sounds exactly like a lot of boys calling out and chaffing one another.

So extraordinary was this whole scene, that on first witnessing it we were overcome with astonishment, and it seemed to us almost impossible that the little creatures whose antics we were watching were actually birds and not human beings. Seemingly reluctant as they had been to enter the water, when once there they evinced every sign of enjoyment, and would stay in for hours at a time.

As may be imagined, the penguins spent a great deal of time on their way to and from the water, especially during the earlier period before the sea-ice had broken away from the ice-foot, as they had so far to walk before arriving at the open leads.

As a band of spotless bathers returning to the rookery, their white breasts and black backs glistening with a fine metallic luster in the sunlight, met a dirty and bedraggled party on its way out from the nesting ground, frequently both would stop, and the clean and dirty mingle together and chatter with one another for some minutes. If they were not speaking words in some language of their own, their whole appearance belied them, and as they stood, some in pairs, some in groups of three or more, chattering amicably together, it became evident that they were sociable animals, glad to meet one another, and like many men, pleased with the excuse to forget for a while their duties at home, where their mates were waiting to be relieved for their own spell off the nests.

After a variable period of this intercourse, the two parties would separate and continue on their respective ways, a clean stream issuing from the crowd in the direction of the rookery, a dirty one heading off towards the open water; but here it was seen that a few who had bathed and fed, and were already perhaps half-way home, had been persuaded to turn and accompany the others, and so back they would go again over the way they had come, to spend a few more hours in skylarking and splashing about in the sea.

In speaking of these games of the penguins, I wish to lay emphasis on the fact that these hours of relaxation play a large part in their lives during the advanced part of the breeding period. They would spend hours in playing at a sort of "touch last" on the sea-ice near the water's edge. They never played on the ground of the rookery itself, but only on the sea-ice and the ice-foot and in the water; and I may here mention another favorite pastime of theirs. Small ice-floes are continually drifting past in the water, and as one of these arrived at the top of the ice-foot, it would be boarded by a crowd of penguins, sometimes until it could hold no more. This "excursion boat," as we used to call it, would float its many occupants down the whole length of the ice-foot, and if it passed close to the edge, those that rode on the floes would shout at the knots of penguins gathered along the ice-foot who would shout at them in reply, so that a gay bantering seemed to accompany their passage past the rookery.

Arrived at the farther end, some half a mile lower down, those on the "excursion boat" had perforce to leave it, all plunging into the tide and swimming against this until they came to the top again, then boarded a fresh floe for another ride down. All day these floes, often crowded to their utmost capacity, would float past the rookery. Often a knot of hesitating penguins on the ice-foot, on being hailed by a babel of voices from a floe, would suddenly make the plunge, and all swim off to join their friends for the rest of the journey, and I have seen a floe so crowded that as a fresh party boarded it on one side, many were pushed off the other side into the water by the crush.

Once, as we stood watching the penguins bathing, one of them popped out of the water onto the ice with a large pebble in its mouth, which it had evidently fetched from the bottom. This surprised me, as the depth of the sea here was some ten fathoms at least. The bird simply dropped the stone on the ice and then dived in again, so that evidently he had gone to all the trouble of diving for the stone simply for the pleasure of doing it. Mr. J. H. Gurney in his book on the gannet, says they (gannets) are said to have got themselves entangled in fishing-nets at a depth of 180 ft. and that their descent to a depth of 90 ft. is quite authentic, so that perhaps the depth of this penguin's dive was not an unusual one.

The tide at the open water leads where they bathed ran a good six knots, but the Adelies swam quite easily against this without leaving the surface.

In the water, as on the land, they have two means of progression. The first is by swimming as a duck swims, except that they lie much lower in the water than a duck does, the top of the back being submerged, so that the neck sticks up out of the water. As their feet are very slightly webbed, they have not the advantages that a duck or gull has when swimming in this way, but supplement their foot-work by short quick strokes of their flippers. This they are easily able to do, owing to the depth to which the breast sinks in the water.

The second method is by "porpoising."

This consists in swimming underwater, using the wings or "flippers" for propulsion, the action of these limbs being practically the same as they would be in flying. As their wings are beautifully shaped for swimming, and their pectoral muscles extraordinarily powerful, they attain great speed, besides which they are as nimble as fish, being able completely to double in their tracks in the flash of a moment. In porpoising, after traveling thirty feet or so underwater, they rise from it, shooting clean out with an impetus that carries them a couple of yards in the air, then with an arch of the back they are head first into the water again, swimming a few more strokes, then out again, and so on.

Perhaps the most surprising feat of which the Adelie is capable is seen when it leaps from the water onto the ice. We saw this best later in the year when the sea-ice had broken away from the ice-foot, so that open water washed against the ice cliff bounding the land. This little cliff rose sheer from the water at first, but later, by the action of the waves, was undercut for some six feet or more in places, so that the ledge of ice at the top hung forwards over the water. The height of most of this upper ledge varied from three to six feet.

Whilst in the water, the penguins usually hunted and played in parties, just as they had entered it, though a fair number of solitary individuals were also to be seen. When a party had satisfied their appetites and their desire for play, they would swim to a distance of some thirty to forty yards from the ice-foot, when they might be seen all to stretch their necks up and take a good look at the proposed landing-place. Having done this, every bird would suddenly disappear beneath the surface, not a ripple showing which direction they had taken, till suddenly, sometimes in a bunch, sometimes in a stream, one after the other they would all shoot out of the water, clean up on to the top of the ice-foot. Several times I measured the distance from the surface of the water to the ledge on which they landed, and the highest leap I recorded was exactly five feet. The "take-off" was about four feet out from the edge, the whole of the necessary impetus being gained as the bird approached beneath the water.

The most important thing to note about this jumping from the water was the accuracy with which they invariably rose at precisely the right moment, the exact distance being judged during their momentary survey of a spot from a distance, before they dived beneath the water, and carried in their minds as they approached the ice. I am sure that this impression was all they had to guide them, as with a ripple on the water, and at the pace they were going, they could not possibly have seen their landing place at all clearly as they approached it, besides which, in many cases, the ledge of ice on which they landed projected many feet forwards from the surface, yet I never saw them misjudge their distance so as to come up under the overhanging ledge.

During their approach they swam at an even distance of about three or four feet beneath the surface, projecting themselves upwards by a sudden upward bend of the body, at the same time using their tail as a helm.

Their quickness of perception is shown very well as they land on the ice. If the surface is composed of snow, and so affords them a good foothold, they throw their legs well forward and land on their feet; but should they find themselves landing on a slippery ice-surface, they throw themselves forward, landing on their breasts in the tobogganing position.

The Adelies dive very beautifully. We did not see this at first, before the sea-ice had gone out, because to enter the water they had only to drop a few inches; but later, when entering from the ice terraces, we constantly saw them making the most graceful dives.

At the place where they most often went in, a long terrace of ice about six feet in height ran for hundreds of yards along the edge of the water, and here, just as on the sea-ice, crowds would stand near the brink. When they had succeeded in pushing one of their number over, all would crane their necks over the edge, and when they saw the pioneer safe in the water, the rest followed.

When diving into shallow water they fall flat, but into deep water and from any considerable height, they assume the most perfect positions and make very little splash. Occasionally we saw them stand hesitating to dive at a height of some twenty feet, but generally they descended to some lower spot, and did not often dive from such a height; but twelve feet was no uncommon dive for them.

The reluctance shown by each individual of a party of intending bathers to be the first to enter the water may partly have been explained when, later on, we discovered that a large number of sea-leopards were gathered in the sea in the neighborhood of the rookery to prey on the penguins. These formidable animals used to lurk beneath the overhanging ledges of the ice-foot, out of sight of the birds on the ice overhead. They lay quite still in the water, only their heads protruding, until a party of Adelies would descend into the water almost on top of them, when with a sudden dash and snap of their great formidable jaws, they would secure one of the birds.

It seemed to me then that all the chivvying and preliminaries which they went through before entering the water arose mainly from a desire on the part of each penguin to get one of its neighbors to go in first in order to prove whether the coast was clear or not, though all this maneuvering was certainly taken very lightly, and quite in the nature of a game. This indeed was not surprising, for of all the animals of which I have had any experience, I think the Adelie penguin is the very bravest. The more we saw of them the fonder we became of them and the more we admired their indomitable courage. The appearance of a sea-leopard in their midst was the one thing that caused them any panic. With dozens of these enemies about they would gambol in the sea in the most light-hearted manner, but the appearance of one among them was the signal for a stampede; but even this was invariably gone through in an orderly manner with some show of reason, for, porpoising off in a clump, they at once spread themselves out, scattering in a fan-shaped formation as they sped away, instead of all following the same direction.

As far as I could judge, however, the sea-leopards are a trifle faster in the water than the Adelies, as one of them occasionally would catch up with one of the fugitives, who then, realizing that speed alone would not avail him, started dodging from side to side, and sometimes swam rapidly round and round in a circle of about twelve feet diameter for a full minute or more, doubtless knowing that he was quicker in turning than his great heavy pursuer; but exhaustion would overtake him in the end, and we could see the head and jaws of the great sea-leopard rise to the surface as he grabbed his victim. The sight of a panic-stricken little Adelie tearing round and round in this manner was a sadly common sight late in the season.

When the sea-ice had gone out, leaving open water right up to the ice-foot, a ledge of ice was left along the western side of the rookery, forming a sort of terrace or "front," with its sides composed of blue ice, rising sheer out of the water to a height of some six feet or more in places. From this point of vantage it was possible to stand and watch the penguins as they swam in the clear water below, and some idea was formed of their wonderful agility when swimming beneath the surface. As they propelled themselves along with powerful strokes of their wings, they swerved from side to side to secure the little prawn-like euphausia which literally swam everywhere in the Antarctic seas, affording them ample food at all times. Their gluttonous habits here became very evident. They would gobble euphausia until they could hold no more, only to vomit the whole meal into the water as they swam, and so lightened, start to feast again. As they winged their way along several feet beneath the surface, a milky cloud would suddenly issue from their mouths and drift slowly away downstream, as without the slightest pause in their career they dashed eagerly along in the hunt for more.

When a penguin returned to his mate on the nest, after his jaunt in the sea, much formality had to be gone through before he was allowed to take charge of the eggs. This ceremony of "relieving the guard" almost invariably was observed.

Going up to his mate, with much graceful arching of his neck, he appeared to assure her in guttural tones of his readiness to take charge. At this she would become very agitated, replying with raucous staccato notes, and refusing to budge from her position on the eggs. Then both would become angry for a while, arguing in a very heated manner, until at last she would rise, and standing by the side of the nest, allow him to walk onto it, which he immediately did, and after carefully placing the eggs in position, sink down upon them, afterwards thrusting his bill beneath his breast to push them gently into a comfortable position. After staying by him for a little while, the other at length would go off to bathe and feed.

When the chicks began to appear all over the rookery, a marked change was noticed in the appearance of the parents as they made their way on foot from the water's edge to the nests. Hitherto they had been merely remarkable for their spotless and glistening plumage, but now they were bringing with them food for the young, and so distended were their stomachs with this, that they had to lean backward as they walked, to counterbalance their bulging bellies, and in consequence frequently tripped over the inequalities of the ground which were thus hidden from their gaze.

What with the exertion of tramping with their burden across the rookery, and perhaps on rare occasions one or two little disputes with other penguins by the way, frequently they were in some distress before they reached their destination, and quite commonly they would be sick and bring up the whole offering before they got there. Consequently, little red heaps of mashed-up and half-digested euphausia were to be seen about the rookery. Once I saw a penguin, after he had actually reached the nest, quite unable to wait for the chick to help itself in the usual manner, deposit the lot upon the ground in front of his mate. When this happens the food is wasted, as neither chick nor adult will touch it however hungry they may be, the former only feeding by the natural method of pushing his head down the throat of a parent, and so helping himself direct from the gullet.

When the chicks are small they are kept completely covered by the parent who sits on the nest. They grow, however, at an enormous rate, gobbling vast quantities of food as it is brought to them, their elastic bellies seeming to have no limit to their capacity; indeed, when standing, they rest on a sort of tripod, formed by the protuberant belly in front and the two feet behind.

To see an Adelie chick of a fortnight's growth trying to get itself covered by its mother is a most ludicrous sight. The most it can hope for is to get its head under cover, the rest of its body being exposed to the air; but the downy coat of the chick is close and warm, and suffices in all weathers to protect it from the cold.

Some way back I made allusion to the way in which many of the penguins were choosing sites up the precipitous sides of the Cape at the back of the rookery. Later I came to the conclusion that this was purely the result of their love of climbing. There was one colony at the very summit of the Cape, whose inhabitants could only reach their nests by a long and trying climb to the top and then by a walk of some hundred yards across a steep snow slope hanging over the very brink of a sheer drop of seven hundred feet onto the sea-ice.

During the whole of the time when they were rearing their young, these mountaineers had to make several journeys during each twenty-four hours to carry their enormous bellyfuls of euphausia all the way from the sea to their young on the nests—a weary climb for their little legs and bulky bodies. The greater number who had undertaken this did so at a time when there were ample spaces unoccupied in the most eligible parts of the rookery.

Large masses of ice were stranded by the sea along the shores of the rookery. These fragments of bergs, some of them fifteen to twenty feet in height, formed a miniature mountain range along the shore. All day, parties of penguins were to be seen assiduously climbing the steep sides of this little range. Time after time, when halfway up, they would descend to try another route, and often when with much pains one had scaled a slippery incline, he would come sliding to the bottom, only to pick himself up and have another try.

Generally, this climbing was done by small parties who had clubbed together, as they generally do, from social inclination. It was not unusual for a little band of climbers to take as much as an hour or more over climbing to the summit. Arrived at the top they would spend a variable period there, sometimes descending at once, sometimes spending a considerable time there, gazing contentedly about them, or peering over the edge to chatter with other parties below.

Again, some half a mile from the beach, a large berg some one hundred feet in height was grounded in fairly deep water, accessible at first over the sea-ice, but later, when this had gone, surrounded by open water. Its sides were sheer except on one side, which sloped steeply from the water's edge to the top.

From the time when they first went to the sea to feed until the end of the season, there was a continual stream of penguins ascending and descending that berg. As I watched them through glasses I saw that they had worn deep paths in the snow from base to summit. They had absolutely nothing to gain by going to all this trouble but the pleasure they seemed to derive from the climb, and when at the top, merely had a good look round and came down again.

Lee Wulff

THE ATLANTIC SALMON

The salmon, as an *alevin*, still carrying part of the egg sac from which she had developed, was one of similar thousands that had been spawned by a pair of oversize Atlantic salmon. She looked like all the other millions and millions of alevins that were emerging from the small pockets of water beneath the stones and coarse sand of the Humber's spawning beds, yet not more than one in five thousand was like her. Taking unto herself the characteristics of her forebears she was destined to be one of the monarchs of the river, if she survived.

She continued to draw nourishment from the dwindling

egg sac until it was completely absorbed. Then, less than an inch long but with muscles hardened and hunger active, she became a *parr* and was on her own. Her food was any living thing that came within her reach and yet was small enough to fit within her mouth. Her fear, an instinctive thing, was of anything that moved and was big enough to eat her. She lived and grew between the two forces, ruthlessly striking those living things upon which she could prey and fearfully fleeing those she could not master. Hers was a world in which death was everywhere and mercy did not exist.

The great bulk of the thousands of fertile eggs of that single spawning pair had survived to the parr stage. They had become a pale cloud of scattered bits of motion on the streambed. Life for them was not yet a matter of individual ability but purely a matter of circumstance. A passing trout chanced to see their movement. The squaretail turned savagely into them. Almost as a unit they darted into tiny caverns of water between the stones for protection, but the trout, by opening his mouth quickly, drew in water from the crevices among the stones and sucked in the small parr, too. He devoured thirty before tiring of the game.

An eel working upstream from the shallow lake below passed through them and took her toll of lives. A salmon parr of three years, a fish five inches long, found them and ate dozens. But, of course, the salmon was not one of those to die.

As days went by she worked her way farther and farther from the shoal of her birth and became more and more of an individual, more dependent upon her own quickness and foresight for food and safety. She was at home then in the twisting currents that made up the flow around her. Her body was covered with its full complement of scales, each growing, ring by ring, as she expanded. Those below were iridescent silver and those of her back were dark brown. Down each side the brown pigmentation extended in a series of narrow transverse bands called finger marks, and overlaying both the silver and brown of her sides was a pattern of bright vermilion spots. Another pattern of black spots was slightly less noticeable in the solid brown of her back. Her tail was definitely forked.

One day she saw an insect settle to the undulating mirrorlike surface above her and struggle to rise into flight again. Before her mind had fully realized her opportunity, instinct had started her on a wild rush upward. Her jaws closed on the midge and she bent her body downward for her dive. But her speed was too great for that maneuver. Instead, she carried on up into the air in her first clean leap. From then on she leaped often in her surface feeding.

She spent her first winter under the ice in Birchey Pond, that wide but shallow lake half a mile downstream, clinging to the shore waters where she could dart quickly into spaces between the ice and lake bed too narrow for the preying trout to follow. There, in the greater mass of the past summer's dead vegetation, she found more insect life and better feeding than was to be found in the river itself. The temperature dropped almost to freezing and a strange lethargy possessed her. Even though food was scarcer than in the summer she cared little and hunted for it less. Her scale rings came closer together for the time element between them was always the same and her body was growing little.

One gray day of winter when she had been idly swimming along just under the ice, a great black shape moved toward her. She swam from it frantically and not until she turned away from its course did she realize that she had not been pursued. She had seen a spent salmon, black, thin, and ugly, a long black skeleton of a fish that had lost more than half its weight through spawning and starvation. Like all salmon, restless fish that they are, the slink wandered the lake aimlessly, waiting unknowingly for the freshening water of spring that would start another change within him. When that change came he would turn silvery again and be possessed of a sudden hunger and would follow the spring floods down to the sea a second time.

When the slink's coat brightened and he headed downstream to the sea, the salmon also lost her lethargy. Food was everywhere and she fed ravenously again, her speed and fierceness remarkable for her two-inch length. Her second year was a continuation of her first and in it she almost doubled her length. She liked the lake and moved constantly through its quiet waters. Occasionally an urge moved her into the current where the river flowed in, holding a place in the current for a while but each time the urge faded and she returned to cruise the still waters. Most of the Humber's parr were clinging to the current where they swam no more, no less, than she did to hold their stream positions, but they found less food and grew more slowly.

The warming water of her second spring filled her with a new desire. It was nothing so simple as her need for food or even her selection of the still water as her feeding ground. It was an intangible force that drew her to the outlet of the lake, where the brown water poured wildly over the lip to race through the rapids below. Other parr were gathering there, too, in schools, taking a place in the current and holding it.

Only occasionally did they feed on the passing food. In the end they swam less swiftly than the current and let it carry them backward over the brink into the rough water and, still facing upstream and swimming slowly, they let it take them gradually downstream.

Almost all the other parr were four years old, but they had been spawned by smaller fish. The salmon was precocious and more developed for her age, though smaller than the rest.

The group was tumbled about in the rough water below the Birchey Forks, where the main flow from Aidies Lake comes in, but they all re-formed in groups and moved downstream. There was little hesitation among them now. In the swiftest water they turned to face upstream and let themselves be carried down by swimming forward slowly. Otherwise they turned and swam down with the flow. They kept formation perfectly through the dark and boiling water of the Smooth Rapids, but Bear Reef with its pounding rapids scattered them wildly. Still, there was little change in the content of the group when it re-formed and drifted on.

At the lip of the Big Falls they hesitated. The pull of the swift current was alarming to them and the heavy vibration set up by the falling water was new and strange. It was more than an hour before the first stragglers let themselves be swept back over the high barrier. By twos and threes the rest followed. The water swept them down through the fifteen-foot drop and released them in a swirling mass of foam and water. Twisting and turning, they were carried on to a smoother flow. Her old group scattered, the salmon joined another school, and continued with them toward the sea.

Forty-five miles below their starting point they reached Deer Lake. The school of parr moved steadily westward following the northern shore. Within the salmon her urge to migrate was diminishing. She broke formation frequently to feed. Halfway to the outlet of the lake she swung away from the others, all four-year-olds, and let them travel on without her. The pull of habit to revert to her lake feeding of the year before outweighed everything else and there she spent her third summer. Fall found her almost six inches long.

She began to feed on smaller fish more than before. Small trout, one-year-old parr, and a lesser number of sticklebacks made up much of her diet. Still no insect dared touch the water within her sight. Her rise would be swift, instinctive, certain. It was as if she were a trap that was always set and the sight of a moving insect was the thing that released it. Unless the insect was lifeless or well beneath the surface her rise would carry her on out into the air.

When winter cooled the water and food became scarce again, the old familiar lethargy returned. This time she fought it. In the conflict of decisions the new one won out. She found the outlet and she moved on downstream alone. The Big Rapids that would have unbalanced her the spring before not once threw her over on her side. Downstream she swam, through the narrowing gorge, over the Shellbird Island shoal where the bare, rock face of Breakfast Head towered a thousand feet above her. There was no heed in her for the cars that moved by on the highway or the trains that puffed along the river's southern shore. Twelve miles below Deer Lake she felt the tides and tasted the strong brine that thinned the brown of the Humber water.

For days she moved in and out with the tide, feeding well and growing more rapidly than she had ever grown before. At the end of a week her change in color was complete. A coating of shining silver clothed the lower two-thirds of her body covering the parr markings and red spots of her freshwater coat. Her back had turned from brown to bluish gray, but the scattered black spots still showed through it slightly. Nature had prepared her for the sea. Now she was a *smolt*. From above she looked like a part of the blue-green water of the bay; from below she was a hazy silver shadow against the sky. Leaving the brackish water she moved slowly toward the open sea.

Three years later the salmon is far out at sea and deep down in water that is never brighter than at twilight. She was stirred to another migration. Six years old now, this was the third time since going to sea that she'd felt that annual restlessness. In her fourth spring the desire had passed quickly. In her fifth she had traveled fifty miles before turning back to the bountiful feeding of the ocean's floor. Like her parents and their parents before them she had held to the sea for three full years and grown steadily. It was incongruous to see her now and think of her early years in the river. After three years of stream life she had come to sea, a parr six inches long and less than three ounces in weight. In an equal time at sea she had become a magnificent salmon and her weight was forty-three pounds. Under a compulsion she could no longer resist she started threading her way upward through the underwater valleys.

Following the slope of the sea floor, she worked toward shore, still feeding, but restless, pausing little. One night she struck a strange current in which the fresh water had reduced the salinity of the sea. All the next day she followed it, swimming near the surface, feeling the strangeness of the motion of the waves. At night she turned away again, leaving the vaguely haunting traces of fresh water. Thirty-six hours later she struck another current of mixed waters and again she turned to nose up into it. It drew her between a group of ragged islands and on into a great bay. The water pattern was familiar to her now and as she moved deeper into the bay she was joined by other salmon and they traveled in a school.

Where the currents mixed she held to the major one, tasting in it the bitterness of peat and the remembered taste of limestone in their certain blending. Until then she had not realized clearly that she was headed for the river, but like a farm boy who leaves home at twenty to return at forty and finds the air laden with old scents in a mixture that could nowhere else be duplicated, the big salmon surged ahead, swimming strongly. She ceased to feed entirely, covering the fifteen-mile length of the Humber Arm against the tide in three hours. Then she spent another ten at the river's mouth, moving in and out of the fresh water to soften the shock of the change.

With her in formation swam more than twenty other salmon. Most of them were *grilse*, or salmon of only one year's sea feeding. They were slim fish with forked tails and a weight of only four or five pounds. Others were mature salmon of two years of sea feeding, sturdy fish with broad, square tails, weighing from ten to twenty pounds. Only two were salmon that had spawned before and were returning from their second pilgrimage to the sea. She was the only three-year maiden in the school, the only one with three straight years of the bounty of the sea behind her, and she dwarfed them all.

As they moved in and out of the fresh water, first one and then another of the school leaped, a maneuver the big salmon had almost forgotten during her years of sea feeding. She leaped, too, unknowing practice for the barriers that lay between her and her spawning bed. On the moonlit floodtide the school entered the river and swam steadily against the easy flow. They covered the lower river as a unit, finding no difficulty in traveling through its lazy waters. The whole school followed her leadership as she paralleled the north shore through Deer Lake's seventeen-mile length. Five miles above the lake they struck the first bad water at the Cache Rapids.

A few grilse were missing when the school re-formed at the head of the white water and pushed ahead. At dawn they rested, tasting the bitterness of the brown water and feeling the change that had started within them. Their stomachs were shrinking and their teeth were starting to recede into their gums. Within them spawn and milt were beginning to form. Beneath the urgency that had drawn them back to their rivers was an underlying nausea that made all food distasteful and was to remain with them till long after spawning. They rested on the shoal at Harriman's Steady. Settling to the bottom in scattered groups of two and three, they sought out eddies where they could hold their positions in the soothing flow with the least exertion. Two canoes passed over them and they paid them no more heed than they would have to a pair of lazy blue sharks at sea, for few fish could outswim them. As the canoes passed a few feet over their heads they moved aside, and then moved back.

At the Big Falls less than half the original school was

462

together, but there they joined thousands of salmon who were about to try to leap the high rock face or were resting up from previous failures. The big salmon was much larger than all but a few of the rest. Her stream days of constant leaping were farther behind her and her bulk was so great that even her superior strength would be taxed to the utmost to make the leap successfully.

The first day she did not try to leap the falls, but she did make a few practice leaps, moving into the air after a run of only half a dozen feet. Other salmon were going over regularly but the new-old sensation of running water and rocky streambed was still unusual for her in her new size. The river was high and she took a position in the main current near the center of the pool, moving forward a little as the hours passed until she felt the buffeting and vibration of the falling water.

On the second evening she tried a leap, but it was short and when she fell she struck viciously against the rock well behind the curtain of falling water. The blow stunned her and she lay still on the bottom for a moment before righting herself. Even as she twisted up and moved into the flow, two eels had been closing in, fresh from tearing out the gills of a grilse that had not revived in time. All through the night she rested near the tail of the pool.

The third evening found her back in the foamy water beneath the falls. This time her leap was good. From an eight-foot start she broke into the air and traveled twelve feet upward, head forward, to touch the brown water at the lip of the falls and spurt up through it to conquer the fifteen-foot wall of rock.

Morning found her lying in Taylor's Pool not far from her birthplace. She had hesitated briefly at the Birchey Forks, swimming first to one and then the other. Unerringly she chose the Birchey water and crossed the lake to the upper river. Here at Taylor's Pool she passed on and then turned back, halted by the slight variation in the water above the mouth of Taylor's

Brook. She settled comfortably into a lie out there under four feet of water and rested. Overhead, the water moved smoothly in an apparently even flow, but just ahead of the big salmon a rock jutted up from the streambed to mar the steady sweep of water. Behind it the water eddied and swirled and in that churning flow the salmon rested with little effort and yet enjoyed the soothing motion of the water against her sleek and restless body.

Her teeth were almost completely absorbed into her flesh and they were sore. The spawn within her was growing, too, adding to the nausea that she found in the fresh water. She still looked much as she had in the sea, but here as well a change was taking place. The silver of her scales was beginning to tarnish, a tarnishing as yet so slight that it could only be noticed in comparison with another salmon taken directly from the sea.

She knew the need for conserving her strength to survive the three-month period of starvation that lay between her and the spawning time . . . and for the long months that lay beyond, if she were to live through them, too. Yet the restlessness of all salmon was a part of her, causing a conflict in her mind. Sometimes, when the idleness palled too much, she lifted to the surface to take a floating leaf into her mouth and then eject it. Again, especially in the evening or at night, she would make a sudden spurt forward, lifting a high wave on the still surface as if to reassure herself that her great strength was still complete. Once, as a small white butterfly passed over her in its erratic flight, she lifted, more in instinct than from plan of mind, in a great surge that took her six feet into the air to capture it. And then the fisherman's fly came along . . .

Joseph Wood Krutch

THE GREAT
CHAIN OF LIFE

THE ANIMAL'S FIRST NEED: PROTEIN
AND ORIGINAL SIN

Is a Volvox any less remarkable than a bird, or a bird any less remarkable than a man? Of course it is—in a sense. But miracles cannot be compared. One is quite as incomprehensible as the other and if man did not exist a Volvox or a robin would be as difficult to "explain" as man himself.

To compare the three is like trying to compare infinites. You cannot say how many times greater one is than another because they are larger than anything that can be imagined. Everything that lives is incommensurate with everything that does not. It has characteristics no nonliving thing even hints at, and in that sense life is an absolute.

Of all these characteristics, the most indescribable is consciousness. To *be* and to know that one *is* is the ultimate privilege, and the ultimate burden, of Man. But whether or not consciousness is an essential and universal characteristic of living things we cannot possibly know. All that lives may or may not be at least dimly aware of itself. But whether it is or is not, there are other characteristics—like the ability to grow and to reproduce—which are possessed by the smallest, the simplest, and the humblest. And perhaps the most fundamental of all is the ineluctable necessity for food. Because the inanimate does not necessarily either change, or grow, or reproduce, it is self-sufficient. But even the potentially immortal amoeba must nourish itself or die.

Of Volvox, we admitted that we could not say decisively whether it was plant or animal, though between many of even the one-celled organisms the distinction is already quite definite. But what, after all, is this distinction which we so take for granted? Since we are to be concerned from now on almost exclusively with the animal rather than the plant, we should no doubt ask. And what the answer will finally come down to is: Food.

A delightful and still popular nonsense book is called *How to Tell the Birds from the Flowers*. This, you may say, is not usually very difficult. But if you generalize the question a little further, if you ask how to tell the animals from the plants, it is not so easy. In fact, biologists are usually hard-pressed when it comes to finding satisfactory definitions for such fundamental things or making fundamental distinctions in terms that will stick. In the case of Volvox they give up. As regards the higher creatures the best they have ever been able to do is to say: "Animals are compulsory protein feeders; plants are not."

This may seem pretty farfetched, or at least pretty irrelevant to what we have in mind when we think of an animal. "I love animals" means something. "I love compulsory protein feeders" is nonsense. Nevertheless the definition really is foolproof and it is the only one that is. "Compulsory" has to be put in because of such insect-eating plants as the common sundew, the pitcher plant, and the Venus's-flytrap. All of them consume protein though they can get along without it. No animal can.

What this means in plain language is that all animals must eat something which is or was alive. It may be either a plant or another animal but only plant or animal matter contains protein and without it they cannot live. No animal, therefore, can be innocent as a plant may be. The latter can turn mere inorganic chemicals into living tissue; the animals cannot. All of them must live off something else. And that, perhaps, is the deepest meaning of Original Sin.

The soul—if there is any such thing—is unknown except as it inhabits an animal body. And the animal body must be nourished by what is, or has recently been, living a life of its own. Hence the first necessity for every animal is the necessity of finding something to eat. He may or may not require shelter to protect himself and he may or may not have to discover or provide some sort of home for his offspring. But find something to eat he must.

Of the temperamental Madam Fremstad it is said that when she sat down to a certain dinner she flung the roast on the floor with an indignant exclamation: "Pork before Parsifal—Bah!" To a bird, on the other hand, worms are a legitimate part of the joy of life and he sees no incongruity in the fact that they are turned into song.

Most birds accomplish this necessary business in direct,

relatively simple ways. Nearly everybody has seen the robin tugging away at an earthworm who holds on for dear life—but usually in vain. The only mysterious part of the business is the robin's gift for knowing where the worms are and to this day a dispute still rages over the question whether he has been listening for his underground breakfast when he cocks his head and then digs, or merely getting one eye into a position where it can look straight down.

The warblers search indefatigably for insects on leaves and blossoms; nighthawks and whippoorwills fly with huge wide-open mouths to funnel in gnats, flies, and moths on the wing. Your woodpecker makes the bark fly from dead limbs in order to get at what lurks beneath, though he also, especially in spring, drums vigorously just for the sake of making an interesting noise. And because he is a percussionist rather than a singer, he will be grateful for the novel timbre of a tin roof if you have one.

The sapsucker who drills little circles of holes around your apple tree to drink sap seems a bit more ingenious than most other birds, and certain woodpeckers who drive acorns into prepared recesses in the bark of a tree in such a way that the squirrels will not be able to get at them before the birds dig them out again in winter seem a little more so. Another woodpecker, one who uses a small stick for the same purpose, is astonishing. But on the whole birds tend to be mere hunters of small game or collectors of wild fruits who live off the country like primitive nomads.

Moreover, despite Western man's aversion to insects, their food seems reasonable and proper. But if every animal must eat something organic, it seems to be also true that everything organic can be eaten by something. Cockroaches thrive on the ancient paste beneath wallpaper and they as well as various other insects revel in the glue of book bindings. "Poisonous" is a meaningless term unless you specify "to whom." Thus the common chipmunk of the East nibbles with impunity the Deadly Amanita or Destroying Angel mushroom, one of the most potent poisons in the world. The so-called drugstore beetle can live for years sealed into a bottle of such violent poisons as aconite and belladonna.

Some creatures, including man, are almost omnivorous and in a pinch can be nourished on almost anything organic, though man cannot get any good out of the grass (not even if he cooks it) on which the ox grows great. Other animals are absurdly restricted, like the Australian koala "bear" who has caused zoo keepers endless trouble because he can't eat anything except mature eucalyptus leaves—young ones poison and will ultimately kill him. Most whales, including the species that grows to be the bulkiest and heaviest of all the animals of sea or land, nourish themselves preposterously on microscopic or nearly microscopic plankton floating in the ocean, though one sort—the killer whale—seems to go against nature by slaughtering almost any animal that comes his way.

Literally nothing is wasted. Everything nourishes something else until the bacteria finally get hold of it and return it to the soil after breaking it down once more into inorganic compounds which no animal could eat but which plants can again transform into protein. Hamlet shared the ancient delusion that decently buried men become the food of maggots, though actually "the devouring worm" cannot reach them. But he had, in general, the right idea when he exclaimed: "A man may fish with the worm that hath eat of a king, and eat of the fish that hath fed of that worm."

On the physical plane no doctrine could be truer than that of reincarnation. Every living body, including our own, has lived many times before. Humble plants seize upon merely chemical elements or compounds and organize them into more complex combinations which thereupon begin to live or, as some would say, to be used by life—which is not the same thing. The animal that eats them raises these compounds to even more advanced levels of complexity and they are inspired with more complex forms of life. The robin who visits his nestlings every five minutes is busy turning bugs into birds.

Omnivorous man may eat the animal and thus build his flesh from its flesh. If he is at last returned directly to earth then what was a man is resolved quite rapidly into plant food again. If his body is burned, then it returns by only one step to the merely inorganic. But if by any chance he is, as was once more common, devoured by some carnivorous beast, then at least one more step downward interposes. But by whatever stages he returns to chemical simplicity the upward phase will presently begin anew. After each fall there is a rise. During millions of years the level achieved at the end of the rise was higher than any that had ever been reached during any million years before man evolved. And nobody knows whether or not the upper limit has been reached.

Thus the robin pulling a worm from the earth or a whale straining plankton from the sea is not merely satisfying an animal need. He is lifting protein from one level to another. Perhaps, as some would say, the higher form of life is "nothing but" the more highly organized compounds that constitute the body of the bird, the whale, or the man. Perhaps, as others would insist, the more complex body is merely somehow necessary to the higher life which informs it. But in either case the individual could not start where the plant started. Some living thing, either plant or animal, had to manufacture for him those compounds that are the simplest he can use.

Only man seems to have made much of an art or ritual out of eating. And that fact is more surprising than might at first sight be supposed. After all he was long preceded by other creatures in the elaboration of various other necessary or useful activities. Thousands of years before he organized games for exercise, animals had been engaged in play markedly formalized. They had also elaborated family life and given to the business of caring for their offspring emotional concomitants that point plainly toward the human. Their ability to make sounds of some sort may have been at first merely a warning to enemies or even an involuntary cry of rage, but long before man learned to speak, animals had seized upon their simple gift and elaborated it into song or some other system of sounds ritually performed. As for the sex instinct, it began many eons ago to generate activities almost as protean as those of man himself. During courtship animals sing, dance, waft perfumes, display colors, and indulge in innumerable forms of mere showing off. In this department there was little left for human beings to invent. But the art, the philosophy, and the ethics of eating are all almost his alone.

Among the lower vertebrates—frogs and snakes—eating is apparently a distasteful if not positively painful process, and some of them succumb to it only at long intervals. Predators, on the other hand, make it an occasion of savage fury. But few if any animals can be said to *dine* in the sense that they may be said to play, sing, make love, or practice the arts of homemaking and domesticity. And since, except for the domesticated animals, few will eat more than is good for them, it seems a reasonable conclusion that they eat to live but are hardly capable of living to eat.

Voltaire called man the only animal who drinks when he is not thirsty and he might have added, "Or eats when he is not hungry." Perhaps that is one of the reasons why he is also the only one who has developed in magic, in manners, or in religious dogma scruples and taboos in connection with eating—more of them indeed than in connection with any other natural process except the sexual. Thus the gourmet and the ascetic generate one another. To live to eat suggests, by reaction, that eating itself is sinful; and while the worldly try to see how much food they can consume, the monk not only tries to see how little he can survive on but regards the indispensability of that little as evidence of his depraved state. We think it indecent to eat much when we mourn, and when we celebrate we think it obligatory to provide more food than is good for us.

Some individuals as well as some civilizations have felt the impulse to compromise by abstaining at least from what Bernard Shaw calls "the practice of consuming the corpses of animals," but the most solemn mystery of the Christian religion (and it is not in this respect unique) is the moment when we practice theophagy. The robin, nevertheless, is neither ashamed of his worm nor inclined to elaborate either the acts or the emotions which accompany its acquisition. And what is true of him seems to be true of most animals. Some birds do offer food as a demonstration of affection, but that is as far as it goes.

The earliest men, like most of the so-called higher animals, did not make a living; they simply found it. Like many animals they could hunt for food and perhaps store it up, but they did not breed the animals they slaughtered and did not plant or cultivate the nuts and berries they collected. No vertebrate before Neolithic man was herdsman or gardener.

Perhaps this fact seems too expected to be worth stating. And so it would be if a certain paradox which we will meet again and again did not present itself. No *vertebrate animal* is more than hunter and storer but many *lower* creatures are. Many of the domestic arts had been elaborately evolved by insects and become complex "cultures" millions of years before man or any of his direct ancestors invented the simplest techniques to make his manner of getting a living very different from that of the mere beast of prey.

Spiders spun silk long before even man himself could do better than wrap himself in skins. Bees concentrated honey from nectar long before he had even built a fire. For that matter, even the birds who shared with him the technical backwardness of the vertebrates built nests while he could still do no more than seek shelter in ready-made caves.

So far as the arts and sciences are concerned, man was long what many of his fellow mammals still are: members of a very much retarded race. Whatever other characteristics he and they were developing these characteristics did not include any that resulted in more ingenious ways of living or of making a living. It must have been only a few thousand years ago that men developed a language capable of conveying any practically useful information as precise as that used by the bee when he tells his fellows how to find a good pasture he has discovered. And it was only yesterday that man invented navigational techniques as ingenious as those used by these same bees who steer by the polarized light from the sun.

If, as the anthropologists believe, man has been a social animal for not more than a million years then certain insects discovered the advantages of a cooperating group something like thirty million years before he did and our progenitors might profitably have heeded an injunction "go to the ant," not so much to learn industry as to learn the agriculture and the animal husbandry man had not yet dreamed of.

Nearly everybody knows that ants "keep cows." To your sorrow you may see, almost anywhere in the temperate zone, aphids overgrazing your favorite flowers and see the ant-dairymen milking their sweet juice. With a little patience you may even see the ants actually putting their milch cows out to pasture and you may curse them for their ingenuity. But it is somewhat less well known, though more astonishing, that other species of ants practice agriculture very elaborately, ingeniously, and successfully. No Paleolithic man about to reach the point where he would begin to plant gardens and thereby qualify himself to be raised to the rank of *neo* instead of *paleo* ever investigated the insect's method. But if he had been capable of doing so he might have advanced a few thousand years at one jump on the basis of what he could have learned about soil preparation and planting. The paradox of the socially retarded mammal ought to be, but so far as I know has not been, investigated.

Most of the books, either technical or popular, which deal with such things usually choose as their example of insect technology one of the tropical agriculturists. But it is not necessary to go outside the United States to find ant farmers who have mastered very advanced techniques of agriculture. The arid Southwest where I now live is the section of our country where they have colonized most successfully. On any late summer walk I am pretty sure to come across both the communities of these farmers and other communities that specialize in making a living in quite a different way.

In certain areas—especially those where there are a good many clumps of a certain hard prickly grass which springs up after a summer rain and is by now ripe and dry—I see rings of yellow chaff surrounding the nest entrances, in and out of which ants are streaming either with ripened grain or with bits of chaff to be piled outside. This is a busy harvest time and the ants seem to know very well what they are doing. Solomon said not only, "Go to the ant, thou sluggard; consider her ways, and be wise." He added that the ant "gathered her food in the harvest." And Solomon was taken to task by later naturalists who did not know that any ants actually did harvest seeds and who either blamed him for nature-faking or apologized for him by suggesting that the bit about harvesting was one of those convenient "late additions" to the ancient text.

It was not until 1829 that a British army man stationed in India first noticed and confirmed the fact that Solomon was in this instance a better naturalist than anyone in the centuries intervening since the heyday of his moralizing. But apparently even wise Solomon did not know that other species had made the great step from the mere harvesting of wild plants to genuine agriculture.

Neither his harvester ants nor mine are more than *mere* harvesters. They neither plant nor tend the grass from which they gather the grain. Theirs is a kind of operation which was, perhaps, not beyond the capacity of late Paleolithic man. But other ants have gone far beyond what either he or most ant species are capable of.

If I go into some area slightly more deserty than that which the harvesters favor, I will come across imposing piles of sand, six inches or more across, two or three inches high, and sloping inward toward the nest entrance so that the latter is at the bottom of a crater shaped like an inverted cone and hence designed to funnel any rain that may fall into the underground chambers. Instead of the harvesters' pile of chaff, scattered green leaves—often those of the resinous creosote bush—may be seen here and there.

These leaves do not look very nutritious, and for ants they would not be. Yet the ants are carrying them below ground as purposefully as their harvesting cousins carried the edible grass seeds. A broken trail of dropped leaves leads to a shrub perhaps fifteen feet away. Ants have climbed it, bitten off leaves, carried them to the ground and then into the nest. In fact they do all this on so large a scale that in tropical countries related species do enormous damage to orchards.

Neither these tropical species nor those of my desert are going to eat the leaves, though for many years it was assumed that they would. The leaves will be used indirectly. And this fact makes it all the more remarkable that these farmer ants should know their usefulness. It is, after all, one thing to recognize food and to take the next step of gathering it for future rather than for present use. Even a secondary use for such food might be accidentally discovered. But a great deal more fore-sight and insight of some sort, conscious, or by now at least completely unconscious, is required to gather something quite inedible which can nevertheless be used to produce food. Be-tween the two there is all the difference between the nomad who gathers nuts to eat or to carry with him and the settled gardener who has learned that food can be grown, not merely found. And this actually is the difference between the har-vesting ants and those who nip leaves from a creosote bush. Those leaves are to be used by gardeners who, in some sense, recognize their value as humus and as plant food.

I have never dug into one of their nests, but I have seen specimens of what you find if you do—amorphous dirt-brown masses of rather spongy-looking material one might be hard put to guess the nature of. Actually it is composed of more or less decayed leaves. And if it has been dug up at the right time it will be penetrated everywhere by the white threads of a fungus. This "mushroom" is the only food of the ants, who not only prepared their mushroom bed but planted the fungus itself where it would grow.

All the ants who have adopted this way of making a living belong to the New World; all belong also to a single tribe called *Atta*; and all the members of this tribe practice mushroom culture whether, like most of them, they live in the tropics, invade the Southwest, or, in the case of one small species, may get as far north as New Jersey. The first observers assumed that the leaves were eaten. As a matter of fact the account published by the first to grasp the truth was neglected and it was not until a German student in the eighteen-nineties elab-orately described what he had discovered in the tropics that accurate knowledge of the agricultural ants begins. Since then many entomologists, including the great American student Wil-liam Morton Wheeler, have worked out all the details and even succeeded in following the development of laboratory-raised specimens from egg to established community.

There is nothing hit-or-miss about the methods employed. Almost purely instinctive though the procedure is assumed to be—at least by now—it has all the appearance of being as intelligently purposeful as any of the steps taken by men to grow any difficult crop.

To begin with it must be remembered that mushrooms are a very difficult crop indeed. They require a very special soil and very rigidly controlled conditions of humidity and temperature. To preserve a pure culture or "stand" of a given kind is more difficult still because the air is full of fungus spores; where one kind flourishes others are very likely to gain a foothold. Yet each member of the Attid tribe does grow a special mushroom that furnishes its only food and does keep the culture pure—no "weeds" grow in their gardens. Moreover the crop is artificially controlled in such a way that it produces only the threads, which act like roots, and the tiny above-surface globules, which the ants eat. Remove the agriculturists from a nest and the garden soon goes wild, assuming a different vegetative form and producing less of the edible material.

Suppose you start your observations with a thriving colony of, say, the species upon whose aboveground activities I have been from time to time casting an eye—although I see nothing except the procession carrying creosote or mesquite leaves from the plants through the entrance leading to the nest. When Wheeler dug up a small colony of this very species he found that the chamber with garden was more than three feet below the surface.

Down there a group of workers are receiving the leaves, cleaning them carefully, and then chewing them into a pulp moistened with saliva. When a small pellet has been properly prepared it is added to the outer edge of the already growing garden, so that the garden becomes larger and larger as time goes on. Meanwhile other workers are keeping the growing mushrooms in good condition and feeding the newly hatched, wormlike larvae with bits of the mature mushrooms, which furnish the only food they or the adult will ever eat.

All this is remarkable enough. But how does a garden get started? Where do the spores to be planted come from?

To answer that question one must know first that most ants, like the honey bee, come in at least three (among the ants often more) different forms. There is the queen who lays eggs, there are sterile females who do most of the work ("maiden aunts," Thoreau called them), and there are the males who are good for nothing except that once in their lives some of them impregnate a virgin female.

The female ant leaves the nest only once. She and the young males both have wings and sometime during the summer, at what looks like a prearranged signal, the youths and maidens pour out of the nest, leap into the air, and begin a dance (What pipes and timbrels, what wild ecstasy!). I have not yet seen this happen at any of the nests of the mushroom-growers, but not long ago I did see the nuptial dance performed by the members of a colony of a certain less interesting red species.

From a distance you might mistake the dancing swarm for gnats, though the individuals are many times gnat size. Approach and you will see that mating couples, still joined, are dropping to the ground. Soon they separate and the life of the male is nearly over. But the female of the mushroom-growing species has many difficult things to do before she can become a queen in her own right and spend the rest of her life laying eggs that will be fertilized by the sperm she stored up during her one mating experience. Almost immediately she tears off the wings, for which she has no further use. Soon she will dig a little cavity and retire into it, never to emerge again into the light. There she will lay the eggs from which a generation of workers will develop to enlarge the nest, care for the next generation, and so on.

All this seems to involve the solution of many problems even for the majority of ant species which collect food only where they can find it. But what of the Atta queen who must start a garden planted with a particular "seed" and begin to grow a difficult crop?

Well, the new Atta queen "foresaw" all the difficulties and made certain preparations, despite the fact that she had never before been outside the nest in which she was born.

The most important of these wise preparations was this: before she left on her wedding journey she bit off a piece of the fungus garden, and instead of swallowing it carried it as a sort of dowry at the back of her mouth cavity. Then, when she found herself widowed on her wedding day but safely underground in the tiny little excavation she had made, she removed the bit of fungus garden from her mouth, placed it carefully on the ground, and laid a few eggs beside it. Soon she must grow enough food to feed not only herself but also her young. There are yet no helpers to bring in leaves to fertilize the garden. What does she do?

What she does is rather indelicate, but not more so than what is involved in certain methods adopted by the wise Chinese. She fertilizes the incipient garden with her own excrement. And that also she does "deliberately." She breaks off minute fragments from the small growing mass, holds them under her anal vent, deposits a drop of liquid feces upon them, and then puts them back—near, but not on, the original tiny speck of culture from which they were torn. In this way she manages to raise just enough produce to feed the first hatched larvae. Because food is scarce they develop into stunted workers. But presently they are collecting leaves and soon the garden

479

is thriving. As is so commonly the case with those who establish new colonies, the first year is the hardest.

Now how on earth did it ever come about that a mere insect had developed a highly complicated gardening technique un-told millennia before man or any other vertebrate animal had taken the first blundering steps in even the general direction of agriculture? Henri Fabre would have answered: "Simply because God built the necessary system of instinctive actions into one of the smallest of his creatures." Most entomologists would reply almost as dogmatically: "Because in the course of millions of years accidental variation and blind chance resulted in useful actions which had sufficient survival value to fix the habits upon the race which happened to practice them and gradually to bring the techniques to perfection."

If you object that this seems improbable, they will reply that given time enough nothing is in this sense improbable. Everything that could happen will, in time, actually happen. Nevertheless, if it still seems to you that something more than blind chance helped the process along, you can make the or-thodox account seem still more difficult to swallow by pointing out that the improbable thing must have happened not merely once, but three times. Some termites and at least one beetle also grow mushrooms. In both cases the techniques must have been developed quite independently of one another and of the ants. Not once but three times agriculture was invented by an insect.

Wheeler, who recognized the difficulties, says finally: "These insects in the fierce struggle for existence, everywhere apparent in the tropics, have developed a complex of instinctive activities which enables them to draw upon an ever-present, inexhaustible food-supply through utilizing the foliage of plants as a substratum for the cultivation of edible fungi."

But Wheeler, who besides being a notable taxonomist and

a very distinguished observer was also a man of considerable philosophical humor, must have realized that as *explanation* the sentence just quoted is on a par with Polonius' explanation of madness. For to define true evolution, what else is it but— evolving?

PARENTHOOD: WITH LOVE AND WITHOUT

Most people are more interested in young animals than in grownups, and at any zoo the mother with her baby attracts the largest crowd. Parental concern is a touch of nature which even those usually indifferent to their fellow creatures recognize as making them kin. There is nothing else which suggests so strongly that the animal is living in an emotional world very much like our own.

Yet many animals get along without any parental care whatsoever and many parents are completely indifferent to their offspring. All except the very simplest one-celled creatures have to have a father and a mother—or at least a pair of great-grandparents somewhere in their past. But only the begetting is indispensable. Some are protected in youth and then educated; others are on their own from the moment of birth or hatching. And the difference is by no means always a difference between higher and lower or even a characteristic of the class to which a given creature belongs.

Most fishes and reptiles get little attention from their parents, yet the common stickleback of the ponds builds a nest faithfully guarded by the male. All young mammals must nurse, and to that minimum extent the mother must concern herself with them. But there is a vast difference between the minimum attention some must get along with and the long period of protection and of education accorded to others. Mammals receive less care than some birds and, with the exception of man,

481

no other creature makes such elaborate preparation for the welfare of its offspring as many insects do.

In some sense the fox and the rabbit, like the cat and the bird, seem to love their young. On the other hand it can hardly be imagined that the toad can love the tadpoles it will never see. Yet when the toad returns to the water to lay its eggs it seems, from the outside, to be exercising a degree of foresight far beyond any that could be expected of any mammal other than man. And by comparison with many insects the frog does not do anything remarkable. The instinct upon which both depend enables them to perform feats only man, among creatures even partly dependent upon intelligence, can approach.

Just as the toad "knows" that its eggs must hatch in water, so the Monarch butterfly "knows" that the caterpillars which are its children will have to find themselves on a milkweed when they emerge; and the cabbage butterfly "knows" that its eggs must be laid on the leaves of some member of the cabbage family—even though neither butterfly has any special need for milkweed or cabbage at any other period of its adult life. Yet the butterfly will never see its children and would not recognize them if it did.

That there is a difference between the "wisdom" of the caterpillar and that of a bird is clearly implied by the difference between their conduct when something untoward takes place. Try to take the eggs from a robin's nest and it will attempt to drive you away, exhibiting unmistakable signs of distress. Scrape the eggs of a Monarch from the leaf of a milkweed and even though it has not yet finished the laying process, it will pay no attention whatsoever. It "knows" that they should be put on that plant. It does not "know" anything else about them.

Obviously, then, "mother love" sometimes accompanies parental care but is not an indispensable condition for it, and the paradox is this: there are two very successful ways of helping the young to survive. One is to be aware of their existence and to take more or less intelligent steps, as the cat or the bird or the squirrel does, to meet situations as they arise. The other is not to "know" in our sense what you are doing or how to modify an almost invariable routine. Yet from the standpoint of survival it is by no means certain that the second does not work quite as well as the first. Insects seem to survive rather more successfully than most mammals.

What then is mother love good for? A difficult question indeed. And one hard to answer on any of the biologist's usual assumptions, which seem to make the insect more "sensible" just because it has no sense.

Not all insects have, to be sure, developed any form of parental care. Take for example the common walkingstick of New England, which has at some time or other surprised most country dwellers when they have happened to see it. No other insect of our temperate region has managed to perfect so successfully the sly device which consists in protecting itself from enemies by looking like something inedible. But it takes no care whatever of its eggs, much less of its children. The eggs are dropped casually to the ground, in such numbers that they sometimes patter off the leaves of trees with a sound like raindrops, and, as I know from experiment, these unprotected eggs sometimes lie two years before they hatch. Still, such carelessness is the exception rather than the rule. Pick out any six-legged creature and the chances are that he will do something remarkable when the time comes to prepare for that future generation which, in the majority of cases, it will never live to see.

Take for example, the wasps—unpopular creatures, to be sure, but endowed with what looks like remarkable foresight and most admirably concerned with the welfare of their posterity. They are likely to appear as unwelcome visitors when fruit or sweets are put out at a picnic, and they are also often seen flying through the air with some caterpillar, some insect, or some spider clasped between their legs. The sweets they are consuming on the spot and it is their favorite food. Usually, at least, the living victim they are flying away with is not for themselves but for children who are not yet even eggs. It will be quite a while before these offspring will be ready to eat the game just taken by their thoughtful parents.

There are many kinds of wasps which follow many different customs. Some are sociable (among themselves) as bees also are; some are solitary. Some build the large familiar nest out of paper manufactured by chewing bits of wood into pulp; others prefer adobe structures. But like related tribes of human beings they have many "culture traits" in common, and the habit of laying up a provision of fresh meat for the exclusive use of their children is widespread among the "solitaries."

For our example we might as well choose the "mud dauber." Nearly everybody has seen its nest plastered against a wall in a barn, a garage, or even within the protected entryway of a country house, where it is still less welcome. Not everybody has seen the maker at work preparing the tube or stocking it with provisions. Still fewer have ever inquired what it is all about, though the story is one of those most often told in books of natural history.

This solitary mud dauber, unlike the social ants, or bees, or wasps, is not a creature whose whole life is a succession of incredible acts. He (or in this case she) was born as a wormlike grub, and without leaving the nest she made herself a cocoon from which she emerged as a fully grown winged insect. From then until the time came to prepare for the next generation she has led a simple life: feeding on flowers or fruits, sleeping where she happened to find herself, and assuming no responsibilities—unless you count as a responsibility being ready to receive a mate when he comes along at the proper time. But when her eggs begin to ripen she abandons her thoughtless life and exhibits what looks like remarkable foresight and remarkable solicitude.

First she collects, one after another, little pellets of mud —almost any sort will do if it is sufficiently damp and sticky. She adds them one to one until she has constructed several neat little tubes, each of which has required thirty or forty separate loads, and when they have dried she undertakes to lay up in them provisions for the children who are not yet even eggs. Just what kind of game she will seek depends upon the species to which she happens to belong. Some solitary wasps specialize in flies, others in beetles, others in caterpillars, still others in grasshoppers, cockroaches, or what not. But none ever deviates from its hereditary preference, and so the common New England mud dauber takes only spiders.

She may need as many as ten or fifteen for each infant and it will take assiduous searching to find enough suitable ones to provision a whole nest. But when she has located a victim she knows precisely how to handle it, and this is the most remarkable part of the whole business. She paralyzes it with a sting and then carries the victim to the mud nursery. Presently she will lay a number of eggs on the spiders and when the egg hatches the wormlike larva will start to feed upon the stunned but often still-living victim.

Our common mud dauber goes in for rather small spiders, which are easily handled, but some of her cousins are bolder —notably the large steel-blue and orange solitary of the Southwest, who attacks the great hairy tarantula much larger than herself and usually wins the duel with an adversary himself well supplied with both venom and the fangs with which to inject it. But our common New England species is no less remarkable than any of her relatives so far as the most surprising part of the performance is concerned. All have mastered the same extraordinary technique. The poison they inject through their sting into the victim penetrates one of its nerve ganglia; the prey is reduced to impotence, though still living, and is carried unresisting to the cell prepared for it.

Here we come face to face with one of the most hotly disputed incidents in all the vast repertory of insect marvels. It was Henri Fabre, perhaps the greatest of all modern observers, who first studied in detail the paralyzing technique, and he regarded his discoveries as the most remarkable of his career. The French species he selected happened to be a caterpillar-hunter. Fabre announced, first, that his wasp expertly sought out with its sting a particular ganglion of the victim's nervous system and, second, that the caterpillar was paralyzed but not killed, because its body must remain alive and fresh until the larva was ready to eat it. Wasps had no icebox but they nevertheless knew how to keep game from spoiling. Moreover, he thought, all this was one more proof of his favorite thesis: God must have built into the insects an automatic mechanism capable of automatically performing all the complicated acts they had never had the opportunity to learn and whose purpose they do not understand.

Since Fabre's time other observers have attempted to check his results, some because of an impartial desire to know, some perhaps because they wanted to reject his thesis. Some prefer to believe that insects did, at one stage of evolution, actually learn things for themselves. Others insist that on the contrary everything in their behavior is explicable on the basis of natural selection, which has gradually given the appearance of purpose to actions originally random. In any event the two Peckhams, who made more than half a century ago the classic observations on American solitary wasps, came to the conclusion that on this continent at least not all of Fabre's statements would stand up. The wasp does paralyze its victim. Often that victim does stay alive until it is eaten. But sometimes it dies and, so far as they could tell, a dead, partly decomposed spider was just as acceptable to the wasp larvae as a fresh one. Therefore, they were inclined to think, Fabre was overinterpreting when he maintained that the greatest marvel of all was the "foresight" exhibited by the hunter who paralyzed but did not kill his victim. Perhaps, they suggested, the purpose of the sting is merely to subdue the spider or caterpillar. Perhaps it does not matter whether it dies or lives on in a comatose state. Perhaps the fact that it often does live on is a mere accident.

This dispute does not, however, greatly concern us at the moment. Take the less wonderful version of the wasp's behavior and it is still wonderful enough and quite sufficient to establish our present point. Once in its life the wasp makes elaborate preparations for the survival of the offspring it will never see. Somehow or other the ability to repeat the performance in the future is transferred to the egg. Somehow or other this ability will be built into the larva and then, somehow or other, survive the almost complete destruction of the larva's body in the course of the sickness that will fall upon it when, without leaving the mud nest, it has spun the cocoon from which it will finally emerge as an adult wasp.

There are many insects whose preparations for the next generation are even more elaborate and occupy much more time. The social wasps, for instance, lay up no stores of living game because they or their firstborn are going to stay around the nest and bring fresh, carefully masticated food to feed the larvae as they grow. This is one of the things the adults are about when you see them swarming about the round paper nests in your apple tree or about the open combs attached to the roof of your porch. But the very fact that the mud dauber is only briefly seized by a concern for the future generation and that his other habits are relatively uncomplicated makes him the most convenient illustration of the thing we are concerned with,

namely the fact that parental and, in this instance, purely pre-
natal care of a very effective and essential kind can be given
by a creature as remote from us as the solitary wasp.

Here, surely, is an extreme example of doing something
well without—so at least it would seem—having the slightest
idea why you are doing it. And for that reason we can hardly
feel that there is in the wasp's behavior any touch of nature
which makes us kin. We may, like Fabre, see in his behavior
the glory of God, or, like some evolutionists, a machine which
seems to imitate purpose so cunningly as to arouse in some
the suspicion that what is called purpose even in human beings
is also only a simulacrum. Certainly, however, we would hesitate to talk about mother love in connection with a solitary
wasp. Yet we still have to admit that if it's survival value you
are after then mother love doesn't work any better. Young
wasps seem to survive just as abundantly as young foxes, or
kittens, or human beings. And so we must ask again whether
mere "survival value" can account for the existence in other
animals of mother love as an emotion.

Because the wasps and all the other insects are presumably
incapable of love or of any other emotion or thought, their
techniques, no matter how ingenious or how successful, leave
us cold.

The wasps are expert but alien. There is something about
their blind foresight (an impossible but inescapable phrase)
which is repulsive and terrifying. They belong to that part of
the universe which operates beyond our comprehension and
almost beyond our sympathy. They are precisely what we don't
want to be. Theirs is a form of life which might exist forever
and still be without significance for us. They go their way, we
go ours. And between us there is no communication. We are
capable of love; they presumably are not.

How different by comparison is the often blundering but

conscious solicitude of warm-blooded animals. As the family cat approaches her term she searches restlessly for a suitable place to give birth to her babies—and often settles finally upon what seems to be a highly unsuitable one. As likely as not she will herself conclude that she made a mistake and will move the helpless creatures to another, sometimes to another and another. The wasp was confident, unhesitant, and efficient. The cat is fussy, apprehensive, and uncertain. But in that uncertainty and fussiness we recognize something of ourselves. In the wasp we recognize nothing.

The wasp children will get no education and they will need none. Like us, the kittens will need it and like us they are more or less at the mercy of a mother's whims. She will wash their faces whether they need it or not and disregard all protests. She will bring them insects to play with and, later, a mouse. She will even sometimes snatch the mouse away again, acting exactly as the mother lion is said to act, and momentarily defend it with a growl just to teach her young that they must expect to face serious competition later on. When she takes them out of doors she will run along a gently sloping branch, encourage them to follow her, and then, on the sink-or-swim principle, try to nudge them off until they have learned to hold on with extruded claws.

What we find most engaging is not so much the wisdom of her actions as the apparent motives behind them. The very fact that, unlike the wasp, she does not quite know what to do about anything, the fact that her intentions are far beyond her competence helps us to recognize in her something akin to ourselves. And it may be even her greatest follies that seem most human.

Like human beings, some other warm-blooded creatures are quite capable of being so stirred up emotionally that they forget the occasion of their excitement and lose sight of their original intentions—as was plainly the case with a pet duck whose antics once amused me. Becoming a mother for the first time at the age of twelve years, she was frantically solicitous over the safety of her offspring. She had no trust even in the human beings who for twelve long years had never offered her anything but kindness, and she would fly at them in a fury if they approached too close. But her absurdity went far beyond that. If one persisted in coming near she would presently, in a very excess of anger, forget who it was she was mad at and start attacking the ducklings themselves.

Yet to say that this behavior revealed an impossible gulf between the mind of a duck and the mind of a man is to reveal an extraordinary ignorance of what the mind of man is like. How often has some fond father, angry because something went wrong at the office, come home and punished for nothing the very children for whom his concern had made some unfavorable turn in his business so disturbing. Nothing is more human than striking at the object nearest at hand when under pressure of blind anger.

Konrad Lorenz, the great Austrian observer, tells the story of one of his ravens who fell in love with him. Not a man to consider personal inconvenience where there was something to be learned, he consented to receive the masticated worms which the affectionate bird insisted upon feeding him. But when the raven—a male—tried to entice him into a nesting hole he was physically unable to accept the invitation. Obviously the raven's insight into the situation was highly imperfect. Either he could not perceive, or was determined to disregard, the fact that a man is too large to enter a raven-sized nest.

No wasp would be capable of such folly—not because it has more insight, but because it does not need any. Perhaps, as many entomologists would be willing to grant, it is not the pure automaton Fabre tried to believe it. Possibly it is endowed with a dim consciousness and some power to adapt to a new situation. But it rarely depends upon either. The raven and the man are alike in that each depends much upon emotion, intelligence, and insight—all of which are, nevertheless, often inadequate to the situation.

A few weeks ago I happened to be walking along the sandy shore of a Pacific inlet where little snowy plovers were nesting. Several fluttered along the ground just out of my reach, putting on the broken-wing act that, as all hunters know, various birds have learned. Are the plovers aware or not aware that their ruse may serve to lead a predator away from the nest? Here, in other words, is an instance where it is impossible to say whether the act is blindly instinctive or accompanied by some understanding of its purpose.

Neither can we be sure in the case of the opossum or of the common, harmless little hog-nosed snake, both of whom "play dead" very convincingly. Mechanistically inclined biologists insist that they are merely paralyzed by fright and neither knows what he is doing, nor can avoid doing it. Indeed one experimenter has recently claimed that the possum can be "cured" of his tendency to hysterical fright by shock treatments like those now administered to human victims of certain mental disorders! But there is really no way of being sure how much, if at all, the plover or the snake or the possum knows the usefulness of his behavior. All that we can be sure of is that certain useful actions can be performed either with or without intellectual awareness and emotional involvement.

Field observers who have made the all-too-few classic studies of the higher mammals in freedom have shown conclusively that some of them not only protect their young but educate them in ways clearly implying a kind of awareness related to our own. In England, for instance, Tregarthen found the otter a most competent as well as a most solicitous mother. She plays with her children and also punishes their bad behavior. Unlike insects, otter babies are not born with a set of instincts to guide them along the paths their species should tread. They have to be taught to swim, to catch fish and frogs and rabbits. There seems little doubt that instruction is given and manners are learned.

Yet otters never become any more perfect in the art of being successful otters than the orphaned wasps do in the art of being wasps. Instinct works more surely than either habit or intelligence. Yet we call the otter a "higher" animal than the wasp because the way in which it goes about the business of living is more like ours. It has intellectual awareness and, what is perhaps even more important, its actions are accompanied by emotions, whereas the awareness and the emotions of an insect must be exceedingly dim, if they exist at all.

An anecdote told originally by the English naturalist Hammerton has often been repeated by those who would dismiss as foolish sentimentality any concern over the emotional distress of even the higher animals. It seems that a certain cow of his acquaintance was so grief-stricken by the death of her calf that she refused to eat or give milk. The calf was skinned, stuffed with hay, and given back to her. She licked it affectionately, regained her appetite, and again gave milk. But the best was yet to come. Finally she wore a hole in the calf's skin, and when the hay came out she munched it contentedly.

Now the cow, like most animals which have been protected by man but not accepted into the human family as the dog and cat have, is no doubt a very stupid animal. But that is not the point, since it is the reality of the emotion, not the presence or absence of any clear insight into the circumstances surrounding it, with which we are concerned. The cow's intellectual understanding of the situation was obviously unbelievably dim. Her distress in the face of a mystery, nevertheless, was real and might be compared with that of a man to whom the sense that there is tragedy at the heart of the universe is as overwhelming as his inability to grasp intellectually the cause or meaning of the situation.

An experiment biologists are fond of citing is very relevant here. They say—I have never tried the experiment—that if you snip off the long slender abdomen of a dragonfly and present it to the mutilated creature's mouth it will eat this half of its body with complete unconcern, and undoubtedly it would take quite as gladly the body of its mate or its child. Faced with its natural food, insect and cow alike did the natural thing—which was to eat; and the cow was as incapable of being put off by the fact that the food seemed to come from an unnatural source as the insect is of being put off by the fact that it has just been mutilated. But there is no reason to suppose that the dragonfly would be capable, as the cow was, of mourning the loss of its young. And that is what makes, from the human point of view, the immeasurable difference.

When, ever so long ago, the insect clan took the turning that led it down the road of ever-elaborating instincts and the very remote ancestors of the mammal took the other, which has led ultimately to man, each committed itself to momentous choices—one of which was, of course, between dependence upon a mind that could judge a situation and dependence upon a fixed pattern of behavior that could not easily be varied. But this is not by any means the whole story.

Somehow or other awareness means not only intellectual grasp but also emotional involvement. Both either first came into being or at least first became a conspicuous part of a living creature's existence a very long time ago, though not when life itself began. And from the human standpoint emotional involvement is quite as important as intellectual grasp. Even the animals with whom we live most intimately, the dog and the cat, bewilder us when we try to understand their minds. They seem sometimes so intelligent, so understanding; at other times so incapable of grasping a situation that seems to us over-whelmingly obvious. We never know quite what to make of them when we consider them as intellectually our kind. Often we wonder whether in our sense they can think at all, and a great gulf opens between us. But it is clear enough that they share our emotions even though they cannot share our thoughts. And it is not merely that they are glad or sad. We see them also jealous, hurt, sometimes ashamed. And here again the touch of nature which makes us kin is not intellectual but emotional.

Where did it come from when we find it either in them or in ourselves? If it seems to have less "survival value" than the insect's efficient instinct, then how can any "survival of the fittest" explain it?

THE NEED FOR CONTINUITY:
MORE LIVES THAN ONE

At the beginning of the fifth chapter of *Alice in Wonderland* Alice has an important conversation with a caterpillar. Thinking of her own recent experiences, she complains that it is very confusing to change size and shape. The Caterpillar—brusque as all Wonderland creatures are—replies: "It isn't."

"Well, perhaps you haven't found it so yet," said Alice; "but when you have to turn into a chrysalis—you will some day, you know—and then after that into a butterfly, I should think you'll feel it a little queer, won't you?"

"Not a bit," said the Caterpillar.

"Well, perhaps your feelings may be different," said Alice; "All I know is, it would feel very queer to me."

"You!" said the Caterpillar contemptuously, "Who are you?"

Having reached this impasse they proceed to explore several others that do not concern us. But this one does. Like all other mammals we human beings take the continuity of our corporeal forms for granted. Except in Wonderland babies do not turn into pigs, or vice versa. If they did we might never have made the important assumption that our souls, our *personae*, or our egos, are similarly continuous. And if that assumption did not exist there would be no basis for that whole universe of ethical ideas without which men would not be men. Who would dare hold a butterfly responsible for what he did as a caterpillar?

The fact remains that for a vast number of all the different kinds of animal creatures on the earth today at least two bodies and two lives—sometimes several—are taken as a matter of course. And in very many cases neither the two bodies nor the two lives resemble one another in any way whatsoever.

The legless, vegetarian, gill-breathing tadpole grows legs, develops lungs, completely reorganizes its digestive tract, and crawls out onto the land to live the rest of its life as a meat-eating toad who never need go near the water again until something—it can hardly be memory—tells it to lay eggs in the water from which it came.

Who that has seen an ethereal Luna moth fluttering his great delicate wings on the windowpane and looking as though he had indeed come from the moon would ever guess that he was only a few weeks before the sluggish but ravenous green worm so repulsive to most people? Or that the harmless lacy-winged antlion, who looks so much like a miniature dragonfly, was once a flat dark-colored little bug lying hidden at the bottom of a sand pit waiting to seize with murderous pincers the ant who tumbled into his treacherous trap? The wisest child of such parents cannot possibly know his own father—or for that matter the father his own son. Neither could any human being guess which children came from which adults. To this day it sometimes turns out that some creature long ago baptized by science with a specific name is merely the young form of some other separately named and classified.

Even with us, children go through a sufficiently bewildering experience in growing up. They become aware of desires unknown before and they are often painfully embarrassed by minor physical changes, such as the breaking of the voice or the swelling of the breasts. Yet by comparison to the life of a frog or a butterfly the corporeal development of a mammal is extremely uneventful and unimaginative. If variety is the spice of life we have very little of it, and we make such a fuss over this little that it is hard to imagine what the problems of adolescence would be if young people fell into a deep sleep at fourteen and then came to with wings.

Your butterfly has to be literally born again. He returns to something like an embryo and then he grows up different —almost as though he were correcting some early mistake.

What, aside from the excitement of living two lives, is the point of that? What does the insect gain, and how in the course of what we glibly accept as "evolution" did he ever develop such a design for living?

Even the most self-confident biologist is likely to answer these questions less readily than he does most others, and it would be a pity to take even the best answers before comprehending fully the phenomenon. What visibly happens is something one would never believe if one did not see it. And the physiological process behind the visible happenings is no less remarkable. Before we even raise the question why, let us consider what does take place.

Many a lover of nature, many a professional biologist even, continues his boyhood habit of collecting a few larvae or chrysalises in summer and of tending them until the day comes when a dead worm bursts out of his mummy case to flit away on wings his humble form never seemed to promise. "Creeping" and "flying" are established in our vocabulary as the very symbols of the most contemptible and the most glorious forms of motion. Yet the caterpillar who went to sleep ignorant of anything except creeping wakes up to waft himself nonchalantly away on the most beautiful wings either nature or the human imagination has ever been able to imagine. No wonder that even the Greeks, un-otherworldly as they were, could not help being led to think of an airy soul leaving the gross dead body behind. "Psyche," or "butterfly," was their word for "soul."

Suppose we select for observation one of the most familiar and widespread of American butterflies—the large red-brown and black Monarch, which few dwellers even in the depths of cities have missed seeing. It ranges over the entire United States and in recent years has migrated, perhaps on American ships, to England, to Australia, and the Philippines. Country dwellers often see Monarchs gathered into large flocks in late summer ready to begin a southward migration very surprising for a butterfly, because most kinds either perish, leaving only eggs or chrysalises behind or, in some species, hide the winter out almost motionless in some shelter.

During the summer each female Monarch had sought out some member of the milkweed family and glued to its leaves a number of tiny eggs. Seen under a hand lens they are distinctively pretty—greenish in color, conical in shape, and neatly ribbed—but they are also too small to be noticed often except by those who look for them. Presently the egg hatches into a tiny caterpillar, which immediately begins the only business of its young life—immoderate eating. If the pasture holds out it never leaves the plant upon which it was hatched and does nothing besides eat, except on the several occasions when it pauses briefly to shed the skin grown too tight to be longer endured.

Presently it reaches full growth and it is then that the casual country walker usually notices it first. It is fat, soft, and to most people repulsive. But it continues to eat ravenously, so that sections of the leaf disappear visibly as the creature moves its head up and down the edge, taking great bites as it goes. Cylindrical feces almost as big around as the caterpillar itself fall to the ground one after another in rapid succession and do not mitigate the general impression of grossness. Moreover the caterpillar has made itself as conspicuous as possible by ringing the green of its body with black and yellow stripes. Monarch larvae have a bitter taste detested by birds and every member of the race is making sure that any inexperienced bird which tries one will not find it difficult to remember what a Monarch larva looks like.

Pick him up (if you are not too squeamish) and you will see that, crowded together near the front end, are the six normal-looking legs all butterfly larvae have which correspond to the obligatory six legs of all adult insects. But there is nothing else about him remotely suggesting a butterfly or indeed an insect of any kind. These six close-together legs are not sufficient for his length, and yet there is an absolute rule of nature that no butterfly can have even in the larval state more than six. You will therefore see at the hind end a double row of fleshy little stumps that clasp the leaf upon which the creature is feeding and serve as substitutes for legs.

If you decide to raise him in captivity for purposes of observation and should this be your first experience, then you will presently be filled with despair. Your caterpillar seems sick. He has grown very sluggish and, incredibly, his incredible appetite fails, no matter how fresh the leaves of his favorite— in fact his only—food you may provide for him. But do not be alarmed. Your caterpillar is about to lose his life in order that he may gain it. Beneath his skin, invisible to any observer, drastic changes have been going on. He no longer eats because he could no longer digest if he did. Moreover, a tough membrane enclosing his whole body is forming just beneath the skin and it is making him so stiff that he can no longer move freely.

Soon he will take up a position either on a twig of the food plant or perhaps upon some other nearby support, natural or man-made. From a special gland at his rear end a little gluey substance will be secreted, and as it hardens it will attach the tip of his body firmly to the support. No butterfly caterpillar spins enough silk to wrap himself, as many moths do, within a cocoon; but the Monarch glue is liquid silk. Soon after he is firmly attached he will let go with all his legs and hang head downward. This is the last step before the caterpillar becomes a chrysalis.

The transformation from sick caterpillar to quiescent chrysalis is less often observed than the emergence of the adult because it happens so quickly. As a matter of fact there isn't much to see. Everything except the one final event takes place beneath the outwardly unchanged skin. Suddenly this skin splits near the rear end and, tissue-thin, drops to the ground. Where a moment ago there was only a worm there is now the mysterious chrysalis. But the Monarch was a good choice for observation because his chrysalis is the prettiest of any American butterfly.

The color is a pale, luminous, leaf-green; the shape somewhat ovoid but broken gracefully into two sections differently curved and as right as an amphora. Probably it follows "the laws of dynamic symmetry" (if there are any such things). It is chastely ornamented with a row of small gilded dots raised slightly above the surface where they half encircle the chrysalis, just at the line where the curvature changes.

"A green coffin with golden nails" someone once called it, and the phrase is accurate as such pretty phrases usually are not. The dots really are not yellow but precisely the color of gold leaf. If they serve any practical function I have never heard it suggested what that function might be. They can hardly be explained by any of the methods commonly used to reduce the beautiful in nature to the merely utilitarian. There is no "sexual selection" to be made at that phase of the Monarch's life cycle. They are certainly not conspicuous enough to serve, like the caterpillar's stripes, as the warning "I taste bad." Nor can the gilding be explained away as the necessary mechanical result of structure as the beautiful designs on the invisible diatom sometimes are. Possibly some substance necessary to the caterpillar but no longer usable just happens to be golden and just happens to be excreted along the line so gracefully placed in relation to the design of the whole. At least we will let it go at that.

Keep the green coffin under observation for a while and presently the yellow-brown of the developing wings will be seen through the thin transparent outer skin. Then one day the skin will rupture and a rather sorry-looking creature, rumpled and feeble, will somehow catch hold of a support with its six legs and rest motionless for several hours. Its abdomen is disproportionately fat, its wings crumpled into little disorderly packets like the not-yet-unfolded petals of a filmy poppy. Gradually they expand and pass from shapelessness to shape. As they do so the fat body grows slenderer, because the fluid that once distended it has been forced through the veins of the wings to expand and stiffen them.

As they grow, the whole insect begins to gain in strength and confidence. An hour ago it had been almost as helpless as a premature baby and hardly able to cling to its support. In an hour more the wings will begin to flutter and the Monarch will sail away—not to the very brief life of many moths, but perhaps to flit until the end of summer and then, with crowds of companions, to start the migration that will carry it hundreds of miles away to some southern clime. Next spring a few with battered wings will make their way north again. Are these northward bound individuals retracing their previous journey, or are they members of a generation begot by those who went south? The question has been long in dispute. Butterflies cannot be banded as easily as birds!

In any event, few other things looking so fragile stand up for so long under the buffetings of life in this world. In so far as it is a machine, it runs not on gasoline but on the energy supplied by the sugar it gets when it sips the nectar of flowers with the long tongue that can probe so deep. Perhaps, in so far as it is more than a machine, it runs on its own tiny portion of whatever else it is that makes man and beast determined to live.

Anyone who has thus watched the progress from caterpillar to finished butterfly has seen incredible things happen. If he has any capacity for wonder he may have been too stunned to ask any questions or to want to probe any deeper. But sooner or later it may occur to him that a great deal has been kept hidden, that what he has seen is merely a series of suddenly revealed transformations. The skin of the chrysalis split to release the still crumpled but fully formed adult. For all he has been able to see, the two transformations were almost as sudden as the metamorphoses of the fairy tale. Some enchanter waves a wand and a man is a beast or a beast is a man. Yet we know that nature does not work that way. Nothing grows except through a series of steps. And these steps have been hidden from the observer. Twice the veil was rent but each time he was presented with a *fait accompli*. What actually went on within the coffin with the golden nails? By what steps does nature make a caterpillar into a butterfly?

Those questions are not wholly unanswerable—at least in the same limited sense that we can answer how nature makes a chicken out of an egg. And the answer is similar though somewhat more complicated, because in this case nature has first to make an egg (or at least an embryo) out of a chicken.

When our caterpillar looked sick he really was sick. He had been struck with what might well be mistaken for some sort of suppurating infection. His muscles and his organs were beginning to dissolve into what looked like a sort of pus. He had just life enough left to perform the final rites which accompany his entombment within the chrysalis before he was returned almost to death.

If we had been watching a fly—a common housefly for instance—the degenerative process would have continued until the maggot had been all but completely dissolved into a creamy liquid seemingly as featureless as the yolk of an egg. Had we been watching instead one of the presumably primitive insects in which the adult is not as different from the infant as is the case with the fly or the butterfly, comparatively little destruction of the already organized living material would have taken place. But here in the case of our Monarch the amount of degeneration which must occur before the larva can be born again—and born better—is between the two extremes.

Once the skin has been shed and the creature—or what is still left of it—ceases to move, the destruction of the original organs and the fashioning of new ones goes on apace. Free-moving cells, much like the phagocytes or white blood corpuscles in the human body, absorb and carry away the disintegrating material, and at the same time new organs begin to form. Ever since the day when the caterpillar was hatched from the egg it has carried within its body certain little groups of cells which were useless until now. They are the buds, if the term be permitted, from which the butterfly's organs will develop, and these organs grow on the material the phagocytes have been carrying away from the parts of its dead self. From these buds a butterfly began to form as the caterpillar was dissolved. No new material save perhaps air and moisture is available any more than anything is available to an egg closed within its shell. What is more remarkable, perhaps, is that almost nothing is left over. The material in one caterpillar is just sufficient to make one butterfly!

"Almost nothing"—the qualification is interesting. If you have kept your butterfly under continuous observation from the time it ruptured the skin of the chrysalis until the moment when it took wing you probably observed, at some moment not long before the last, that one or two drops of liquid fell from its rear end to the ground. "What," you may have asked, "is that?" It was Nature's miscalculation; or rather, her margin of safety. Being sure to have enough—and she cannot predict just how much was going to be lost by evaporation from the chrysalis—she had one or two drops left over.

Different phenomena strike different people with sudden amazement. And it so happens that the drop or two which fall from the rear end of a newly emerged butterfly has long seemed to me among the most astounding things I have ever observed. Whatever told the creature so much in the course of its already eventful life tells it one thing more: "You won't need that now."

The legends of many peoples are full of changing-into-something-else not unlike what the butterfly takes for granted. Ovid's *Metamorphoses*—by no means an entomological work—is one of the most enduringly popular books ever written. The impossible possibility that a man or even a beast might turn into some wholly different creature seems to fascinate something buried deep in human nature. Did it arise, one wonders, independently as one of the never-to-be-fulfilled dreams of which presumably only man is capable, or was it suggested to him by the fact that so many other creatures do change out of all recognition? The question is not likely ever to be answered, but if it could be it might throw some light on another important one: Is what we call "the imagination" limited—as an old theory of psychology held that it was—to the recombination of materials supplied it by the senses, or is it capable of genuine creation; is it able to body forth before the mind's eye what never was even in its constituent parts; what, in some cases, never could be?

But why and how did so many creatures acquire the power to turn and the habit of turning into something else, when that is the very last thing possible to so many others? It can hardly be just because some feel more strongly than others that variety is the spice of life. Or at least that is not the kind of explanation to satisfy our habits of mind. We look for and we often find "explanations" in terms of mechanism and function which are temporarily satisfying. And though in the case of insect metamorphosis they are not so pat as they are in certain others, they do exist.

Aristotle knew that a butterfly was first a worm and then a chrysalis. He describes the sequence briefly in that great compendium of information and misinformation called the *History of Animals*. It was certainly part of the current lore which he summarized and there is no knowing for how long some men at least had been familiar with the phenomenon. But Aristotle says nothing about the how or the why and even today biologists are likely to be unusually tentative in their discussion. Nevertheless, ever since the theory of evolution inspired them with the hope of explaining everything in evolutionary terms, the problem has been repeatedly approached.

The most confident statement which most would be willing to make is a negative one. What was once generally supposed to be one of the master keys won't unlock this particular secret. Whatever the explanation may be, it is not that a modern butterfly has to be a caterpillar first because it was once nothing but a caterpillar and must therefore, in the course of its development, go through its evolutionary history. It is not, in other words, because "Ontogeny recapitulates phylogeny," or, in plain English, because the development of an individual briefly summarizes the evolutionary development of the species to which it belongs.

That may explain why, for instance, a human embryo has gill slits like those of an embryo fish and has also at a later stage of embryonic development a good deal more of a tail than the miserable, too easily broken little projection at the end of the spinal column which is all the adult can boast of to represent the imposing appendage his monkey-like ancestors found useful. It may even explain why children like to play Indian at a certain stage in their lives. Perhaps at that age they really are Indians. But it won't explain what happens to a Monarch butterfly. And there are many reasons why it won't.

In the first place, when ontogeny really does recapitulate phylogeny—and the rule is not as invariable as was once supposed—it doesn't do more than suggest rather than realize the earlier structure of the species. The gill slits of the human embryo never become functional; the future human being never breathes water with them. And if the caterpillar were a mere recapitulation of something in the racial history of butterflies it wouldn't be the perfect functioning organism it is.

In the second place, the fossil record of insects forbids the assumption that the immediate ancestors of the butterflies were caterpillars in the same sense that the immediate ancestors of the seals were land-dwelling creatures and the immediate ancestors of toads were water animals which later took to dry land. The fact that seals return to the land to give birth to young which have to be taught to swim and the fact that toads, on the other hand, return to the water to lay the eggs which hatch into water-breathing tadpoles really does illustrate a sort of recapitulation and the fossil record confirms what the habits would suggest.

But there is nothing to correspond to all this in the case of the butterfly. The oldest fossil insects belong in the class with those modern kinds which hatch out as something very like miniature adults and undergo only minor changes. As the paleontologist follows the story forward through time he finds that as more and more "modern" types appear the changes taking place in the course of the individual's life history become more and more radical. In other words the complete metamorphosis, instead of being something brought down from the very remote past, is something developed fairly late in the course of evolution. In fact it seems pretty safe to say that the modern caterpillar, the modern chrysalis, and the modern adult butterfly have all evolved into their present complexity more or less simultaneously, not one after the other. Hence the caterpillar as well as the winged adult are both "modern." Of course it may be true that all animals, including man, go back ultimately to the famous "wormlike creature." But the butterfly did not come from it much more directly than the mammal did and the caterpillar is not a mere intermediate stage between the two.

This leaves us with a fact almost as staggering as the fact of evolution itself. We have to recognize not only that life began somehow and slowly developed the orderly processes by which an egg becomes a child and the child grows up into an adult but that somewhere along the line certain living creatures also developed the power, not implicit in the other system of development, to change radically their whole structure and all their habits at the midway of their mortal life: the power to start out as caterpillars and to end up as butterflies. This fact the biologist simply has to accept, and if he wants to go on asking questions he must turn to those less ultimate in their implications.

Granting the fact that metamorphosis did arrive relatively late, the layman (or even the neo-Lamarckian) might be tempted to put such a question in the form: "Well then, why should any creature have wanted to live more lives than one, to submit itself to so drastic a break in the continuity of even its bodily structure and its habits of life?" A more orthodox formulation would be: "What purpose is served? What advantage in 'the struggle for life' does such a course of development confer?"

You can say, if you like, that all insects have their skeletons on the outside and that therefore they can grow only if they molt these skeletons from time to time. You can add that for some reason the skin or skeleton of functional wings is never molted so that flying insects can't grow once they have become capable of flying. Ergo, butterflies must be something other than butterflies until they have reached their maximum size.

Rather more simply, you can also argue that when winter or some other unfavorable season has to be endured there is an advantage in passing it during a resting period like that of the hibernating woodchuck or an erstwhile caterpillar now withdrawn into its chrysalis. Or you can say that certain advantages may result from a sort of division of labor which lets the caterpillar devote most of his time to accumulating nourishment while the butterfly is left free to concentrate upon the business of mating and egg laying. But it has to be admitted that many creatures manage quite successfully to survive the winter without turning into something else and to eat as well as make love during the same life stage. The explanations are satisfactory only in the limited sense that it is satisfactory to say that a nighthawk has a large mouth so it can catch insects on the wing and a woodpecker a long bill so that it can dig them out of dead trees. Either method works and we still don't know why one was adopted in one case, the other in another. Perhaps there is no reason except that more different kinds of creatures can be supported on this earth if every possible opportunity for making a living is exploited. Nature abhors monotony as much as she abhors a vacuum.

No doubt it would be very exciting to a human child if he could look forward to sprouting wings somewhere well this side of the grave. Still there are compensating advantages in not being a butterfly, and one of them is that we don't have to sacrifice continuity: that mentally and spiritually the human child is father to the man, that even the kitten is father to the cat, in a way that a caterpillar cannot possibly be father to a butterfly.

Notoriously the insects, even if not quite the pure machines they are sometimes called, learn precious little in the course of their lives and depend very largely upon inherited instincts to solve the problems of living. Even the "highest"—the ants and the bees—can profit extremely little from experience. Nearly all the "progress" evolution has made possible for them is in the direction of more and more complex patterns of instinctive behavior, not in the direction of increased intelligence as we understand the term.

It may very well be that the discontinuity in their lives is at least partly responsible for this fact. Even in a caterpillar inherited instincts are, to be sure, somehow carried safely through that near-destruction of his organization which takes place in the chrysalis. At least the butterfly "knows" that its eggs must be laid on the food plant which the larva, not the adult, must have. But it is hard to imagine that anything *learned* by an individual could similarly survive.

Once, many millions of years ago, the insects were by all odds the most up-and-coming of living creatures. They had a long head start on all the rest of us, and some think that in certain respects—notably in respect to social organization—they are ahead of us still. But if you measure progress in terms of consciousness, adaptability, or the power to learn from experience, then radically different creatures passed them long ago.

It has often been pointed out that the line the primitive ancestors of the insects first took imposes certain limits upon future development. It worked and it still works amazingly well within the limits, but the limits cannot be passed. If, for instance, you are going to have your skeleton on the outside and get oxygen through a simple system of ramifying tubes, then you can't get very big and a man-sized insect is mechanically unthinkable.

At first sight the vertebrate scheme—an internal skeleton exposing the soft vital parts to all the dangers of any physical environment—may seem to be much less advantageous. But the compensating advantages have proved overwhelming. One of them is that you can grow without molting. And if it really is because insects molt that they have developed the habit of metamorphosis then that habit itself may be responsible for their failure to develop what we call minds and for their consequent dependence upon instinct. Your butterfly does not have to be educated. He is born knowing everything he will ever need to know. But that means also nearly everything he ever can know.

All things considered, it is perhaps just as well that human adolescents don't retire into a chrysalis at fourteen. Forgetting everything they had learned up to that age would be too large a price to pay even for wings.